T0207439

Lecture Notes in Computer Science $\qquad$ 13986

Founding Editors

The series Lecture Notes in Computer Science (LNCS), including its subseries Lecture Notes in Artificial Intelligence (LNAI) and Lecture Notes in Bioinformatics (LNBI), has established itself as a medium for the publication of new developments in computer science and information technology research, teaching, and education.

LNCS enjoys close cooperation with the computer science R & D community, the series counts many renowned academics among its volume editors and paper authors, and collaborates with prestigious societies. Its mission is to serve this international community by providing an invaluable service, mainly focused on the publication of conference and workshop proceedings and postproceedings. LNCS commenced publication in 1973.

Gisele Pappa · Mario Giacobini · Zdenek Vasicek
Editors

Genetic Programming

26th European Conference, EuroGP 2023
Held as Part of EvoStar 2023
Brno, Czech Republic, April 12–14, 2023
Proceedings

Springer

Editors
Gisele Pappa ⓘ
Universidade Federal de Minas Gerais
Belo Horizonte, Minas Gerais, Brazil

Mario Giacobini ⓘ
Università degli studi di Torino
Turin, Italy

Zdenek Vasicek ⓘ
Brno University of Technology
Brno, Czech Republic

ISSN 0302-9743   ISSN 1611-3349 (electronic)
Lecture Notes in Computer Science
ISBN 978-3-031-29572-0   ISBN 978-3-031-29573-7 (eBook)
https://doi.org/10.1007/978-3-031-29573-7

This Springer imprint is published by the registered company Springer Nature Switzerland AG
The registered company address is: Gewerbestrasse 11, 6330 Cham, Switzerland

Preface

The 26th European Conference on Genetic Programming (EuroGP 2023) took place at the Brno University of Technology (BUT), Brno, Czech Republic, April 12–14, 2023. The conference was held in a hybrid model to allow both in-person and online attendance.

Genetic Programming (GP) is a unique branch of evolutionary computation that has been developed to automatically solve design problems, in particular computer program design, without requiring the user to know or specify the form or structure of the solution in advance. It uses the principles of Darwinian evolution to approach problems in the synthesis, improvement, and repair of computer programs. The universality of computer programs, and their importance in so many areas of our lives, means that the automation of these tasks is an exceptionally ambitious challenge with far-reaching implications. It has attracted a very large number of researchers and a vast amount of theoretical and practical contributions are available by consulting the GP bibliography[1].

Since the first EuroGP event in Paris in 1998, EuroGP has been the only conference exclusively devoted to the evolutionary design of computer programs and other computational structures. In fact, EuroGP represents the single largest venue at which GP results are published. It plays an important role in the success of the field, by serving as a forum for expressing new ideas, meeting fellow researchers, and initiating collaborations. It attracts scholars from all over the world. In a friendly and welcoming atmosphere authors present the latest advances in the field, also presenting GP-based solutions to complex real-world problems.

EuroGP 2023 received 38 submissions from around the world. The papers underwent a rigorous double-blind peer review process, each being reviewed by multiple members of an international Program Committee.

Among the manuscripts presented in this volume, 14 were accepted for full-length oral presentations (38% acceptance rate) and 8 as short talks. In 2023, papers submitted to EuroGP could also be assigned to the "Evolutionary Machine Learning Track". Among the 38 submissions, the authors of 9 papers indicated their papers fit the track, with 4 accepted for full-length oral presentation and one as short talks. Authors of both categories of papers also had the opportunity to present their work in poster sessions to promote the exchange of ideas in a carefree manner.

The wide range of topics in this volume reflects the current state of research in the field. The collection of papers covers interesting topics including developing new variants of GP algorithms for both optimization and machine learning problems as well as exploring GP to address complex real-world problems.

Together with three other co-located evolutionary computation conferences (EvoCOP 2023, EvoMUSART 2023, and EvoApplications 2023), EuroGP 2023 was part of the Evo* 2023 event. This meeting could not have taken place without the help of many people. The EuroGP Organizing Committee is particularly grateful to:

---

[1] http://liinwww.ira.uka.de/bibliography/Ai/genetic.programming.html.

- SPECIES, the Society for the Promotion of Evolutionary Computation in Europe and its Surroundings, aiming to promote evolutionary algorithmic thinking within Europe and wider, and more generally to promote inspiration of parallel algorithms derived from natural processes.
- The high-quality and diverse EuroGP 2023 Program Committee. Each year the members give freely of their time and expertise, in order to maintain high standards in EuroGP and provide constructive feedback to help the authors to improve their papers.
- Nuno Lourenço (University of Coimbra, Portugal) for his dedicated work with the submission system.
- João Correia (University of Coimbra, Portugal) and Francisco Chicano (University of Málaga, Spain) for their great work on the Evo* publicity, social media service, and website.
- Sérgio Rebelo (University of Coimbra, Portugal), João Correia (University of Coimbra, Portugal) and Tiago Martins (University of Coimbra, Portugal) for their important graphic design work.
- Our invited speakers, Marek Vácha and Evelyne Lutton, for their inspiring and enlightening keynote talks.
- Finally, we express our continued appreciation to Anna I. Esparcia-Alcázar (Universitat Politècnica de València, Spain), from SPECIES, whose considerable efforts in managing and coordinating Evo* helped towards building a unique, vibrant, and friendly atmosphere.

April 2023

Gisele Pappa
Mario Giacobini
Zdenek Vasicek

Organization

Program Chairs

Gisele Pappa   Universidade Federal de Minas Gerais, Brazil
Mario Giacobini   University of Torino, Italy

Publication Chair

Zdenek Vasicek   Brno University of Technology, Czech Republic

Local Chairs

Jiri Jaros   Brno University of Technology, Czech Republic
Lukas Sekanina   Brno University of Technology, Czech Republic

Publicity Chair

João Correia   University of Coimbra, Portugal

Conference Administration

Anna I. Esparcia-Alcazar   Evostar Coordinator

Program Committee

R. Muhammad Atif Azad   Birmingham City University, UK
Wolfgang Banzhaf   Michigan State University, USA
Heder Bernardino   Universidade Federal de Juiz de Fora, Brazil
Anthony Brabazon   University College Dublin, Ireland
Stefano Cagnoni   University of Parma, Italy
Mauro Castelli   Universidade Nova de Lisboa, Portugal
Antonio Della Cioppa   University of Salerno, Italy
James Foster   University of Idaho, USA

Jin-Kao Hao   University of Angers, France
Erik Hemberg   ALFA Group at MIT-CSAIL, USA
Malcolm Heywood   Dalhousie University, Canada
Ignacio Hidalgo   Universidad Complutense de Madrid, Spain
Ting Hu   Queen's University, Canada
Domagoj Jakobovic   University of Zagreb, Croatia
Ahmed Kattan   Umm Al-Qura University, Saudi Arabia
Krzysztof Krawiec   Poznan University of Technology, Poland
Andrew Lensen   Victoria University of Wellington, New Zealand
Nuno Lourenço   University of Coimbra, CISUC, DEI, LASI,
                                   Portugal
Penousal Machado   University of Coimbra, CISUC, DEI, LASI,
                                   Portugal
James McDermott   National University of Ireland, Ireland
Eric Medvet   University of Trieste, Italy
Quang Uy Nguyen   Le Quy Don Technical University, Vietnam
Michael O'Neill   University College Dublin, Ireland
Una-May O'Reilly   Massachusetts Institute of Technology, USA
Fabrício Olivetti De França   Universidade Federal do ABC, Brazil
Tomasz Pawlak   Poznan University of Technology, Poland
Stjepan Picek   Radboud University, Netherlands
Peter Rockett   University of Sheffield, UK
Conor Ryan   University of Limerick, Ireland
Lukas Sekanina   Brno University of Technology, Czech Republic
Moshe Sipper   Ben-Gurion University of the Negev, Israel
Lee Spector   Hampshire College, USA
Ivan Tanev   Doshisha University, Japan
Ernesto Tarantino   ICAR-CNR, Italy
Leonardo Trujillo   Instituto Tecnológico de Tijuana, Mexico
Leonardo Vanneschi   Universidade NOVA de Lisboa, Portugal
Man Leung Wong   Lingnan University, Hong Kong, China
Bing Xue   Victoria University of Wellington, New Zealand
Mengjie Zhang   Victoria University of Wellington, New Zealand

Contents

Short Presentations

Long Presentations

A Self-Adaptive Approach to Exploit Topological Properties of Different GAs' Crossover Operators

José Ferreira[1], Mauro Castelli[1], Luca Manzoni[2], and Gloria Pietropolli[2]([⊠]) [ID]

[1] NOVA Information Management School (NOVA IMS), Universidade Nova de Lisboa, Campus de Campolide, 1070-312 Lisboa, Portugal
{m20180422,mcastelli}@novaims.unl.pt
[2] Dipartimento di Matematica e Geoscienze, Università degli Studi di Trieste, Via Alfonso Valerio 12/1, 34127 Trieste, Italy
lmanzoni@units.it, gloria.pietropolli@phd.units.it

Abstract. Evolutionary algorithms (EAs) are a family of optimization algorithms inspired by the Darwinian theory of evolution, and Genetic Algorithm (GA) is a popular technique among EAs. Similar to other EAs, common limitations of GAs have geometrical origins, like premature convergence, where the final population's convex hull might not include the global optimum. Population diversity maintenance is a central idea to tackle this problem but is often performed through methods that constantly diminish the search space's area. This work presents a self-adaptive approach, where the non-geometric crossover is strategically employed with geometric crossover to maintain diversity from a geometrical/topological perspective. To evaluate the performance of the proposed method, the experimental phase compares it against well-known diversity maintenance methods over well-known benchmarks. Experimental results clearly demonstrate the suitability of the proposed self-adaptive approach and the possibility of applying it to different types of crossover and EAs.

1 Introduction

Evolutionary computation (EC) [2], a subfield of artificial intelligence, leverages computing power to model global optimization strategies that mimic natural evolution and can be applied to several domains [13,19,24]. Genetic algorithms [8] belong to the family of EC and handle a population of candidate solutions represented as a sequence of genes. GAs are stochastic optimizers, minimizing or maximizing an objective function while exploring the underlying search space. This space can be better understood considering a geometrical and topological view of the evolutionary process. In particular, candidate solutions can be described as points in a geometric space within a dynamical system (expressed as the whole GA search process) that changes these points as time (generations) goes through. To describe a candidate solution, one can characterize it genotypically by studying its genetic information or phenotypically by studying its fitness,

G. Pappa et al. (Eds.): EuroGP 2023, LNCS 13986, pp. 3–18, 2023.
https://doi.org/10.1007/978-3-031-29573-7_1

meaning its ability to address the optimization problem at hand. The merging of these two dimensions establishes a fitness landscape in which all the possibilities within the genotypical domain, and consequent fitness outcomes, can be envisioned. Depending on the effect they produce in the underlying space, it is possible to identify two main categories of search operators used by a GA to act on a point/solution: (1) geometric operators (that result in a convex search), and (2) non-geometric operators (that result in a non-convex search) [16]. A convex search will contract the hypervolume of the hypercube that represents the fitness landscape observable by the search process, while a non-convex search has, usually, a non-zero probability of expanding it. In this context, it is crucial to understand the idea of a topology associated with a problem. Using an algorithm to solve a problem means essentially applying a strategy for searching that topology in a (hopefully) optimized fashion. Existing literature proposed different search operators in the context of GAs [11]. Each genetic operator produces a specific effect on the search process, and it is challenging to determine which operators are more effective in addressing a specific problem and which operator is more effective in a given phase of the search process. As a consequence, existing works proposed to dynamically modify the probability of using genetic operators based on some criteria [4,12,22]. In this work, we propose a self-adaptive approach to exploit the properties of geometric and non-geometric crossovers to achieve a more effective search. We expect this method can help overcome (or at least reduce) the problem of premature convergence of the population, one of the main limitations of GAs. In particular, we rely on a self-adaptive technique which, based on the current stage of the search process, decides whether to use the non-geometric crossover to (possibly) increase the population's convex hull. This work differs from the existing methods that, in the majority of the cases, simply modify the probabilities of crossover and mutation based on the status of the search process but without considering topological information concerning the genetic operators. In particular, by adapting the search operators used in GAs, we will leverage the continuous need to apply either geometric or non-geometric crossover in different phases of the search process characterized by specific space topology conditions [3].

The remaining part of the manuscript is organized as follows: Sect. 2 reviews some concepts concerning the geometric properties of genetic operators and convex search; Sect. 3 links this study to the existing literature; Sect. 4 presents the proposed self-adaptive approach; Sect. 5 outlines the experimental settings; Sect. 6 analyzes the results achieved, while Sect. 7 summarizes the main findings of the paper and suggests future research avenues.

2   Fundamental Concepts

This section presents important concepts and definitions related to the geometrical properties of genetic operators. To frame the discussion, it is essential to recall some ideas developed in the geometric framework that unified various EAs [15]. This framework analyzes the working principles of the genetic operators from a mathematical perspective.

2.1   Crossover

Let S be the space of all possible solutions and the image set $\mathrm{Im}[OP]$ the set of all possible offspring produced by a recombination operator OP with non-zero probability.

A recombination operator OP belongs to the geometric crossover class $\mathcal{G}$ [15] if there exists at least a distance d under which such a recombination is geometric:

$$OP \in \mathcal{G} \iff \exists d : \forall p_1, p_2 \in S : \mathrm{Im}\left[OP\left(p_1, p_2\right)\right] \subseteq [p_1, p_2]_d .$$

On the other hand, a recombination operator OP belongs to the non-geometric crossover class $\overline{\mathcal{G}}$ if there is no distance d under which such a recombination is geometric:

$$OP \in \overline{\mathcal{G}} \iff \forall d : \exists p_1, p_2 \in S : \mathrm{Im}\left[OP\left(p_1, p_2\right)\right] \setminus [p_1, p_2]_d \neq \emptyset.$$

The geometric crossover leads to the creation of offspring lying on the segment that connects the parent individuals in the space. As pointed out by Moraglio, there are three properties [17] that arise from using a geometric recombination operator:

- *Property of Homology.* It states that the recombination of one parent with itself can only produce the parent itself.
- *Property of Convergence.* It states that the recombination of one parent with its offspring cannot produce the other parent of that offspring unless the offspring and the second parent coincide.
- *Property of Partition.* It states that two recombinations, the first of parent a with a child c of a and b, and the second of parent b with the same child c, cannot produce a common grandchild e other than c.

If any crossover operator fails to meet any of these properties, it is, by definition, non-geometric.

In this paper, we use two recombination operators: *one-point crossover* (geometric) and *extension ray crossover* (non-geometric). Among the (several) existing crossover operators, we decided to rely on these simple operators for presenting the proposed self-adaptive approach. This choice mitigates the causal relationship between the use of more complex types of crossover operators and the results achieved in the experimental phase. One-point crossover [23] is a mask-based crossover for binary strings that produces offspring in the segment between the two parents. On the other hand, extension ray crossover [18] extends the segment passing through both parents, thus producing offspring outside this segment.

The other operator typically adopted in the GA framework is the mutation. Anyway, we will not discuss its property in further detail, as the study of this paper is focused on the crossover operator (we do not use the mutation in the search process).

2.2   Convex Combination, Convex Hull, and Convex Search

This section reports the concepts necessary to fully understand the role of geometric and non-geometric operators in the search process and discusses diversity maintenance strategies in convex search. A *convex combination* is a combination of vectors where all coefficients, their multiplicative factors, are non-negative and sum up to 1. A set S closed under convex combinations is a convex set. In this case, any $a, b \in S$ implies $\overline{ab} \subseteq S$. The *convex hull* of this set is the boundary of its convex closure, also called a convex *polytope*.

Geometric crossover leads to convex outcomes and reduces the convex hull of the present generation's pairs of parents. In other words, the distance between two children will be smaller than the one between their parents. Intuitively, this formulates a concept of convex search given that executing selection and crossover multiple times over any number of generations will lead to a search space reduction [15]. Figure 1 illustrates a hypothetical spatial evolution from one generation to the subsequent, showing the modification of the global convex hull produced by the usage of selection and geometric crossover.

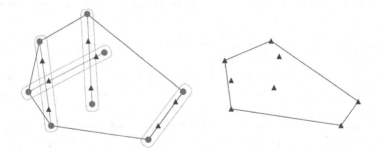

Fig. 1. Convex hull reduction through a generation. Gray dots are the vertex of the polytope which represents the solution space. Red triangles are the individuals created through geometric crossover. (Color figure online)

A fundamental problem researchers encounter while creating new search strategies is the one of premature convergence. It exists where space exploration leads to final stages where neither the global optima nor acceptable local optima are present or will be present because while diminishing the volume of the population's convex hull, these solutions are left out. Diversity maintenance strategies often focus on creating or maintaining distance between members of the population, and they represent an attempt to counteract premature convergence. However, these strategies will only have an effect relative to the continuous volume-decreasing convex polytope. As a consequence, there will be diminishing returns when it comes to these strategies. In particular, these methods (abstracting from the vast number of different implementations) artificially increase the chances of future genetic material propagation of specific individuals in the name of diversity. This is achieved by augmenting the fitness of genotypically remote individuals and/or by artificially reducing the fitness of genotypically close individuals in subpopulations. Phenotype-agnostic methods also exist, focusing only on promoting or demoting certain individuals with the help of genotypic distances. Each method (or a combination of methods) leads to the same outcome - the increase of targeted individuals' probabilities regarding mating/survival in the population. Cross-generational Probabilistic Survival Selection (CPSS) reduction [20,21] and fitness sharing [6] are examples of these strategies.

Examples of strategies that will disrupt this pattern of diminishing returns include: 1) the spawning of new random solutions, as they have a non-zero probability of landing outside the global population's convex hull, or 2) non-geometric crossover since it is a local non-convex transformation with a non-zero probability of creating offspring outside the global convex hull [9].

3 Related Works

Maintaining the diversity of individuals in the population received greater attention in recent years. In particular, a wide variety of methods to enhance diversity have been developed, and for a detailed review of these techniques, the reader can refer to [7]. In this section, we recall the most commonly-used methods, including Diversity Control oriented Genetic Algorithm (DCGA) and the Self-adaption Genetic Algorithm (SA). In the experimental phase, the performance of these techniques will be compared against our proposed method to assess its usefulness in improving the GA search process.

Fitness sharing is the most frequently used technique in literature. Here, population diversity is maintained via introducing a diversity function, which ensures the mitigation of unbridled head-to-head competition between widely disparate points in the solution space [6]. Another popular method is *deterministic crowding*, where the diversity issue is solved by forcing every offspring to compete with one of its parents and eventually replace it if the offspring is not worse [5]. Most existing diversity-maintenance mechanisms – as the two examples aforementioned – require problem-specific knowledge to set up specific parameters properly. A clear example is DCGA [21]. In particular, in this method, the selection criterion exploits the distance between a candidate individual and the best-performance individual and uses it (based on a probabilistic function) to produce a higher selection probability for a candidate solution with a larger distance. Hence, to ensure the effectiveness of the method, the probability function must be defined properly.

In [10], authors proposed SA to control the diversity of the population without explicit parameter setting. A self-adaptation mechanism is proposed: for controlling diversity, two measures are introduced: the *difference function*, which computes the degree of dissimilarity, and the *contribution*, which monitors the effect of the recombination.

4 Methodology

This section discusses the proposed method. Firstly, Sect. 4.1 provides details on how to perform diversity maintenance dynamically. Subsequently, Sect. 4.2 provides an in-depth description of how to couple such dynamic diversity maintenance with a combination of geometric and non-geometric crossovers.

4.1 Dynamic Diversity Maintenance

As stated above, one of the principal weaknesses of GA is the premature convergence to a solution, sometimes causing the consequent stuck of the algorithm in a local optimum. One of the recurrent ideas to overcome this issue is the maintenance of a certain level of diversity among candidate solutions. This work proposes a *self-adaptation mechanism* to control and guarantee diversity in the population and, simultaneously, to avoid the time-consuming task of setting specific hyperparameters of the GA.

As proposed in [10], a successful diversity maintenance strategy consists of looking at the population as a society divided in multiple groups. In this case, a *group* is represented by candidate solutions that share similar chromosomes. This approach promotes recombination between parents of different group and, in the meantime, disincentives crossover among individuals belonging to the same group. To formalize this concept, two quantities must be introduced: a *preference type*, which affects the characteristics of diversity in mating, and *contribution*, which measures the merit of each preference type. In this work, we rely on the same idea to determine the most suitable crossover operator to be used in the different generations of the search process.

Mating. The preference type τ is a parameter that indicates the preference of an individual to recombine with another one based on their degree of diversity. It is a positive quantity ranging between $0 \leq \tau \leq \tau_{max}$, where τ_{max} is the maximum preference type. Higher values of τ will lead, intuitively, to offspring which differ from the parents, thus encouraging diversity among members of the population.

τ is used to compute the *difference function* $\mathcal{D}$ as follows:

$$\mathcal{D}(\tau, d_i) = 0.5 + \frac{\tau}{\tau_{max}}(d_i - 0.5) \tag{1}$$

where d_i is the difference between the first selected individual x_1 and a candidate mate y_i, calculated as follows:

$$d_i = \frac{h(y_i, x_1)}{\ell} \tag{2}$$

where h is the Hamming distance between two individuals, and ℓ is the length of a chromosome.

At this point, we have all the ingredients to define how the mating is performed (i.e., for a maximization problem). Once the first individual x_1 is provided, then its recombination mate x_2 is selected in the following way:

$$x_2 = \arg\max_{i \in s_t}[f(y_i)\mathcal{D}(\tau, d_i)] \tag{3}$$

where f is the fitness function and s_t is the tournament size. Thus, a candidate who has a higher value of $\mathcal{D}$ has more chance to be selected as the second parent. Let us remark that when $\tau = 0$, the probability of selection does not depend on d, and the mating is just a fitness-based selection.

Diversity Control. Considering that the degree of diversity of the population is controlled by the preference type, it is necessary to define a procedure that correctly updates the value of τ according to the population's needs. The idea is to associate a parameter at each possible value of τ, called *contribution*, which quantifies how solutions with a given preference type produce better quality offspring. Contribution depends on t (the training epoch) and τ, and it is defined as the ratio between successful and total crossover:

$$Contribution(\tau, t) = \frac{\#SuccCross(\tau, t)}{\#Cross(\tau, t)} \tag{4}$$

At generation $t + 1$, the probability of choosing τ will be equal to its contribution at generation t. Thus, the more a preference type is associated with the creation of good-quality individuals, the higher will be the probability for it to be reused. In this work, we defined a crossover to be successful if produces at least an offspring with fitness equal to or better than both parents. This is different with respect to the approach described in [10], where authors defined as positive crossover a recombination in which the fitness of the resulting offspring is superior to the one of both the parents. In fact, in the (rare) scenario where the offspring and its best parent share the same fitness value and do not coincide, the introduction of a new solution may produce a positive effect on the search process. An example is the presence of a plateau – a part of the space in which all points have the same fitness score – in the fitness landscape.

More precisely, the diversity control procedure works as follows:

1. Randomly generates the initial population of individuals and evaluates their fitness.
2. Initialize the contribution equally for each preference type.
3. Select an individual and its partner with the aforementioned mating procedure. Recall that the probability of choosing τ is equal to its contribution.
4. Create two new individuals by crossover and evaluate their fitness.
5. Repeat step **4** and **5** for the whole population.
6. Compare the fitness of the offspring with the one of their parents. Update the contribution values consequently.
7. Repeat step **3–6** until a termination criterion is satisfied.

In this work, to maintain all the preference type values throughout the search process, we impose a minimum threshold of 10% for each contribution.

4.2 Self-adaptive Crossover

In Sect. 2.1, we introduced two different definitions for crossover:

- *Geometric crossover*, which, if employed alone, decreases the size of the population convex hull, thus being a diversity reducer.
- *Non-geometric crossover*, which has the ability to behave as a diversity enhancer method.

By applying the mating routine described in Sect. 4.1, our objective is to obtain a method that allows us to self-adapt the choice of crossover operators in the GA algorithm depending on the specific stage of the search process. The idea is that most recombination will still be geometric, but occasionally non-geometric crossover will be applied to avoid the situation of premature convergence where the global optima may not be contained in the convex hull. Intuitively, this combination can be described as a phenomenon of *conspansion*, i.e., material contraction during space expansion. Contraction is a consequence of the geometric crossover, while expansion is a consequence of the non-geometric one.

As mentioned above, the procedure described in Sect. 4.1 can be adequately modified for the selection of the most performing type of crossover. Specifically, non-geometric crossover will be chosen if the parent x_1 has preference type $\tau = \tau_{max}$, otherwise geometric one will be used. In fact, a high preference type indicates the need to augment the diversity in the population. Therefore, when τ assumes the highest possible value, the crossover technique that is a diversity enhancer must be selected.

To summarize, when the convex search starts leading to negative effects on population phenotype, the contribution parameter values associated with preference types related to geometric crossover start to decrease. Non-convex search, as a consequence, will be selected with more probability, as the share of contribution of the preference type τ_{max} (i.e., the one linked to non-geometric crossover) will increase. On the other hand, if the non-geometric crossover causes a worsening in the individuals' fitness, the geometric crossover will be preferred by reducing non-geometric crossover contribution.

Let us emphasize the fact that the expansion of the global convex hull is not ensured at each step. Firstly, as the choice of which crossover to use is non-deterministic, it can simply not occur during a generation. Secondly – and more important – the individuals produced by non-geometric crossover are not necessarily outside the global convex hull.

We propose and investigate two variants for the self-adaptive crossover introduced, namely P and P'.

P'

1. Tournament selection, size= 3
2. Eliminate tournament winner x_1 from population
3. Difference function tournament, size=length of the population
4. Eliminate winner x_2 from population
5. Return children (y_1, y_2)

P

1. Tournament selection, size= 3
2. Difference function tournament, size= 3
3. Return children (y_1, y_2)

The main difference between these two variants lies in the selection technique. In particular, we want to study how different approaches in the choice of the second parent affect the algorithm. In fact, once the first parent x_1 is fixed, P randomly selects only a limited set of candidates and computes the difference function over this set. On the other hand, P' computes the difference function for all the individuals of the population to find the perfect fit x_2.

5 Experimental Settings

Table 1. Definitions and optimum values (minimum) of the CEC 2017 benchmark functions.

	No.	Functions	Opt.
Unimodal functions	1	Shifted and Rotated Bent Cigar	100
	2	Shifted and Rotated Sum of Different Power	200
	3	Shifted and Rotated Zakharov	300
Simple multimodal functions	4	Shifted and Rotated Rosenbrock	400
	5	Shifted and Rotated Rastrigin	500
	6	Shifted and Rotated Expanded Schaffer F6	600
	7	Shifted and Rotated Lunacek Bi-Rastrigin	700
	8	Shifted and Rotated Non-Continuous Rastrigin	800
	9	Shifted and Rotated Levy	900
	10	Shifted and Rotated Schwefel	1000
Hybrid functions	11	Zakharov; Rosenbrock; Rastrigin	1100
	12	High-conditioned Elliptic; Modified Schwefel; Bent Cigar	1200
	13	Bent Cigar; Rosenbrock; Lunacek bi-Rastrigin	1300
	14	High-conditioned Elliptic; Ackley; Schaffer F7; Rastrigin	1400
	15	Bent Cigar; HGBat; Rastrigin; Rosenbrock	1500
	16	Expanded Schaffer F6; HGBat; Rosenbrock; Modified Schwefel	1600
	17	Katsuura; Ackley; Expanded Griewank plus Rosenbrock; Schwefel; Rastrigin	1700
	18	High-conditioned Elliptic; Ackley; Rastrigin; HGBat; Discus	1800
	19	Bent Cigar; Rastrigin; Griewank plus Rosenbrock; Weierstrass; Expanded Schaffer F6	1900
	20	HappyCat; Katsuura; Ackley; Rastrigin; Modified Schwefel; Schaffer F7	2000
Composition functions	21	Rosenbrock; High-conditioned Elliptic; Rastrigin	2100
	22	Rastrigin; Griewank; Modified Schwefel	2200
	23	Rosenbrock; Ackley; Modified Schwefel; Rastrigin	2300
	24	Ackley; High-conditioned Elliptic; Griewank; Rastrigin	2400
	25	Rastrigin; HappyCat; Ackley; Discus; Rosenbrock	2500
	26	Expanded Schaffer F6; Modified Schwefel; Griewank; Rosenbrock; Rastrigin	2600
	27	HGBat; Rastrigin; Modified Schwefel; Bent Cigar; High-conditioned Elliptic; Expanded Schaffer F6	2700
	28	Ackley; Griewank; Discus; Rosenbrock; HappyCat; Expanded Schaffer F6	2800
	29	15; 16; 17	2900
	30	15; 18; 19	3000

The set of functions used, described in Table 1, is the *CEC 2017* function suite [1] for single-objective real-parameter numerical optimization. The suite is composed of unimodal, multi-modal, hybrid, and composition functions that are shifted, rotated, and non-separable. Their characteristics of noise and ruggedness make them excellent candidates for studying the effectiveness of the proposed approach, as they require different degrees of diversity in the population to be solved efficiently. The search space is $[-100, 100]^D$. $D = 10, 30$ are investigated in this experimental phase.

Table 2. Experimental settings. All the values of the hyperparameters coincide for $D = 10, 30$, except for the length of the chromosome, which is 200 in the former case and 600 in the second one.

Parameter	Value
Population size	400
Length of chromosome	{200,600} bits
Number of generation	200
Number of independent run	30
Crossover probability (Pc)	100%
Mutation rate (Pm)	0%
Tournament size	3

To assess the performance of the proposed method, we considered a GA with a population size equal to 400 and a search process that runs for 200 generations. Thus, the total number of fitness evaluations for each experiment is equal to $MaxFES = 80000$, i.e., the product between these two quantities. In this experimental phase, we decided to concentrate our attention only on crossover operators, whereas mutation is not allowed, to fully understand how the introduction of non-geometric crossover can impact the overall performance of the algorithm. As the algorithm is stochastic, 30 runs have been performed for each benchmark function. Further details concerning the implementation of the GA are reported in Table 2.

The results obtained by the two variants P and P' of the self-adaptive crossover are compared with:

- The vanilla GA.
- Two variants of DCGA: DCGA1 and DCGA2. In particular, we considered the following DCGA parameters [21] values: for DCGA1, $c = 0.01$, and $a = 0.19$, while for DCGA2, $c = 0.234$, $a = 0.5$.
- SA, the self-adaptive GA outlined in Sect. 4.1.

This is a relatively broad group of techniques as it considers vanilla GA, variants of DCGA (which uses a static strategy to maintain diversity), and finally self-adaptive algorithms: SA (where the crossover operator is fixed) and P and P' (where there is a choice between geometric and non-geometric operators).

Regarding the self-adaption mechanism, we use $4 - (\tau \in 0, \ldots, 3)$ – preference levels both for SA, P, and P' (Figs. 4 and 5).

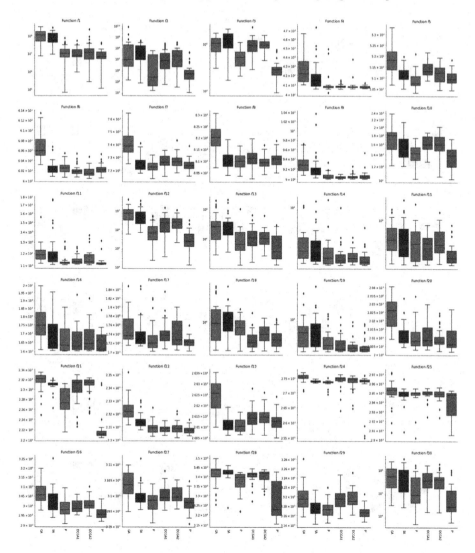

Fig. 2. Median of the fitness over the 30 independent runs for the considered benchmark problems – $D = 10$.

6 Experimental Results

As stated in Sect. 4.2, the goal of this study is to compare the performance of our algorithm against a wide range of well-known GA-based methods. The experimental results, computed over 30 independent runs, are reported through box-plot for $D \in \{10, 30\}$ in Fig. 2 and Fig. 3. Experimental results show that the proposed method generally outperforms the other competitors in the vast majority of the benchmark functions. Specifically, at least one algorithm between

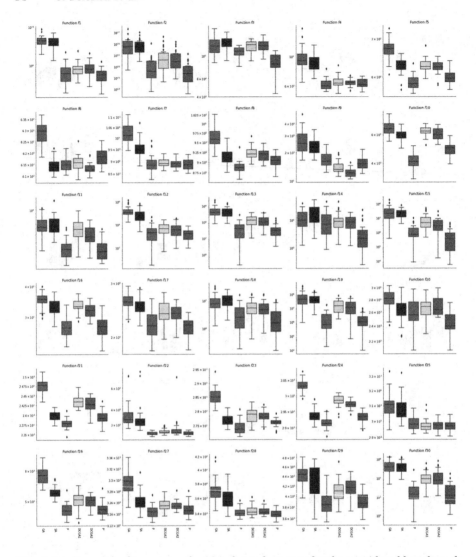

Fig. 3. Median of the fitness over the 30 independent runs for the considered benchmark problems – $D = 30$

P and P' leads to an improvement of the fitness in 25 functions of the 10-th dimensional case and in 27 functions of the 30-th dimensional one. The fitness gap between our techniques and the other methods taken into account increases together with the dimension of the problems, suggesting that our algorithm is particularly suitable for solving challenging optimization problems in higher dimensions. On the other hand, when the results of P and P' are compared, we obtain that for $D = 10$ P' seems to achieve better fitness values, while for $D = 30$ the two methods show comparable performances.

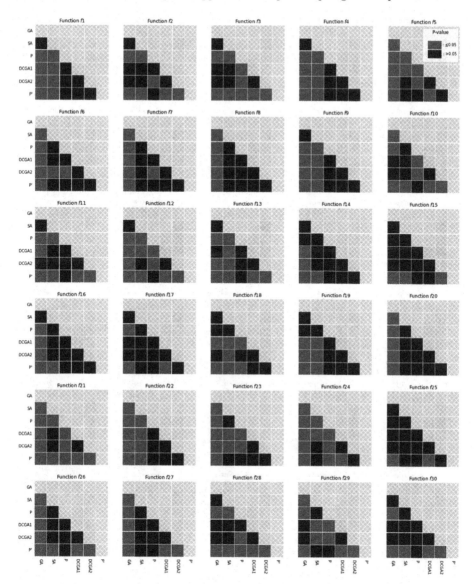

Fig. 4. P-values returned by the Mann-Whitney U test with the Bonferroni correction for each of the 30 functions ($D = 10$). Green denotes a p-value for which the alternative hypothesis cannot be rejected. Red denotes a p-value for which the null hypothesis (i.e., equal median) cannot be rejected. (Color figure online)

To investigate whether our method significantly outperforms the others, the Mann-Whitney U statistical test (computed considering a significance level $\alpha = 0.05$ and the Bonferroni correction [14]) results are displayed in Fig. 2 for $D = 10$, and in Fig. 3 for $D = 30$. Based on these results, it is possible to confirm that the

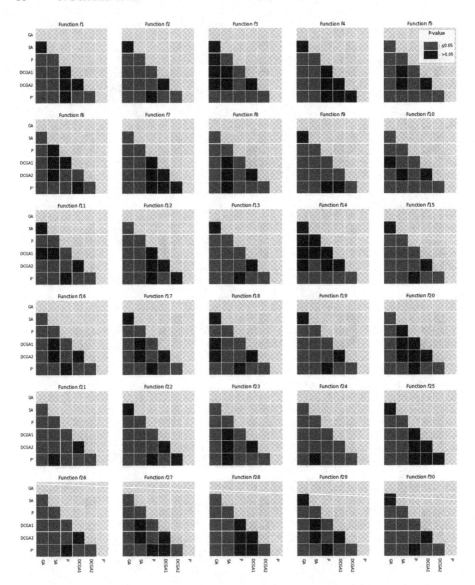

Fig. 5. P-values returned by the Mann-Whitney U test with the Bonferroni correction for each of the 30 functions ($D = 30$). Green denotes a p-value for which the alternative hypothesis cannot be rejected. Red denotes a p-value for which the null hypothesis (i.e., equal median) cannot be rejected. (Color figure online)

proposed technique can produce better performance with respect to the other competitors – and with a statistical significance – in the vast majority of the considered benchmarks.

Again, Mann-Whitney U statistical tests confirm that we obtain better results in $D = 30$ w.r.t. $D = 10$, confirming the hypothesis that the proposed self-adaptive method improves the search for the optimal value when the dimension – and thus, the difficulty of the problem – increases.

7   Conclusions

GAs are a popular technique in the EAs family. Despite their success in different domains, they suffer from a premature convergence problem, where the final population's convex hull might not include the global optimum. Population diversity maintenance strategies are fundamental to counteract this problem, but the typical GA convex search still reduces the population's convex hull over the generations. Thus, the idea of this work is the definition of a self-adaptive method for counteracting the reduction of the convex hull produced by the application of geometric crossover and, at the same time, preserving population diversity. The paper proposed a self-adaptive method for using geometric and non-geometric crossover operators in different stages of the search process based on the information provided by the current candidate solutions and accommodating the ever-evolving necessities of the underlying search space/problem topology. To assess the performance of the proposed approach, an extensive experimental phase was performed considering the CEC 2017 benchmark suite and comparing our proposal against the vanilla GA and popular diversity maintenance techniques. Experimental results clearly show the superior performance of the proposed method and the advantage provided by using both geometric and non-geometric crossover in the search process. In the future, we plan to test the proposed approach with different EAs and crossover operators.

References

1. Awad, N., Ali, M., Liang, J., Qu, B., Suganthan, P., Definitions, P.: Evaluation criteria for the cec 2017 special session and competition on single objective real-parameter numerical optimization. Technical report (2016)
2. Bäck, T., Fogel, D.B., Michalewicz, Z.: Handbook of evolutionary computation. Release **97**(1), B1 (1997)
3. Castelli, M., Manzoni, L., Gonçalves, I., Vanneschi, L., Trujillo, L., Silva, S.: An analysis of geometric semantic crossover: a computational geometry approach. In: IJCCI (ECTA), pp. 201–208 (2016)
4. Castelli, M., Manzoni, L., Vanneschi, L., Silva, S., Popovič, A.: Self-tuning geometric semantic genetic programming. Genet. Program. Evolvable Mach. **17**(1), 55–74 (2016)
5. Friedrich, T., Oliveto, P.S., Sudholt, D., Witt, C.: Analysis of diversity-preserving mechanisms for global exploration. Evol. Comput. **17**(4), 455–476 (2009)
6. Goldberg, D.E., Richardson, J., et al.: Genetic algorithms with sharing for multi-modal function optimization. In: Genetic algorithms and their applications: Proceedings of the Second International Conference on Genetic Algorithms, vol. 4149. Lawrence Erlbaum, Hillsdale, NJ (1987)

7. Gupta, D., Ghafir, S.: An overview of methods maintaining diversity in genetic algorithms. Int. J. Emerg. Technol. Adv. Eng. **2**(5), 56–60 (2012)
8. Holland, J.H.: Genetic algorithms. Sci. Am. **267**(1), 66–73 (1992)
9. Ishibuchi, H., Tsukamoto, N., Nojima, Y.: Diversity improvement by non-geometric binary crossover in evolutionary multiobjective optimization. IEEE Trans. Evol. Comput. **14**(6), 985–998 (2010)
10. Jassadapakorn, C., Chongstitvatana, P.: Self-adaptation mechanism to control the diversity of the population in genetic algorithm. arXiv preprint arXiv:1109.0085 (2011)
11. Lim, S.M., Sultan, A.B.M., Sulaiman, M.N., Mustapha, A., Leong, K.Y.: Crossover and mutation operators of genetic algorithms. Int. J. Mach. Learn. Comput. **7**(1), 9–12 (2017)
12. Lin, W.Y., Lee, W.Y., Hong, T.P.: Adapting crossover and mutation rates in genetic algorithms. J. Inf. Sci. Eng. **19**(5), 889–903 (2003)
13. Maan, V., Malik, A.: Genetic algorithm application on 3D pipe routing: a review. Recent Innov. Comput. 139–148 (2022)
14. McKnight, P.E., Najab, J.: Mann-Whitney u test. Corsini Encycl. Psychol. 1 (2010)
15. Moraglio, A.: Towards a geometric unification of evolutionary algorithms. Ph.D. thesis, Department of Computer Science, University of Essex, UK (2007)
16. Moraglio, A.: Abstract convex evolutionary search. In: Proceedings of the 11th Workshop Proceedings on Foundations of Genetic Algorithms, pp. 151–162 (2011)
17. Moraglio, A., Poli, R.: Inbreeding properties of geometric crossover and non-geometric recombinations. In: Stephens, C.R., Toussaint, M., Whitley, D., Stadler, P.F. (eds.) FOGA 2007. LNCS, vol. 4436, pp. 1–14. Springer, Heidelberg (2007). https://doi.org/10.1007/978-3-540-73482-6_1
18. Moraglio, A., Togelius, J., Silva, S.: Geometric differential evolution for combinatorial and programs spaces. Evol. Comput. **21**(4), 591–624 (2013)
19. Sharma, S., Kumar, V.: Application of genetic algorithms in healthcare: a review. Next Generation Healthcare Informatics, pp. 75–86 (2022)
20. Shimodaira, H.: DCGA: a diversity control oriented genetic algorithm. In: Second International Conference On Genetic Algorithms in Engineering Systems: Innovations and Applications, pp. 444–449 (1997). https://doi.org/10.1049/cp:19971221
21. Shimodaira, H.: A diversity-control-oriented genetic algorithm (DCGA): performance in function optimization. In: Proceedings of the 2001 Congress on Evolutionary Computation (IEEE Cat. No.01TH8546), vol. 1, pp. 44–51 (2001). https://doi.org/10.1109/CEC.2001.934369
22. Srinivas, M., Patnaik, L.M.: Adaptive probabilities of crossover and mutation in genetic algorithms. IEEE Trans. Syst. Man Cybern. **24**(4), 656–667 (1994)
23. Srinivas, M., Patnaik, L.M.: Genetic algorithms: a survey. Computer **27**(6), 17–26 (1994)
24. Vanneschi, L., Henriques, R., Castelli, M.: Multi-objective genetic algorithm with variable neighbourhood search for the electoral redistricting problem. Swarm Evol. Comput. **36**, 37–51 (2017)

A Genetic Programming Encoder for Increasing Autoencoder Interpretability

Finn Schofield, Luis Slyfield, and Andrew Lensen(✉)

School of Engineering and Computer Science, Victoria University of Wellington,
PO Box 600, Wellington 6140, New Zealand
{schofifinn,slyfieluis,andrew.lensen}@ecs.vuw.ac.nz

Abstract. Autoencoders are powerful models for non-linear dimensionality reduction. However, their neural network structure makes it difficult to interpret how the high dimensional features relate to the low-dimensional embedding, which is an issue in applications where explainability is important. There have been attempts to replace both the neural network components in autoencoders with interpretable genetic programming (GP) models. However, for the purposes of interpretable dimensionality reduction, we observe that replacing only the encoder with GP is sufficient. In this work, we propose the Genetic Programming Encoder for Autoencoding (GPE-AE). GPE-AE uses a multi-tree GP individual as an encoder, while retaining the neural network decoder. We demonstrate that GPE-AE is a competitive non-linear dimensionality reduction technique compared to conventional autoencoders and a GP based method that does not use an autoencoder structure. As visualisation is a common goal for dimensionality reduction, we also evaluate the quality of visualisations produced by our method, and highlight the value of functional mappings by demonstrating insights that can be gained from interpreting the GP encoders.

Keywords: Genetic programming · Autoencoder · Dimensionality reduction · Machine learning · Explainable artificial intelligence

1 Introduction

Dimensionality reduction (DR) involves taking high-dimensional data and producing a representation of it in a space with considerably less dimensions. This is beneficial both as a pre-processing step to improve the effectiveness or tractability of using the data for machine learning, and to gain a better understanding of the data during exploratory data analysis (EDA), for example visualising the data in two or three dimensions. Non-linear dimensionality reduction (NLDR) [11] is of particular interest for data with more complicated relationships that can only be expressed by non-linear relationships. Many existing NLDR methods are able to produce low-dimensional representations of complex datasets while retaining much of the original information [6,17]. However the best performing

G. Pappa et al. (Eds.): EuroGP 2023, LNCS 13986, pp. 19–35, 2023.
https://doi.org/10.1007/978-3-031-29573-7_2

methods have two main drawbacks: they only produce the new data points in low-dimensional space without a means to process new points without retraining, and the way in which they produce the embedding is difficult for a human to directly check and interpret.

Having interpretable models is important for a variety of reasons. One reason is for tasks with human impact such as in medical settings, where there is an ethical obligation to explain any decisions made. Another is as governments move to legislate around the use of artificial intelligence, there is increasing legal obligation for model interpretability. In the case of NLDR, an interpretable model means we can build a clearer picture of how the data in the constructed lower dimensional space relate to the real data in the original space, and as a result have more confidence using the method in settings where this is important. This also makes NLDR a more effective tool for EDA, as we can use both the low dimensional representation and the transformation itself to understand more about the data and the relationships within it.

An autoencoder (AE) is a type of artificial neural network architecture that performs NLDR [5]. It has two main components: an encoder that transforms from the space containing the original data into a space with considerably less dimensions, and a decoder that transforms from the low dimensional space back to the higher one. An AE is an appealing method to use for NLDR because its neural network architecture is capable of representing powerful relationships, it produces a functional mapping that makes it possible to easily transform previously unseen data points into the latent space, and its ability to reconstruct the original data is an appealing metric of success at the DR task. However to achieve high performance, AE architectures usually have far too many parameters to be interpretable without using external methods to provide an explanation.

Genetic programming (GP) is an evolutionary computation (EC) technique where computer programs are evolved over generations [2,18]. GP has inherent potential for interpretability, because it evolves solutions that combine user-selected terminals and functions. GP has recently been demonstrated to be a capable NLDR technique which produces functional mappings [12,14]. Some of these approaches have used a multi-tree GP representation with a custom fitness function for evaluating embedding quality [12–14,22,24], and other work has looked into GP specifically for autoencoding [16,19].

Combining GP and autoencoding is an appealing concept: it has the potential to benefit both from the power of AEs and the inherent interpretability of GP. Existing research into GP for autoencoding has attempted to evolve both the encoder and decoder structure using GP, but has struggled due to the strong dependency between the encoder and decoder, limiting performance [16]. Other attempts instead forgo the architecture, instead using a fully tree-based approach which gives representations that are complex and difficult to interpret [19].

In this work we propose an AE-based approach to NLDR that replaces the encoder component with a multi-tree structure trained by GP but keeps the neural network decoder. We suggest that due to demonstrated suitability of multi-tree GP to NLDR in general, that by replacing *only* the encoder with a multi-tree GP individual whilst retaining the ANN decoder, the value of GP interpretability can be harnessed while avoiding the problems of evolving both the encoder and decoder simultaneously but keeping the benefit of the AE training method and its potentially powerful neural network decoder.

The contributions of this work are summarised by the research goals:

- Investigate existing work relevant to GP for NLDR and autoencoding;
- Propose a novel autoencoding method that replaces the encoder with a multi-tree GP individual while retaining the ANN decoder;
- Evaluate how the method compares at the task of NLDR to conventional AEs and GP-based NLDR;
- Compare how the method performs at visualisation to the baselines; and
- Investigate what potential the method has for interpretability.

1.1 Structure

The rest of the paper is structured as follows. Section 2 discusses related work to provide a context for the method we present. Section 3 outlines our proposed method for creating an autoencoder with a genetic programming encoder. Section 4 describes the experiment design used to test the method, and Sect. 5 the results of running these tests. Further analysis including visualisation examples are presented in Sect. 6, and the paper is concluded in Sect. 7.

2 Background and Related Work

2.1 Non-linear Dimensionality Reduction

While traditional dimensionality reduction techniques such as PCA [7] can be sufficient for producing high-quality embeddings of data, often the underlying structure of a dataset is too complex to be captured by linear combinations and transformations. For these datasets, Non-Linear Dimensionality Reduction (NLDR) techniques are required. These are also sometimes referred to as manifold learning techniques [1].

NLDR techniques can be divided into two classes: mapping and non-mapping. *Mapping* techniques are those which produce the data embeddings in the low-dimensional space, as well as a functional mapping to produce them from the high-dimensional space. *Non-mapping* techniques on the other hand provide only the low-dimensional embedding.

Having access to a mapping has a few key advantages. Firstly, it allows for better interpretation of how the dimensionality reduction has been achieved by being able to identify which of the original features are important to the found embedding. Secondly, it allows for new instances of the data to be placed in the low-dimension space without the need to re-run the DR algorithm again.

A canonical example of a high performing NLDR algorithm is t-distributed Stochastic Neighborhood Embedding (t-SNE) [6]. t-SNE works by constructing a probability distribution over pairs of instances in the original feature space, such that instances close together have a high probability. Then, a second probability distribution is constructed over the instances in the desired low-dimensional space. Lastly t-SNE minimises the Kullback-Liebler (KL) divergence between the two distributions with respect to the locations of the instances in the low dimensional space, resulting in the final embedding. The more recent state-of-the-art Uniform Manifold Approximation and Projection (UMAP) [17] follows a similar process to t-SNE. However, instead of using probability distributions and minimising the KL divergence, UMAP uses fuzzy graph representations of the data in the high dimensional and low dimensional space.

Both t-SNE and UMAP are non-mapping. There have been parametric variations proposed for both that use neural networks to allow for reusable mappings, although they are still extremely complex and difficult to interpret [15,21].

2.2   Evolutionary Computation for Dimensionality Reduction

Various EC techniques have been applied to DR. The most straightforward DR task, feature selection, is simply isolating the most important features. This has been approached by a range of EC techniques, such as bit-string genetic algorithms [10], particle swarm optimisation (PSO) [26], and ant colony optimisation (ACO) [8].

The more complex problem of feature construction involves creating a reduced number of new features using the original features as components for transformation. For this task, the programmatic structure of GP is an obvious candidate. By using the original input features as the GP terminal set, and setting an appropriate fitness function, GP can learn high-performing combinations of features in an explainable way with little constraint on the form of the learned functions. The functional structure of GP trees lends itself to produce not just mappings which are reusable, but also ones that are interpretable.

Genetic programming has been proposed as a potential approach to learn functional mappings for non-linear dimensionality reduction, such as in *Genetic Programming for Manifold Learning* (GP-MaL), which proposed the use of GP for NLDR [12] without the use of an AE architecture. GP-MaL uses a multi tree representation, with w trees to represent w dimensions of the embedding. The fitness function used by GP-MaL is based on the preservation of orderings of neighbours from the original feature space to the embedding. The GP-MaL fitness is somewhat ad-hoc: it uses a particular formulation of neighbour preservation. Later work has proposed other fitness functions, such as that used by the state-of-the-art non-mapping UMAP [17] method [22].

2.3   Genetic Programming for Autoencoding

There are some existing methods that incorporate GP with an AE framework.

The *Genetic Programming Autoencoder* (GPAE) replaced the AE entirely with a linear GP representation [16]. Each individual is comprised of two linear GP programs that represent an encoder and a decoder. Instead of a population-based search, GPAE mutates a single individual using hill-climbing. The multi-tree approach is ruled out due to the inability to share calculations between trees. In fact, given that it is desirable to have minimal redundancy (shared information) between the embedding dimensions, sharing calculations may be a *downside* of GPAE.

Structurally Layered Genetic Programming (SLGP) [19] uses a dual-forest representation, with w trees to construct the embedding and v trees to reconstruct the original input. Representing the decoder as a forest requires a tree for each original feature, which is difficult to train on high-dimensional data. To avoid this, SLGP decomposes the problem to smaller, independent GP runs, each of which considers a subset of features. Thus, the mapping can only take into account combinations of original features which are in the same subset. Learning a tree for each original dimension is also very expensive.

Genetic Programming for Feature Learning (GPFL) is an AE-like approach to feature learning using GP [20]. Feature learning is a tangential task to NLDR that involves learning representations of image data. While GPFL uses a multi-tree representation, it learns the trees sequentially, with each subsequent tree correcting earlier errors. The final individual is a linear combination of the trees. One drawback of GPFL is the indirect model structure. A major potential key benefit to using GP for autoencoding is the ability to produce a clear functional mapping to the low dimension space, which GPFL does not provide.

3   Proposed Method: GPE-AE

After considering the existing work on the subject discussed above, we developed the novel Genetic Programming Encoder for Autoencoding (GPE-AE) method. The overall design of GPE-AE is presented in Fig. 1. Here, an example of learning a 3-dimension embedding of n features is used. Taking the original input dataset with features f, these are used as inputs to a multi-tree GP individual to produce a lower dimension embedding W, where w_i is dimension i of the embedding. W is then used as the input for the ANN decoder, while the original features f are used as training targets. Once the decoder has been trained, it can then output a prediction of the original features f'. This prediction can be used to evaluate the quality of the embedding, and thus the quality of the GP individual.

3.1   GP Representation of Encoder

We use a multi-tree GP representation with each individual being comprised of w trees. Each of these trees represents a functional mapping of the f inputs to a single dimension of the w-dimension space. This is consistent with previous work that showed success with a multi-tree approach [12]. While other work has argued against a multi-tree representation for autoencoding due to its inability to share calculations between trees [16], we argue that for the purposes of dimensionality reduction, separating calculations is actually a strength. In theory, each dimension of embedding should have as little shared information as possible, to ensure they are capturing independent parts of the underlying distribution.

Our GP encoder representation uses 12 functions, as shown in Table 1. In addition to standard arithmetic operators $(+, -, \times)$, we utilise: absolute addition

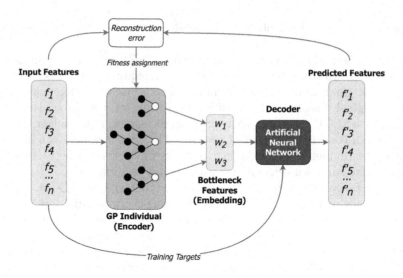

Fig. 1. An overview of GPE-AE, with n features reduced to a 3D embedding.

Table 1. Functions used by GPE-AE. All functions take/produce numeric values.

Category	Arithmetic							Logical			Non-Linear	
Function	$+$	$5+$	$-$	$\lvert+\rvert$	$\lvert-\rvert$	$\times$	$\%$	max	min	if	ReLU	sigmoid
No. Inputs	2	5	2	2	2	2	2	2	2	3	1	1

($\lvert+\rvert$) and subtraction ($\lvert-\rvert$), an addition function which takes 5 inputs (5+), and protected division (%) which returns 1 when the denominator is zero. Absolute arithmetic operators allow for the easier comparison of magnitudes of inputs. The 5+ function allows for more aggressive combination of sub-trees in a more space efficient way. Three logical operators are also included: max, min, and if. if takes three inputs — if the first input is greater than zero, it outputs the second input; otherwise it outputs the third. These allow for more expressive use of the original features beyond arithmetic combinations. To allow for further non-linear transformations, the ReLU and sigmoid functions are used. These are commonly used as activation functions in neural networks, adding the capacity for non-linear learning. Existing GP for NLDR work has also used these [12,14].

The GP terminals used are the original f features of the data, as well as ephemeral random constants (ERCs). ERCs are random values uniformly sampled over the range $[-1,1]$, and remain constant over the evolution once initialised. The use of ERCs allows for scaling and offsetting of features.

Multi-tree GP requires adapting traditional tree-based GP crossover and mutation. In this work, we use *All Index Crossover (AIC)*, where standard crossover is performed on all pairs of trees (with the same index within the multi-tree representation) across each parent. For mutation, we perform standard GP mutation on a randomly selected tree.

3.2   Fitness Evaluation

The fitness of GPE-AE individual I with w trees occurs is evaluated as follows:

1. The features of input data X are used as inputs for I, producing the embedding W with w dimensions.
2. W is used as input to the ANN decoder, with X serving as training targets.
3. Once training of the decoder is complete, a final prediction X' is made using W as the input to the trained model.
4. The reconstruction error is calculated between the original data X and the reconstruction X', which is assigned to I as the fitness.

As in standard AEs, the objective/fitness function is the reconstruction error between the inputs and the predicted outputs. We use root mean squared error (RMSE), as it better penalises large errors [3]. RMSE is defined as:

$$RMSE = \sqrt{\frac{\sum_{i=1}^{n}(\mathbf{x}_i' - \mathbf{x}_i)^2}{n}} \tag{1}$$

for n instances, where $\mathbf{x}_i$ is the ith instance of the input, and $\mathbf{x}_i'$ is the predicted value of instance i (after encoding and decoding).

3.3   Decoder Architecture

The architecture of the ANN decoder requires special consideration. In a conventional AE, it is common to use a "funnel" architecture, where the encoder has hidden layers with a decreasing number of neurons, with the decoder reflecting the encoder. GPE-AE, however, uses a dynamically structured GP encoder. For GPE-AE, we propose the use of a simple multi-layer perceptron for the decoder. The input layer has w inputs, one for each embedding dimension. The output layer has f outputs, one for each of the features of the data in the original space.

In GPE-AE, the role of the ANN decoder is only to evaluate the performance of the GP encoder, and such a decoder needs to be trained for each evaluation (of which there can be many). Therefore, we do not need the decoder to be *perfect*, but merely to be able to reliably distinguish the performance of encoder candidates in a consistent way, and training needs to be relatively short to keep the overall GPE-AE running time computationally feasible. As such, we suggest the use of simpler decoder architectures. This also has the benefit of pressuring the evolution process towards better-structured embeddings: if a reasonably simple decoder cannot sufficiently reconstruct the input, then the embedding is likely ill-formed.

As ANN training is stochastic, we use a hash of the encoder as the seed for training the decoder, ensuring a GP individual always has the same fitness.

4   Experiment Design

To examine the role the complexity of the decoder plays in the performance of GPE-AE, we perform experiments with decoders with either 1, 2, or 3 hidden

Table 2. The NN architectures used by GPE-AE and CAE in the experiments. GPE-AE only makes use of the decoder.

No. Hidden Layers	Encoder Arrangement	Decoder Arrangement
1	[128]	[128]
2	[128, 64]	[64, 128]
3	[128, 64, 32]	[32, 64, 128]

Table 3. GP parameters used for GPE-AE and GP-MaL.

Parameter	Setting	Parameter	Setting
Generations	1000	Pop.Size	100
Mutation	20%	Crossover	80%
Elitism	top 10	Pop. Init	Half-and-half
Selection	Tournament	Tourn. Size	7
Min. Tree Depth	2	Max. Tree Depth	8

layers. The number of neurons at each layer is presented in Table 2. We follow the common "funnel" architecture of AEs, where the decoder incrementally expands the data from the bottleneck to the final reconstruction. As the depth of the decoder increases, so does the number of connections, allowing for more complex decoding structures. However, introducing more connections has a significant impact on the computational cost of evaluating fitness.

We are also interested in how GPE-AE performs on NLDR tasks of varying difficulty. To evaluate this, we perform experiments across a range of embedding sizes (1, 2, 3, 5, and 10), which represent decreasing levels of challenge.

Standard GP parameters used by GPE-AE for all experiments are shown in Table 3. For the decoder, standard neural network parameters are used, as follows. A limit of 100 epochs is used to reduce computational cost, as a NN is required to be trained for each fitness evaluation. As the decoders are reasonably simple, this was considered sufficient. The ReLU activation is used to add non-linearity between hidden layers (as in GPE-AE). The Adam optimiser was used as it performs well in similar problems [9]. A learning rate of 0.001 and mini-batch size of 200 were found to be sufficient in exploratory testing.

For each dataset, embedding size, and number of hidden layers, we perform 30 runs using GPE-AE and both of the comparison methods stated in Sect. 4.1. This accounts for the stochastic nature of the evolutionary process.

4.1   Comparison Methods

To evaluate the performance of our proposed GPE-AE method, we compare it to two relevant baselines: a conventional ANN auto-encoder (CAE) and GP-MaL.

Conventional Auto-Encoder: we perform experiments using the same hidden layer configurations as GPE-AE. The architecture of the encoder mirrors the decoder, as shown in Table 2. By mirroring the decoder, we can be sure that any structure capable of being found by the encoder is capable of being reversed by the decoder. We train the CAE using the same standard NN hyper-parameters as we use for the GPE-AE decoder, and with the same MSE objective function.

GP-MaL: Our motivation behind GPE-AE is to use the AE structure to produce functional mappings for NLDR with GP. Thus, it is also valuable to compare it to another multi-tree GP NLDR method, such as GP-MaL. This allows us to test the hybrid GPE-AE method against both a pure CAE method and a pure multi-tree GP method. Our GP-MaL experiments use the same parameters and terminal/function sets as in GPE-AE.

4.2   Evaluation Measures

As NLDR and autoencoding are unsupervised tasks, there is no "gold-standard" objective measure to compare different methods. For GPE-AE and the CAE, we can directly compare their reconstruction error. However, GP-MaL does not perform reconstruction. While we could use the GP-MaL fitness function to compare all methods, this would then favour GP-MaL, which used it as the optimisation criterion. To avoid this, we propose using the classification accuracy obtained using the low-dimension embedding. This approach has been used in previous GP for NLDR work [12,14], and relies on the assumption that the data labels are important to the structure of the data. We argue that this assumption generally holds, as the ability of the classification algorithm to separate data in a low-dimensional space indicates that important structure within the data has been retained in the embedding.

As this is unsupervised learning, the data labels are not given to GPE-AE or our comparison methods, eliminating bias towards any particular approach. The classification algorithm used for evaluation (after the embeddings have been learned) is the scikit-learn Random forest implementation, using 100 trees. Random forest is an efficient and robust algorithm, making it a good choice for unbiased evaluation [23]. We calculate the classification accuracy using 10-fold-cross-validation on the low dimensional embedding.

Measuring Complexity. To measure the complexity of the models, we can calculate the number of *connections* that the GP trees and the ANN encoder have. As both are directed graphs, we are counting the number of edges in each. For a GP individual, this is $|nodes| - w$, for w roots (trees), as each node except the root have a single parent. For a fully-connected neural network, this is defined by the equation $\sum_{i \in L}^{L-1} L_i L_{i+1}$, where L is the number of layers in the network, and L_i is the number of nodes as layer i. The number of connections approximately represents the complexity or uninterpretability of a model.

4.3   Datasets

The datasets used are presented in Table 4. Clean1 is from openML [25], while the rest are from the UCI Repository [4]. The selected datasets have a range of different dimensionalities, classes and instances to evaluate the performance of GPE-AE across different problems.

Table 4. Datasets used for testing.

Dataset	Instances	Features	Classes
Clean1	476	168	2
Dermatology	358	34	6
Ionosphere	351	34	2
Segmentation	2310	19	7
Wine	178	13	3

5   Results

We first compare the three methods (GPE-AE, CAE, and GP-MaL) in terms of their classification accuracy and complexity (number of connections). These results are shown in Table 5. The results are grouped vertically by the dimensionality of the embedding. For GPE-AE and the CAE, there are three rows per dimensionality, for the three different configurations of hidden layers. For example, GPE 3HL is GPE-AE with a 3-hidden layer decoder, while CAE 2HL is a conventional autoencoder with two hidden layers in both the decoder and encoder. The number of neurons at each layer was presented in Table 2. GP-MaL does not use a decoder architecture, and so has only one row per dimensionality.

Classification Accuracy: The mean classification across the 30 runs for each method on each dataset is presented in the Accuracy column. A Wilcoxon significance test was performed with a p-value of 0.05. The tests were performed using each configuration of hidden layers, with the GPE-AE and CAE methods being tested for each configuration and dataset. A "+" next to a GPE-AE accuracy indicates that GPE-AE significantly outperformed CAE with the same hidden layer configuration on a dataset; a "−" indicates GPE-AE performed significantly worse. GPE-AE was also compared to GP-MaL, with a ↑ indicating that GPE-AE significantly outperformed GP-MaL, and a ↓ indicating the opposite.

From our results, GPE-AE was generally better than CAE for the "easier" datasets with 34 or fewer features. This indicates that the GP approach is able to find embeddings with a more separable structure of classes. On the harder problem of the Clean1 dataset with 168 features, the CAE generally outperformed GPE-AE — albeit with a relatively small magnitude of difference. Clean1 has two classes, split 54%:46% — giving a "baseline" accuracy of 54%. All the methods are very close to this when reducing to one dimension, indicating the classifier is only doing slightly better than randomly assigning labels. This is not surprising, as reducing 168 features to a single dimension is inherently a difficult task assuming most of the features are not irrelevant or redundant. GP-MaL often had the highest performance of the three methods, but was only significantly better than GPE-AE in some tests on the Dermatology and Segmentation datasets. Neither of the AE methods show clear trends across different hidden layer configurations. All three methods show clear improvements as embedding dimensionality increases; more dimensions allows for more structure to be retained.

Table 5. GPE-AE compared to conventional auto-encoders (CAE) and GP-MaL.

Method	Clean1		Derma.		Iono.		Segmen.		Wine	
	Acc.	Conn.	Acc.	Conn.	Acc.	Conn.	Acc.	Conn.	Acc.	Conn.
1 Dimension										
GPE 1HL	0.527−	182	0.813+ ↓	**159**	0.850+	**189**	0.653+	**141**	0.914+	216
GPE 2HL	0.544−	302	0.803+ ↓	210	0.871+	237	0.663+	214	0.908	**178**
GPE 3HL	0.546−	132	0.784↓	219	0.865+	192	0.641+	201	0.912	181
CAE 1HL	0.553	21632	0.719	4480	0.714	4480	0.489	2560	0.767	1792
CAE 2HL	0.571	32320	0.712	6592	0.724	6592	0.594	3712	0.811	2560
CAE 3HL	0.581	34336	0.746	8608	0.725	8608	0.561	5728	0.793	4576
GP-MaL	0.582	**128**	0.915	371	0.868	211	0.649	217	0.883	217
2 Dimensions										
GPE 1HL	0.600	374	0.891↓	324	0.883+	273	0.701+	288	0.942	470
GPE 2HL	0.622	276	0.894+ ↓	301	0.886+	244	0.754+	**262**	0.933	342
GPE 3HL	0.582−	**254**	0.872+ ↓	**267**	0.890+	238	0.707+	392	0.916+	352
CAE 1HL	0.591	21760	0.874	4608	0.817	4608	0.573	2688	0.872	1920
CAE 2HL	0.641	32384	0.855	6656	0.823	6656	0.619	3776	0.909	2624
CAE 3HL	0.645	34368	0.852	8640	0.81	8640	0.605	5760	0.911	4608
GP-MaL	0.621	348	0.935	301	0.889	**227**	0.714	325	0.937	**267**
3 Dimensions										
GPE 1HL	0.639−	257	0.909	371	0.896+	326	0.799+ ↓	336	0.938	380
GPE 2HL	0.642−	328	0.909+	425	0.897+	342	0.779+ ↓	299	0.938	**367**
GPE 3HL	0.628−	**106**	0.911+	**305**	0.893+	**277**	0.786+ ↓	**257**	0.940	518
CAE 1HL	0.665	21888	0.910	4736	0.856	4736	0.676	2816	0.928	2048
CAE 2HL	0.676	32448	0.886	6720	0.862	6720	0.709	3840	0.937	2688
CAE 3HL	0.665	34400	0.865	8672	0.860	8672	0.713	5792	0.940	4640
GP-MaL	0.639	301	0.928	450	0.899	348	0.830	320	0.95	419
5 Dimensions										
GPE 1HL	0.692−	323	0.918−	451	0.906	**242**	0.889+	549	0.944+	588
GPE 2HL	0.697−	**242**	0.922+	436	0.903	431	0.888+	**327**	0.931	385
GPE 3HL	0.668−	431	0.897+ ↓	**346**	0.904+	297	0.873+	338	0.936+	547
CAE 1HL	0.707	22144	0.940	4992	0.900	4992	0.806	3072	0.934	2304
CAE 2HL	0.739	32576	0.911	6848	0.895	6848	0.812	3968	0.932	2816
CAE 3HL	0.707	34464	0.882	8736	0.885	8736	0.767	5856	0.927	4704
GP-MaL	0.689	480	0.953	492	0.911	428	0.895	595	0.947	**295**
10 Dimensions										
GPE 1HL	0.746−	410	0.949	808	0.906+	387	0.928+	673	0.956+	604
GPE 2HL	0.747−	**356**	0.933↓	772	0.905−	**378**	0.916+ ↓	**384**	0.964+	832
GPE 3HL	0.717−	491	0.943+ ↓	**452**	0.912−	562	0.911+ ↓	559	0.949+	655
CAE 1HL	0.769	22784	0.946	5632	0.920	5632	0.870	3712	0.938	2944
CAE 2HL	0.775	32896	0.930	7168	0.915	7168	0.856	4288	0.935	3136
CAE 3HL	0.744	34624	0.910	8896	0.905	8896	0.821	6016	0.924	4864
GP-MaL	0.755	577	0.963	754	0.907	457	0.932	839	0.963	**556**

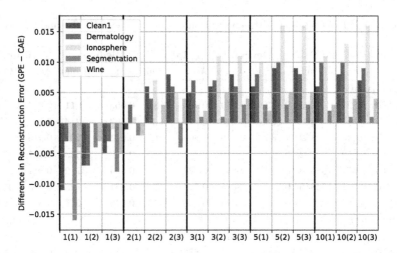

Fig. 2. The differences between the reconstruction error achieved by GPE-AE and CAE. A -ve value means GPE-AE had a lower reconstruction error than CAE. 1(3) represents a dimensionality of 1 with 3 hidden layers in the decoder.

Number of Connections: For GPE-AE and GP-MaL, the average number of connections of the best individuals found across the 30 runs for each method is presented. The number of connections in the neural networks is consistent for all runs of the same configuration, and is presented for comparison. To highlight the effect the complexity of the decoder has on the complexity of the encoder, the smallest average individual size for each test configuration is shown in bold. Even though the CAE designs we propose are fairly simple, they still have a very large number of connections compared to GP, especially on the Clean1 dataset.

Reconstruction Error: The mean difference in reconstruction error between GPE-AE and the CAE is shown in Fig. 2 for each dataset and hidden layer config-uration. The x-axis represents increasing embedding dimensionality from left to right, with the three different hidden layers configurations shown at each embed-ding dimensionality. A negative difference indicates that GPE-AE achieved bet-ter reconstruction error than CAE. There are two clear trends: firstly, that the two methods are very similar in their reconstruction errors (differing by at most ~0.016); and secondly that GPE-AE has lower reconstruction error on the most challenging task of embedding in one dimension, while CAE has lower errors at higher dimensions. This is encouraging for GPE-AE: with many fewer connec-tions, it can achieve a lower reconstruction error with a one-dimensional embed-ding. It does, however, demonstrate the need for further research into improving the performance of multi-tree GP at higher embedding sizes.

6   Further Analysis

A common NLDR task is the reduction of data to two dimensions, explicitly for the sake of visualising the data. To evaluate the suitability of GPE-AE for

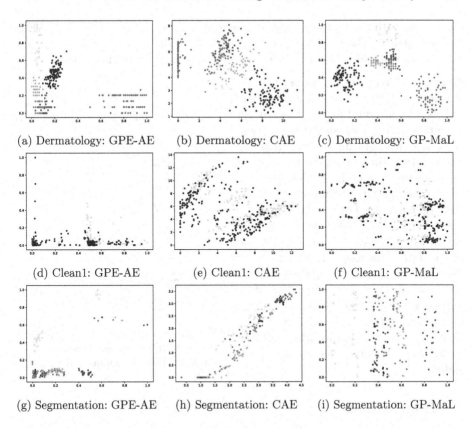

(a) Dermatology: GPE-AE   (b) Dermatology: CAE   (c) Dermatology: GP-MaL

(d) Clean1: GPE-AE   (e) Clean1: CAE   (f) Clean1: GP-MaL

(g) Segmentation: GPE-AE   (h) Segmentation: CAE   (i) Segmentation: GP-MaL

Fig. 3. Visualisations produced by the GP methods and a conventional auto-encoder (CAE) on the Dermatology, Clean1 and Segmentation datasets. The median result of each was chosen for visualisation.

visualisation, we show representative two-dimensional embeddings produced by the three methods in Fig. 3 on Dermatology, Clean1 and Segmentation datasets. These datasets represent a variety of difficulty in terms of number of original features, from 19 for Segmentation to 168 features for Clean1.

On Dermatology, the GPE-AE (and, to a lesser degree, GP-MaL) visualisation is that instances seem to be grouped in rigid "steps" along the y axis. The CAE learns "smoother" mapping functions due to the large number of connections, whereas the GP methods are able to find lower-complexity trees that are sufficient in terms of fitness. For Clean1, none of the methods are able to clearly separate the two classes, which is consistent with the earlier results. This suggests that the relationship between the class distribution and the original feature space is quite complex: it cannot be well-represented in only two dimensions. The Segmentation visualisations are quite different to the earlier ones. The CAE represents the data along a single line, suggesting it is not making "full use" of the two dimensions. GP-MaL splits the yellow and purple classes quite clearly along the x-axis, with the y-axis mostly representing intra-class variation. GPE-AE is able to separate the same classes well, but also pushes a number instances far away from the main clusters — these are perhaps outliers in the dataset.

Evolved Program Analysis: a key strength of GPE-AE over a conventional AE is the interpretable tree-based representation. Figure 4 shows a functional mapping for reducing the Dermatology dataset to a single dimensions. It makes use of a single non-linear operator: a sigmoid function with the input $\min(f9, \max(f21, f13))$. This suggests these features may have some non-linear relationship to the underlying distribution of the data. This mapping uses 12 unique features of Ionosphere's original 34. From this, we can infer that only $\sim$35% of features are required to create a sufficiently good embedding. Random forest classification achieved an accuracy of 0.782 using the embedding produced by this individual.

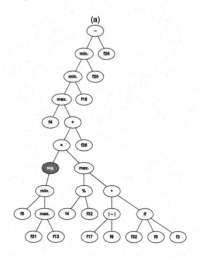

Fig. 4. Dermatology dataset (34 features) reduced to a single feature with GPA-AE using 12 unique features.

7   Conclusions

Autoencoders are a class of unsupervised learning models for learning representations of data. They simultaneously learn a function to encode the data in a low-dimensional space (the encoder) and a function to reconstruct the input data from the encoding (the decoder). Conventional autoencoders use artificial neural networks, which have an opaque structure that makes interpretation of the encoder mappings highly difficult.

Existing work has attempted to replace the entire autoencoder with GP. Some of these represent both the encoder and decoder independently with GP, however these has been difficult to evolve due to the inter-dependency between them. Other approaches have forgone the encoder-decoder architecture entirely by using GP to mimic the reconstructive behaviour of an autoencoder directly. These approaches have used complex and indirect GP representations, which makes interpretation difficult. To address this gap, we proposed the Genetic Programming Encoder for Autoencoding (GPE-AE). GPE-AE retains the ANN decoder, while using a multi-tree GP representation for the encoder. This allows for an interpretable encoding, while still retaining the performance benefits of the ANN decoder.

We have presented the results of experiments to compare GPE-AE to both conventional autoencoders (CAE) and GP-MaL, a similar GP method for NLDR. We found that GPE-AE was competitive with both approaches for producing embeddings which retained the original structure, demonstrating the strength of the approach at finding functional dimensionality reductions. We also have compared two-dimensional visualisations produced by the methods, and assessed how the different approaches can effect these. Finally, we have analysed some selected GP encoders produced by GPE-AE to demonstrate the valuable insights that can be gained by using interpretable AE models.

Future Work: In this work, we kept the structure of the ANN decoder simple and constant for all individuals during the evolution. Future work could explore dynamic decoder structures based on the encoder structure. Simple encoders could make use of simpler decoders, reducing evolution time by reducing decoder training time for simpler solutions where complex decoders are not required.

The datasets used in this week were of relatively low dimensionality, with only one dataset over one hundred dimensions. This was suitable for an initial test of GPE-AE as a proof of concept, but future work could test the method on a wider variety of datasets, especially larger ones with over one thousand features. Performing such experiments would provide more rigorous tests of the suitability of GPE-AE for a wider range of tasks.

Another potential extension of this work is to Variational Autoencoders. The suitability of GP for multi-objective optimisation [14,27] would allow introducing another objective that constrains the shape of the latent distribution.

Finally, this work does not directly take into account the susceptibility of GP to program bloat. It is likely that exploring and applying suitable bloat control methods would lead to more compact and therefore interpretable trees in the GP-based encoders produced by the proposed method.

References

1. Bengio, Y., Courville, A.C., Vincent, P.: Representation learning: a review and new perspectives. IEEE Trans. Pattern Anal. Mach. Intell. **35**(8), 1798–1828 (2013). https://doi.org/10.1109/TPAMI.2013.50
2. Bi, Y., Xue, B., Zhang, M.: Evolving deep forest with automatic feature extraction for image classification using genetic programming. In: Bäck, T., et al. (eds.) PPSN 2020. LNCS, vol. 12269, pp. 3–18. Springer, Cham (2020). https://doi.org/10.1007/978-3-030-58112-1_1
3. Chai, T., Draxler, R.R.: Root mean square error (RMSE) or mean absolute error (MAE). Geosci. Model Dev. Discuss. **7**(1), 1525–1534 (2014)
4. Dua, D., Graff, C.: UCI machine learning repository (2017). https://archive.ics.uci.edu/ml
5. Hinton, G.E., Salakhutdinov, R.R.: Reducing the dimensionality of data with neural networks. Science **313**(5786), 504–507 (2006)
6. Hinton, G.E., Roweis, S.T.: Stochastic neighbor embedding. In: Becker, S., Thrun, S., Obermayer, K. (eds.) Advances in Neural Information Processing Systems 15 [Neural Information Processing Systems, NIPS 2002, 9–14 December 2002, Vancouver, British Columbia, Canada], pp. 833–840. MIT Press (2002)
7. Jolliffe, I.T.: Principal Component Analysis. In: Lovric, M. (ed.) International Encyclopedia of Statistical Science, pp. 1094–1096. Springer, Berlin, Heidelberg (2011). https://doi.org/10.1007/978-3-642-04898-2_455

8. Kashef, S., Nezamabadi-pour, H.: An advanced ACO algorithm for feature subset selection. Neurocomputing **147**, 271–279 (2015)
9. Kingma, D.P., Ba, J.: Adam: a method for stochastic optimization. In: Bengio, Y., LeCun, Y. (eds.) 3rd International Conference on Learning Representations, ICLR 2015, San Diego, CA, USA, 7–9 May 2015, Conference Track Proceedings (2015). https://arxiv.org/abs/1412.6980
10. Leardi, R., Boggia, R., Terrile, M.: Genetic algorithms as a strategy for feature selection. J. Chemom. **6**(5), 267–281 (1992)
11. Lee, J.A., Verleysen, M.: Nonlinear Dimensionality Reduction, vol. 1. Springer, New York (2007). https://doi.org/10.1007/978-0-387-39351-3
12. Lensen, A., Xue, B., Zhang, M.: Can genetic programming do manifold learning too? In: Sekanina, L., Hu, T., Lourenço, N., Richter, H., García-Sánchez, P. (eds.) EuroGP 2019. LNCS, vol. 11451, pp. 114–130. Springer, Cham (2019). https://doi.org/10.1007/978-3-030-16670-0_8
13. Lensen, A., Xue, B., Zhang, M.: Genetic programming for manifold learning: preserving local topology. IEEE Transactions on Evolutionary Computation, pp. 1–15 (2022). early Access
14. Lensen, A., Zhang, M., Xue, B.: Multi-objective genetic programming for manifold learning: balancing quality and dimensionality. Genet. Program. Evolvable Mach. **21**(3), 399–431 (2020). https://doi.org/10.1007/s10710-020-09375-4
15. van der Maaten, L.: Learning a parametric embedding by preserving local structure. In: Proceedings of the Twelfth International Conference on Artificial Intelligence and Statistics, AISTATS 2009, Clearwater Beach, Florida, USA, 16–18 April 2009. JMLR Proceedings, vol. 5, pp. 384–391. JMLR.org (2009)
16. McDermott, J.: Why is auto-encoding difficult for genetic programming? In: Sekanina, L., Hu, T., Lourenço, N., Richter, H., García-Sánchez, P. (eds.) EuroGP 2019. LNCS, vol. 11451, pp. 131–145. Springer, Cham (2019). https://doi.org/10.1007/978-3-030-16670-0_9
17. McInnes, L., Healy, J.: UMAP: uniform manifold approximation and projection for dimension reduction. CoRR abs/1802.03426 (2018)
18. Poli, R., Langdon, W.B., McPhee, N.F.: A Field Guide to Genetic Programming (2008). lulu.com, https://www.gp-field-guide.org.uk/
19. Rodriguez-Coayahuitl, L., Morales-Reyes, A., Escalante, H.J.: Evolving autoencoding structures through genetic programming. Genet. Program. Evolvable Mach. **20**(3), 413–440 (2019). https://doi.org/10.1007/s10710-019-09354-4
20. Ruberto, S., Terragni, V., Moore, J.H.: Image feature learning with genetic programming. In: Bäck, T., et al. (eds.) PPSN 2020. LNCS, vol. 12270, pp. 63–78. Springer, Cham (2020). https://doi.org/10.1007/978-3-030-58115-2_5
21. Sainburg, T., McInnes, L., Gentner, T.Q.: Parametric UMAP embeddings for representation and semisupervised learning. Neural Comput. **33**(11), 2881–2907 (2021)
22. Schofield, F., Lensen, A.: Using genetic programming to find functional mappings for UMAP embeddings. In: IEEE Congress on Evolutionary Computation, CEC 2021, Kraków, Poland, June 28–1 July 2021, pp. 704–711. IEEE (2021)
23. Svetnik, V., Liaw, A., Tong, C., Culberson, J.C., Sheridan, R.P., Feuston, B.P.: Random forest: a classification and regression tool for compound classification and QSAR modeling. J. Chem. Inf. Comput. Sci. **43**(6), 1947–1958 (2003)
24. Uriot, T., Virgolin, M., Alderliesten, T., Bosman, P.: On genetic programming representations and fitness functions for interpretable dimensionality reduction (2022). https://arxiv.org/abs/2203.00528

25. Vanschoren, J., van Rijn, J.N., Bischl, B., Torgo, L.: OpenML: networked science in machine learning. SIGKDD Explor. **15**(2), 49–60 (2013). https://doi.org/10.1145/2641190.2641198
26. Xue, B., Zhang, M., Browne, W.N.: Multi-objective particle swarm optimisation (PSO) for feature selection. In: Genetic and Evolutionary Computation Conference, GECCO 2012, Philadelphia, PA, USA, 7–11 July 2012, pp. 81–88. ACM (2012)
27. Zhao, H.: A multi-objective genetic programming approach to developing pareto optimal decision trees. Decis. Support Syst. **43**(3), 809–826 (2007)

Graph Networks as Inductive Bias for Genetic Programming: Symbolic Models for Particle-Laden Flows

Julia Reuter[1]([✉])[iD], Hani Elmestikawy[2][iD], Fabien Evrard[2][iD],
Sanaz Mostaghim[1][iD], and Berend van Wachem[2][iD]

[1] Institute for Intelligent Cooperating Systems, Otto-von-Guericke-University,
Magdeburg, Germany
{julia.reuter,sanaz.mostaghim}@ovgu.de
[2] Institute for Mechanical Process Engineering, Otto-von-Guericke-University,
Magdeburg, Germany
{hani.elmestikawy,fabien.evrard,berend.vanwachem}@ovgu.de

Abstract. High-resolution simulations of particle-laden flows are computationally limited to a scale of thousands of particles due to the complex interactions between particles and fluid. Some approaches to increase the number of particles in such simulations require information about the fluid-induced force on a particle, which is a major challenge in this research area. In this paper, we present an approach to develop symbolic models for the fluid-induced force. We use a graph network as inductive bias to model the underlying pairwise particle interactions. The internal parts of the network are then replaced by symbolic models using a genetic programming algorithm. We include prior problem knowledge in our algorithm. The resulting equations show an accuracy in the same order of magnitude as state-of-the-art approaches for different benchmark datasets. They are interpretable and deliver important building blocks. Our approach is a promising alternative to "black-box" models from the literature.

Keywords: Genetic Programming · Graph Networks · Fluid Mechanics

1 Introduction

The rapid growth of computational power over the last decades has played an important role in fluid mechanics research, as it enables the direct-numerical-simulation (DNS) of flows of ever-increasing complexity. Within the field of fluid mechanics, the simulation of the so-called *particle-laden flows*, i.e., cases of numerous particles immersed and evolving within a fluid, is particularly challenging. Examples of particle-laden flows can be found in the flow of blood cells in plasma, or in the fluidization of biomass particles in furnaces. Due to the complex particle-particle and fluid-particle interactions, high-resolution simulations

G. Pappa et al. (Eds.): EuroGP 2023, LNCS 13986, pp. 36–51, 2023.
https://doi.org/10.1007/978-3-031-29573-7_3

of such flows are only limited to micro-scale systems with thousands of particles (e.g., see [23,28]). These cannot meet the requirements of real-world applications commonly involving billions of particles. To overcome this issue, volume-filtered approaches solve the flow on a lower resolution [2,7]. To close the information gap owing to the lower simulation resolution, they require information about the fluid-induced force acting on an individual particle $\mathbf{F}_{fluid}$, which depends on the above-mentioned interactions and consequently the number of particles.

Various approaches have been presented in the literature to approximate the value of $\mathbf{F}_{fluid}$, including empirical models [20,22], pairwise-interaction extended models [1,3] and physics-informed artificial neural networks (ANN) [25]. While empirical approaches only predict the mean force, the other methods show promising results to also predict the variations form the mean, but often lack explainability and predict $\mathbf{F}_{fluid}$ only with a certain error. The goal of this paper is to develop interpretable symbolic models for $\mathbf{F}_{fluid}$ which can capture the complex interactions for a large number of particles. Interpretable models for this problem are desirable as they allow for deeper understanding and analysis of the underlying interactions, which remain opaque in approaches up until now.

The identification of interpretable models from experimental and simulation data has recently gained importance to overcome the "black-box" nature of machine-learning algorithms such as ANNs. Genetic programming (GP) is a suitable approach to develop human-interpretable symbolic models from data. Given the problem of predicting $\mathbf{F}_{fluid}$ in particle-laden flows, symbolic models allow a better understanding of the underlying interactions. However, previous work has shown that the straightforward application of GP algorithms on DNS data, for instance to recover the velocity field of the flow around two particles, cannot cope with the complexity of the problem [19]. The complexity of GP algorithms scales exponentially with the number of input variables and functions, thus some pre-processing of the data and/or combination with other model reduction techniques is required to reduce the dimensionality of the problem [4]. In this paper, we present an approach using inductive bias to identify symbolic models for $\mathbf{F}_{fluid}$. Our approach comprises the following two steps:

1. **Inductive bias**: We first train a Graph network (GN) to predict $\mathbf{F}_{fluid}$. This step reduces the problem complexity and makes it tractable for GP.
2. **Symbolic model**: We then employ a GP algorithm to develop symbolic models, which replace the internal ANN blocks of the GN.

Since the underlying pairwise interactions between particles are unknown, we employ two different structures as interaction patterns. Our main contributions are *(i)* the supply of high-resolution DNS data of particle-laden flows at different volume-fractions, *(ii)* an extensible algorithm that combines GN and GP and allows for different underlying structures, and *(iii)* a comprehensive analysis of the resulting equations. Our experiments show promising results for $\mathbf{F}_{fluid}$. Moreover, the symbolic functions are concise and deliver meaningful building blocks.

2   Background and Related Work

Since the problem addressed in this paper is rather complex, we first give an overview about related research in the GP area. We then introduce basic concepts of particle-laden flows as well as related research in a separate subsection.

2.1   Genetic Programming in Physics Applications

Symbolic models (i.e., mathematical expressions) are interpretable and help to understand underlying patterns in data. Under certain conditions, they generalize better and have higher extrapolation capabilities compared to ANNs [10]. Despite all the advantages, identifying symbolic models from data is a non-trivial task. With increased interest in symbolic models for physics applications, efforts have been made to develop algorithms for symbolic regression, which include, but are not limited to [5,9,12,17,21,27,30]. Population-based methods making use of evolutionary algorithms have shown to perform well on these tasks (also known as GP for symbolic regression).

Applications from the physics area are not new to the GP community. Twenty years ago already, Keijzer et al. proposed a dimension penalty as additional objective to evolve equations that are conformal with physical laws [13]. Other approaches include grammar-based GP algorithms, which restrict the search space to pre-defined rules (e.g., see [14,18]), and strongly typed GP [29]. However, as pointed out by Cranmer et al. [10], high-dimensional problems are too complex to be directly approached with GP, due to the combinatorial explosion with increasing number of features and functions. Being a common underlying pattern for many physics applications, they proposed a framework for problems which can be modeled as interacting particles. Since GNs can represent this underlying structure, they first train a GN on the available data, thus induce a bias. The internal parts of the GN are then replaced by symbolic models. In this way, the problem complexity for the GP algorithm is reduced.

Two recent publications address interesting problems at the intersection of GP and fluid mechanics: Zille et al. examined the capabilities of GP algorithms to predict known equations for the flow around a single spherical particle [31]. Reuter et al. [19] extended this approach to two particles. Their work indicates that the problem is too complex to be directly approached with GP. The problem addressed in the present paper has a considerably higher complexity due to the many particles involved.

2.2   Machine Learning for Particle-Laden Flows

Particle-laden flows can be locally characterized based on the particle Reynolds number, Re, and the particle volume fraction, ϕ. Re is a dimensionless quantity characterizing the ratio of inertial effects over viscous effects within the fluid, whereas ϕ is the local fraction of volume occupied by the particles in the mixture.

The identification of an accurate model for the fluid forces acting on particles is a non-trivial task, as shown by the rich recent literature on the matter

[1, 3, 24, 25]. Promising approaches in this area assume so-called pairwise inter-actions between particles [1, 25]. It states that in a flow locally governed by Re and ϕ, the force $\mathbf{F}_{\text{fluid},i}$ acting on a particle i is approximated with a sum of interactions with neighboring particles depending on their relative locations $\mathbf{r}_j$. In this context, the prediction accuracy increases only up to a number of consid-ered neighboring particles between 20 and 30 [1, 25]. Although this assumption introduces a certain error to the models, since only considering first-order inter-action between particles, their predictive abilities are competitive with those of other approaches.

We are mainly interested in the data-driven approaches, which use data from DNS to find an accurate model for $\mathbf{F}_{\text{fluid}}$. Noteworthy publications in this area include [15, 16], which employ multiple linear regression on expansions of spher-ical harmonics. Wachs et al. extract distributions of particle locations within a pre-defined neighborhood from DNS data [24]. These are used to find correlations between the force exerted on a particle and the locations of its neighboring par-ticles. Balachandar et al. were the first to implement an artificial neural network (ANN) for force prediction [3]. The input data comprises the relative particle positions within a neighborhood, as well as Re and ϕ. The ANN severely overfits the training data. A recent publication from Wachs et al. indicates that physics-inspired neural networks (PINN) can overcome the overfitting problem [25]. A main characteristic of this approach is the parameter sharing between neural net-work blocks: Making use of the pairwise interaction assumption, the influence of each neighbor of a particle is calculated by a small ANN, which is shared among all neighbors. The total force on a particle is the linear superposition of the influ-ences of its neighboring particles. Next to the neighbor locations, the predictive features include the local average velocity, which can be approximated from the particle locations. The mentioned approaches impose an underlying form on the model, which is deduced from prior knowledge about the problem. For example, the prior assumptions of the PINN model regarding basic interactions between particles are easy to understand. However, the transformations inside the ANN blocks remain opaque.

Building upon the works of Reuter et al. [19], Wachs et al. [25] and Cranmer et al. [10], our goals are: First, to identify which underlying pattern describes the data best by inducing two variants of bias through GN. Second, to overcome the "black-box" nature of GNs by replacing the network blocks with symbolic models. In this paper, we present an algorithm that fits the nature of the problem at hand.

3 Proposed Methods

As depicted in Fig. 1, the overall algorithm comprises two phases: In the first phase, a GN is trained on the input data. The inductive bias of the GN deter-mines the internal structure, i.e., which particles interact with others and how the influences of multiple neighboring particles are aggregated. This surrogate model facilitates the development of symbolic models, since the general shape

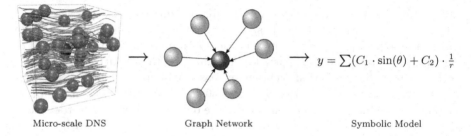

Micro-scale DNS   Graph Network   Symbolic Model

Fig. 1. Symbolic models are generated from simulation data using a GN as surrogate model. Physical particles in the simulation translate to nodes in the GN.

of the equation is determined beforehand. Subsequently, the GP algorithm fits symbolic models to the output of the internal structures of the GN, rather than the actual target variable. The prediction of the target variable is achieved by aggregating the symbolic models, using the same aggregation scheme as previously employed in the GN.

3.1   Graph Networks

Many systems in physics or real-world applications can be represented by graphs, such as spring systems [10] or particles in a particle-laden flow. This motivates the use of GNs to model interactions between objects or particles. GNs are a subtype of graph neural networks (GNN) [6]. They contain network models for each internal structure of a graph.

A particle translates to a node in the GN, which is described by the **node model** Φ^n. A system of q particles is represented by a graph with q nodes n_i, where $i = 1 \ldots q$. A node n_i has incoming and/or outgoing edges from/to the nodes in its neighborhood $\mathcal{N}_i$. The neighborhood can be defined by a number of closest nodes to n_i or all nodes within a certain distance from n_i.

The edges of the graph are represented by the **edge model or message function** Φ^e. The message function $m_{i,j}$ captures pairwise interactions between two nodes n_i and n_j, where $n_j \in \mathcal{N}_i$. The pairwise interaction is determined by the current state of the interacting particles, so that the input to the edge model comprises the features of the two interacting particles. The node model updates the state of a particle as a function of the current state of a particle n_i, as well as the aggregated incoming edge messages. Both Φ^e and Φ^n use shared parameters for all pairwise interactions and node updates.

The global model Φ^g processes all aggregated messages and updated nodes. It computes a global property g for the entire graph. The formal definition of a GN with q nodes is as follows:

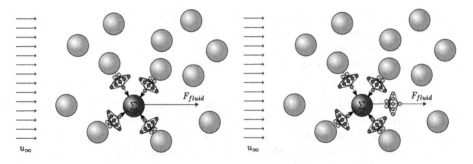

Fig. 2. GN to predict $\mathbf{F}_{\text{fluid}}$ imposed on the red particle in a particle-laden flow, given four particles in its neighborhood $\mathcal{N}_i$: $y = g(x)$ (left) and $y = f(g(x))$ (right). u_∞ is the flow velocity.

$$m_{i,j}(t+1) = \Phi^e(n_i(t), n_j(t)) \tag{1}$$

$$n_i(t+1) = \Phi^n(n_i(t), \sum_{j \in \mathcal{N}_i} m_{i,j}(t)) \tag{2}$$

$$g(t+1) = \Phi^g(n_i(t) \ldots, m_{i,j}(t) \ldots), \text{ where } i = 1 \ldots q, \ j \in \mathcal{N}_i \tag{3}$$

A GN facilitates different ways of predicting a target variable, i.e., different underlying structures. Since the optimal structure of the model to predict a target variable y is unknown, our framework proposes two variants aligned with the structure of the problem at hand:

1. $y = g(x) = \sum_{j \in \mathcal{N}_i} m_{i,j}$: Only the edge model Φ^e is captured. The target variable is the sum of the edge messages received by a node.
2. $y = f(g(x)) = f(\sum_{j \in \mathcal{N}_i} m_{i,j})$: Both the edge model Φ^e and node model Φ^n are captured. The target variable is a function of the summed edge messages. Thus, the summed edge messages are an input to the node model, which predicts the target variable.

Figure 2 depicts these two variants using the example of a particle-laden flow. It becomes apparent that the internal structure of a GN is separable, which means that we can fit separate symbolic models to the outputs of the edge and node models. This facilitates the equation fitting in the next step tremendously.

3.2 Genetic Programming

Genetic Programming is a population-based approach to develop symbolic models from data. The equations are represented as parse trees, which consist of basic operators, functions, features and constants, often referred to as set of primitives. New equations are formed in an evolutionary manner, by applying crossover and mutation on selected equations from the population. With growing interest in symbolic regression for physics applications, the basic GP algorithm got enhanced with techniques from the area of machine learning. An important property is the possibility to include and fit constants in the equations, usually achieved by a regression algorithm on top of the evolution of equations. Other techniques are batch-wise training to process big datasets in a reasonable amount of time, and the use of sophisticated error functions.

Algorithm. The proposed GP algorithm makes use of the training features, which can include raw data as well as pre-processed or transformed data to induce prior knowledge about the problem. In addition, the algorithm employs constants, which are fitted through a regression algorithm. Since no ground truth equation is available, the choice of an appropriate function set is a non-trivial task, with major influence on the result. Preliminary experiments and a coarse function tuning have shown that the set of functions and operators $\{+, *, \sin(\circ),$ $\cos(\circ), \tan(\circ), e^{(\circ)}, \log(\circ)\}$ yields satisfactory results. The fitness function to be minimized is the commonly used mean square error (MSE).

Depending on the underlying structures imposed by the GN, the GP algorithms slightly differ:

1. $y = g(x) = \sum_{j \in \mathcal{N}_i} m_{i,j}$: The symbolic model $\Phi^{e'}$ replaces the message model Φ^e, and is thus fitted to the output of the message function, which was recorded during GN training. Constants in the resulting equations are then refitted to the original target variable to avoid the accumulated approximation error. To this end, we employ the Levenberg-Marquardt algorithm and use the constants found by the GP algorithm as starting values.

2. $y = f(g(x)) = f(\sum_{j \in \mathcal{N}_i} m_{i,j}) = n_i$: The first symbolic model $\Phi^{e'}$ replaces the message model Φ^e and follows the same procedure as in (1). The second symbolic model $\Phi^{n'}$ replaces the node model Φ^n and predicts the target variable, given the influence of the neighboring particles. Thus, it receives the summed influences $\sum_{j \in \mathcal{N}_i} m'_{i,j}$ as function input. To refit the constants, the inner function $\Phi^{e'}$ is plugged into the outer function Φ^n.

Techniques for Physically Meaningful Equations. Physical laws often follow relatively simple equations. Thus, our GP algorithm aims at finding equations of low complexity. At the same time, these equations should be in line with physical laws in terms of units. While approaches like grammar-based GP or a dimension penalty as an additional objective are often effective to avoid unit violations, they are complex to implement and sometimes restrict the search space in an undesirable way. Recent research shows, that relatively simple techniques can also yield satisfactory results [9]. To this end, our algorithm employs a complexity measure, complexity-constrained function inputs as well as certain building rules for the parse trees.

Complexity Measure: To compute the complexity of an equation, each operation, function, feature and constant is assigned a complexity value. The total equation complexity is the sum of the complexity values of the used primitives. The complexity values were determined by a coarse hyperparameter tuning.

Binary operators like addition, subtraction and multiplication, as well as the training features are assigned a complexity value of 1. Constants play an important role in numerous physical laws, such as the gravity constant, to name one. When the number of constants in an equation is unknown, they come with the cost of overfitting the training data if too many of them appear in the same expression. Thus, we assign a higher complexity value of 2 to constants. Unary functions such as $\sin(\circ)$, $\cos(\circ)$, $\tan(\circ)$, $e^{(\circ)}$ and $\log(\circ)$ apply a non-linear transformation to the input. Consequently, they are associated a higher complexity value of 2 compared to the basic operators. The unary operation $\frac{1}{\circ}$ is associated with a complexity of 1.

Complexity-Constrained Function Inputs: Another technique to keep the expressions simple yet effective is to restrict the input of certain operations to a maximum allowed complexity. Our algorithm restricts the input complexity of trigonometric, logarithmic and exponential functions to 8. This means, an expression like $y = \sin(2.0 \cdot x - 3.0)$ with an input complexity of 7 is allowed. $y = \sin(2.0 \cdot x + \log(x) + 3.0)$ with an input complexity of 11 exceeds the limit.

Building Rules: Preliminary experiments have shown that GP algorithms sometimes tend to include multiple nested functions in expressions, for instance $\sin(\cos(\sin(\circ)))$. This behavior is to be avoided, as it can lead to the model overfitting the training data and usually has little meaning in terms of explainability. Consequently, we limit the nesting of trigonometric functions to a maximum of 1, so that $\sin(\sin(\circ))$ is allowed, but further nesting with any trigonometric function is prohibited.

4   Experiment Design

We investigate the viability of the presented approach using benchmark data from the Stokes flow (i.e., $Re = 0$), with four different particle-volume fractions ϕ. For each dataset, separate models are trained for the underlying structures $y = g(x)$ and $y = f(g(x))$.

4.1   Data Generation: Simulation of Particle-Laden Flows

We consider the flow past a stationary array of monodisperse spherical particles in the Stokes regime ($Re \rightarrow 0$), at which the viscous forces dominate. In this Regime, the flow is governed by the Stokes equations as follows

$$\mu \Delta \mathbf{u} - \nabla p = -\mathbf{F}_{\text{fluid}}, \tag{4}$$
$$\nabla \cdot \mathbf{u} = 0, \tag{5}$$

where μ is the fluid viscosity, $\mathbf{u}$ is the fluid velocity, p is the hydrostatic pressure and $\mathbf{F}_{\text{fluid}}$ is the Force acting on the fluid. There is no closed-form solution for

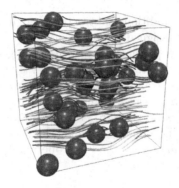

Fig. 3. Random array of stationary spherical particles at $\phi = 0.064$

such equations in complex configurations that include more than a single spherical particle. However, a solution can be built from the superposition of fundamental solutions due to the linearity of the governing equations. The method of Regularized Stokeslets [8] is incorporated to construct the solution of the flow around the array of particles. A single regularized Stokeslet solves the flow driven by locally distributed force ($\mathbf{F}_{\text{fluid}} = \mathbf{g}\phi_\epsilon(|\mathbf{x} - \mathbf{x}_0|)$) in free space, where ϕ_ϵ is an isotropic regularization kernel with compact support over the length ϵ. Each particle is represented by a group of locally distributed forces to achieve the no-slip at the particle surface.

A random array of 30 spherical particles is generated in a unit cube except for one particle which is placed at the center of the cube. Each particle is represented by 300 force markers. The free stream flows in x-direction with uniform velocity $u_\infty = 1$ m/s and the fluid viscosity is $\mu = 1$ kg/(m s). The force $\mathbf{F}_{\text{fluid}}$ has the three components and is computed on the particle located at the center of the cube for 500 random particle arrangements. The scope of this paper is to predict the *streamwise force component* F_{fluid}. We provide benchmark data for each of the following volume fractions $\phi = [0.064, 0.125, 0.216, 0.343]$. Visualization of a sample case is shown in Fig. 3. Each training sample encompasses the following features:

– Relative positions $\mathbf{r}_i$ of the 29 neighboring particles
– Average fluid velocity $\bar{\mathbf{u}}^f$ within the unit cube
– Streamwise force component F_{fluid} exerted by the fluid on the particle of interest

4.2   Data Preprocessing

The raw data generated by the Stokes flow solver from Sect. 4.1 undergoes further transformations before it serves as input to the GN. Initially, the raw particle locations are represented in a three-dimensional Cartesian coordinate system. Preliminary experiments have shown that the GP algorithms perform better when locations are available in spherical coordinates. The GN performance

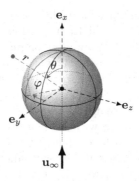

Fig. 4. Spherical coordinate system with radius r, polar angle θ, and azimuthal angle φ.

remains similar for both configurations. Thus, we convert the particle locations to spherical coordinates r, θ and ϕ (see Fig. 4 for exact definition). We assume that it behaves this way because the particle distance r plays an important role in the underlying symbolic model. Furthermore, the trigonometric functions employed in the function set of the GP algorithm are more meaningful with an angle like θ and ϕ as input. This could save the algorithm an intermediate step to compute a dimensionless quantity from the features in Cartesian coordinates.

To increase the number of available data samples, we augment the data by rotations around the axis of the free stream. This has the added benefit of representing symmetries around the free flow direction in our data. In this way, a total of 3,000 samples per training set is available. We split the data with a 3:1 ration into training and test sets. Since the mean force $\langle F_{\text{fluid}} \rangle$ can already be approximated from existing correlations [26], we will predict the deviation from the mean force. This can have the same order of magnitude as $\langle F_{\text{fluid}} \rangle$ itself.

4.3   Algorithm Settings

The features of the neighboring particles, i.e., input features to the edge model, are the relative position from the center particle in spherical coordinates r, θ, φ. The training features of the center particle of interest comprise the local average velocity in x, y and z direction, $\bar{u}_x^f$, $\bar{u}_y^f$ and $\bar{u}_z^f$.

The edge and node models of the GN comprise two fully connected hidden layers with 30 neurons each. We use the hyperbolic tangent as nonlinearity. For both $y = g(x)$ and $y = f(g(x))$ as underlying structures, the output of the edge model is recorded during training of the GN to be used as target features of the GP algorithm. The learning rate with an initial value of 0.002 is adjusted during the training process. The model parameters are optimized by the Adam optimizer. We train the model for 5000 epochs to minimize the MSE as loss function. The GN is implemented using *PyTorch Geometric* [11].

Our GP algorithm is implemented in the *PySR* framework [9]. We run the GP algorithm for 200 iterations, with a population size of 100 individuals. The multi-objective algorithm minimizes the MSE as well as the complexity value of an equation. The best individual from the final Pareto front is identified using a combined measure of accuracy and complexity, as implemented in [9]. The algorithm employs the problem-specific parameters as described in Sect. 3.2, and uses the standard configuration of *PySR* with regards to genetic operators and operator probabilities.

In first trials, we observed that the nested symbolic models $y = f(g(x))$ are more complex than $y = g(x)$, with a tendency to mainly using constants in the outer equations. This can be explained by the two consecutive GP runs for f and g. To keep the comparison fair, we want to allow the algorithm for $y = g(x)$ to use more constants, by reducing the constant complexity to 1. Since the structure of an accurate equation is unknown, and fewer constants can be beneficial for generalization, we still run experiments for $y = g(x)$ and a constant complexity of 2. Considering the four benchmark datasets, this makes a total of twelve experiment instances. The training data and code for this paper are publicly available at https://github.com/juliareuter/flowinGN.

5   Results and Analysis

Since the algorithm comprises two steps, we applied the following procedure: The GN was trained ten times for each experiment variant. We observed similar accuracies for all runs, which are comparable to those of state-of-the-art-approaches [3,25]. We randomly selected one of the ten models as our basis model. In the next step, we employed the GP algorithm to replace this basis model with symbolic models. For statistical comparison, we perform 31 independent realizations of the GP algorithm per experiment instance. The experiments are analyzed regarding the overall algorithm performance, the explainability of the resulting equations as well as validation on unseen data.

5.1   Overall Algorithm Performance

Figure 5 displays the MSE distributions over 31 realizations for each experiment variant. We used the Holm-Bonferroni test to compare the results for each ϕ. The best variants are displayed in bold. For $\phi = 0.064$, no statistically significant difference between the three variants was identified.

We can observe that all experiment variants for all ϕ achieve MSE values of similar magnitude. In general, the nested function $y = f(g(x))$ has a lower spread compared to $y = g(x)$ over 31 runs, i.e., is more reliable to achieve good results. While the medians differ, the best models found by each algorithm have almost the same error value. For most ϕ, no significant difference between the constant complexities $c = 2$ and $c = 1$ for $y = g(x)$ is observable.

5.2   Explainability of Equations

For a more profound analysis of the resulting equations, we select the best and/or most frequently found symbolic model for each experiment variant. The constants of these equations are refitted to the original dataset, since they were

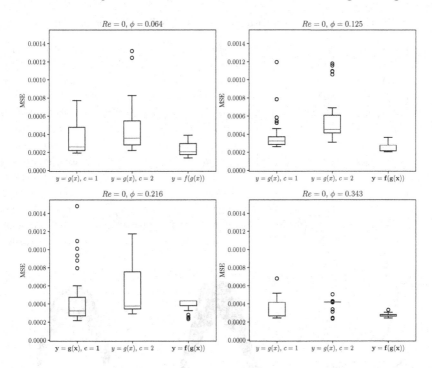

Fig. 5. MSE for different experiment instances over 31 realizations. The variable c indicates the constant complexity for $y = g(x)$. Bold experiments performed best.

trained on the outputs of the GN edge model rather than the target variable. Table 1 shows the refitted equations together with their MSE values on the test dataset. For comparison, the MSE of the GN on the same dataset is displayed.

The equations are concise across all experiment instances. It becomes obvious that the algorithm settings successfully prevented function nesting as well as complex input arguments for trigonometric, logarithmic and exponential functions. Almost all equations are physically meaningful without the use of a dimension penalty or grammar-based approach, only through including prior problem knowledge as constraints. Solely $\sin(r)$ and $\exp(\sin(\theta))$ are unusual terms. While the input comprises six features r, θ, φ, $\bar{u}_x^f$, $\bar{u}_y^f$ and $\bar{u}_z^f$, mainly r and θ are used, twice as well $\bar{u}_x^f$ comes into play.

Having a look at the MSE values, the GP equations perform slightly worse than the GN. The errors of the symbolic models are in the same order of magnitude of 10^{-4} as the GN, but are sometimes about 1.5 times larger. The underlying structure $y = f(g(x))$ performs better for all benchmark datasets. The best equations with $y = g(x)$ as underlying structure show better performance for $c = 1$ than $c = 2$ across all benchmark instances.

For the underlying structure $y = g(x)$, we examined two complexity values for constants, of $c = 2$ and $c = 1$. The constant complexity $c = 1$ resulted in one additional constant for $\phi = 0.125$ and $\phi = 0.216$. The other instances employ the

Table 1. Symbolic models with constants refitted to the original dataset.

ϕ	Experiment	Equation	GP MSE	GN MSE
0.064	$y = g(x)$, $c = 2$	$\sum \left(0.01146r + 0.01146\sin(\theta) - 0.0142\right)\frac{1}{r}$	0.000209	0.000120
	$y = g(x)$, $c = 1$	$\sum \left((0.03448r + 0.03448\sin(\theta) - 0.04238)(-\log(r))\right)$	0.000188	0.000120
	$y = f(g(x))$	$0.0992 \sum \left((r(\sin(\theta) - 0.1312) - 0.1983)(-\log(r))\right) + \bar{u}_x^f - 0.3177$	0.000157	0.000106
0.125	$y = g(x)$, $c = 2$	$\sum \left(0.01397\sin(r) + 0.01397\sin(\theta) - 0.01724\right)\frac{1}{r}$	0.000284	0.000173
	$y = g(x)$, $c = 1$	$\sum \left(0.00839 + (0.01578\sin(\theta) - 0.01644)\frac{1}{r}\right)$	0.000260	0.000173
	$y = f(g(x))$	$0.0597 \sum \left(\left(\sin(\theta) - 0.45368 - \frac{0.12479}{r}\right)e^{-r}\right) - 0.0616$	0.000209	0.000146
0.216	$y = g(x)$, $c = 2$	$\sum \bar{u}_x^f\left(\sin(\theta) - 0.57328 - \frac{0.10557}{r}\right)$	0.000316	0.000206
	$y = g(x)$, $c = 1$	$\sum \left(0.00944 + (0.01932\sin(\theta) - 0.01982)\frac{1}{r}\right)$	0.000247	0.000206
	$y = f(g(x))$	$0.1166 \sum \left(\left(0.17448\sin(\theta) - 0.08318 - \frac{0.01419}{r}\right)\frac{1}{r}\right) - 0.1602$	0.000248	0.000167
0.343	$y = g(x)$, $c = 2$	$\sum \left((0.08249\sin(\theta) - 0.07348)(-\log(r)) + 0.00539\right)$	0.000239	0.000191
	$y = g(x)$, $c = 1$	$\sum \left((0.08749\sin(\theta) - 0.07348)(-\log(r)) + 0.00423\right)$	0.000239	0.000191
	$y = f(g(x))$	$0.3904 \sum \left(\left(0.10982e^{\sin(\theta)} - 0.26635\right)(-\log(r)) + 0.0165\right) - 0.0421$	0.000219	0.000197

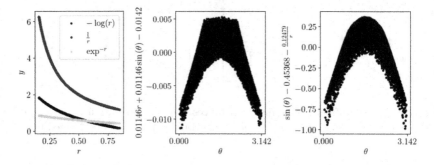

Fig. 6. Insights into frequently used building blocks in the symbolic models.

same number of constants for both complexity values. Comparing the number of constants of the two underlying structures, $y = f(g(x))$ always contains one or two more constants than $y = g(x)$.

Similar patterns across equations of all experiment instances can be identified. Each equation contains a building block that accounts for the distance of a neighboring particle: The terms $\frac{1}{r}$, $-\log(r)$ and $\exp(r)$ scale the influence of a neighboring particle on the particle of interest, i.e., decrease with increasing radius r. Furthermore, each equation contains a larger building block which includes $\sin(\theta)$ and constants or other small terms. We can assume that this building block determines the influence of a particle, which is then scaled with the distance to the center particle. Figure 6 displays the function values of some of the identified building blocks.

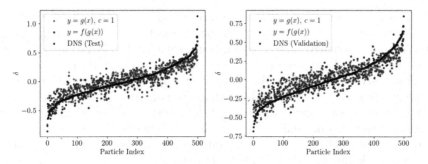

Fig. 7. Normalized predictions of the deviation from the mean force $\langle F_{fluid} \rangle$, i.e., $\delta = \frac{F_{fluid} - \langle F_{fluid} \rangle}{\langle F_{fluid} \rangle}$. Particles are sorted in ascending order by their target value.

5.3   Validation of Symbolic Models

Due to the complex underlying relations, overfitting to training data is a common issue in machine learning for fluid mechanics. Thus, we validate the equations found by the GP algorithm on a dataset with the same values of Re and ϕ, but from a different DNS realization. Figure 7 exemplarily depicts the normalized predictions of the deviation from the mean force $\langle F_{fluid} \rangle$ for $\phi = 0.216$. The left plot shows the predictions for 500 particles from the same realization as the training data, and the right plot from a different realization. The plot as well as the MSE values of 0.000247 (left) and 0.000225 (right) for $y = g(x)$ and 0.000248 (left) and 0.000226 (right) for $y = f(g(x))$ indicate that the equations identified actual underlying patterns and did not overfit the training data. The other benchmarks behave similarly, but are omitted here because of space reasons.

6   Conclusion and Future Work

We presented an approach to develop symbolic models for the fluid force acting on particles in particle-laden flows from simulation data. A GN serves as surrogate model, from which the symbolic models are deduced using the introduced GP algorithm. We include prior knowledge about the problem by employing a complexity measure as well as imposing constraints on the equation generation process. Furthermore, we preprocessed the data to make them manageable for the GP algorithm. Since the shape of the final model is unknown, we examined two underlying structures $y = g(x)$ and $y = f(g(x))$.

   Compared to state-of-the-art approaches, the presented GN achieved similar accuracies [3,25]. The symbolic models consistently perform slightly worse than GN, although errors of both approaches are of the same order of magnitude. A validation on unseen data indicated that our models do not overfit. The underlying structure $y = f(g(x))$ performed best on the provided benchmark instances. We identified building blocks which frequently appear in equations across all benchmark instances. The equations also reveal which features are most influential on the target variable. Altogether, our approach offers a promising, human-interpretable alternative to the hidden transformation in ANN blocks. Building upon the work of Reuter et al., we scaled up from two to thirty particles.

This study confirms the applicability of our approach to the problem at hand. In a next step, we will examine the performance of our approach on the prediction of the other two force components. While most of the equations evolved are physically meaningful, small terms such as $\sin(r)$ violate the unit system. Converting all features to non-dimensional quantities, such as dividing the distance r by the particle radius, can circumvent this issue.

References

1. Akiki, G., Moore, W., Balachandar, S.: Pairwise-interaction extended point-particle model for particle-laden flows. J. Comput. Phys. **351**, 329–357 (2017)
2. Anderson, T.B., Jackson, R.O.Y.: A fluid mechanical description of fluidized beds. I EC Fundam. **6**(4), 524–539 (1967)
3. Balachandar, S., Moore, W.C., Akiki, G., Liu, K.: Toward particle-resolved accuracy in Euler-Lagrange simulations of multiphase flow using machine learning and pairwise interaction extended point-particle (PIEP) approximation. Theoret. Comput. Fluid Dyn. **34**(4), 401–428 (2020)
4. Beetham, S., Capecelatro, J.: Multiphase turbulence modeling using sparse regression and gene expression programming (2021). https://arxiv.org/abs/2106.10397
5. Biggio, L., Bendinelli, T., Neitz, A., Lucchi, A., Parascandolo, G.: Neural symbolic regression that scales. In: International Conference on Machine Learning, pp. 936–945 (2021)
6. Bronstein, M.M., Bruna, J., LeCun, Y., Szlam, A., Vandergheynst, P.: Geometric deep learning: going beyond Euclidean data. IEEE Signal Process. Mag. **34**(4), 18–42 (2017)
7. Capecelatro, J., Desjardins, O.: An Euler-Lagrange strategy for simulating particle-laden flows. J. Comput. Phys. **238**, 1–31 (2013)
8. Cortez, R.: The method of regularized stokeslets. SIAM J. Sci. Comput. **23**(4), 1204–1225 (2001)
9. Cranmer, M.: Pysr: Fast & parallelized symbolic regression in python/julia (2020). https://doi.org/10.5281/zenodo.4041459
10. Cranmer, M., et al.: Discovering symbolic models from deep learning with inductive biases. In: NeurIPS 2020 (2020)
11. Fey, M., Lenssen, J.E.: Fast graph representation learning with pytorch geometric. arXiv preprint arXiv:1903.02428 (2019)
12. Kaptanoglu, A.A., et al.: PySINDy: a comprehensive python package for robust sparse system identification. J. Open Source Softw. **7**(69), 3994 (2022)
13. Keijzer, M., Babovic, V.: Dimensionally aware genetic programming. In: Proceedings of the 1st Annual Conference on Genetic and Evolutionary Computation, vol. 2, pp. 1069–1076 (1999)
14. Mckay, R.I., Hoai, N.X., Whigham, P.A., Shan, Y., O'neill, M.: Grammar-based genetic programming: a survey. Genet. Program. Evolvable Mach. **11**(3–4), 365–396 (2010). https://doi.org/10.1007/s10710-010-9109-y
15. Moore, W.C., Balachandar, S.: Lagrangian investigation of pseudo-turbulence in multiphase flow using superposable wakes. Phys. Rev. Fluids **4**, 114301 (2019)

16. Moore, W., Balachandar, S., Akiki, G.: A hybrid point-particle force model that combines physical and data-driven approaches. J. Comput. Phys. **385**, 187–208 (2019)
17. Rackauckas, C., et al.: Universal differential equations for scientific machine learning (2020). https://doi.org/10.48550/arXiv.2001.04385v4
18. Ratle, A., Sebag, M.: Grammar-guided genetic programming and dimensional consistency: application to non-parametric identification in mechanics. Appl. Soft Comput. **1**(1), 105–118 (2001)
19. Reuter, J., Cendrollu, M., Evrard, F., Mostaghim, S., van Wachem, B.: Towards improving simulations of flows around spherical particles using genetic programming. In: 2022 IEEE Congress on Evolutionary Computation (CEC), pp. 1–8 (2022)
20. Richardson, J.F., Zaki, W.N.: The sedimentation of a suspension of uniform spheres under conditions of viscous flow. Chem. Eng. Sci. **3**(2), 65–73 (1954)
21. Ross, A.S., Li, Z., Perezhogin, P., Fernandez-Granda, C., Zanna, L.: Benchmarking of machine learning ocean subgrid parameterizations in an idealized model. In: Earth and Space Science Open Archive, p. 43 (2022)
22. Schiller, L., Naumann, A.: über die grundlegenden Berechnungen bei der Schwerkraftaufbereitung. Zeitschrift des Vereines Deutscher Ingenieure **77**, 318–320 (1933)
23. Schneiders, L., Meinke, M., Schröder, W.: Direct particle–fluid simulation of Kolmogorov-length-scale size particles in decaying isotropic turbulence. J. Fluid Mech. **819**, 188–227 (2017)
24. Seyed-Ahmadi, A., Wachs, A.: Microstructure-informed probability-driven point-particle model for hydrodynamic forces and torques in particle-laden flows. J. Fluid Mech. **900**, A21 (2020)
25. Seyed-Ahmadi, A., Wachs, A.: Physics-inspired architecture for neural network modeling of forces and torques in particle-laden flows. Comput. Fluids **238**, 105379 (2022)
26. Tenneti, S., Garg, R., Subramaniam, S.: Drag law for monodisperse gas-solid systems using particle-resolved direct numerical simulation of flow past fixed assemblies of spheres. Int. J. Multiph. Flow **37**(9), 1072–1092 (2011)
27. Udrescu, S.M., Tegmark, M.: AI Feynman: a physics-inspired method for symbolic regression. Sci. Adv. **6**(16), eaay2631 (2020)
28. Uhlmann, M., Chouippe, A.: Clustering and preferential concentration of finite-size particles in forced homogeneous-isotropic turbulence. J. Fluid Mech. **812**, 991–1023 (2017)
29. Wappler, S., Wegener, J.: Evolutionary unit testing of object-oriented software using strongly-typed genetic programming. In: Proceedings of the 8th Annual Conference on Genetic and Evolutionary Computation, p. 1925–1932 (2006)
30. Werner, M., Junginger, A., Hennig, P., Martius, G.: Informed equation learning. arXiv preprint arXiv:2105.06331 (2021)
31. Zille, H., Evrard, F., Reuter, J., Mostaghim, S., van Wachem, B.: Assessment of multi-objective and coevolutionary genetic programming for predicting the stokes flow around a sphere. In: 14th International Conference on Evolutionary and Deterministic Methods for Design, Optimization and Control (2021)

Phenotype Search Trajectory Networks for Linear Genetic Programming

Ting Hu[1]([✉]) [ID], Gabriela Ochoa[2] [ID], and Wolfgang Banzhaf[3] [ID]

[1] School of Computing, Queen's University, Kingston, ON K7L 2N8, Canada
ting.hu@queensu.ca
[2] University of Stirling, Stirling FK9 4LA, UK
gabriela.ochoa@stir.ac.uk
[3] Department of Computer Science and Engineering, BEACON Center
for the Study of Evolution in Action, and Ecology, Evolution and Behavior Program,
Michigan State University, East Lansing, MI 48864, USA
banzhafw@msu.edu

Abstract. In this study, we visualise the search trajectories of a genetic programming system as graph-based models, where nodes are genotypes/phenotypes and edges represent their mutational transitions. We also quantitatively measure the characteristics of phenotypes including their genotypic abundance (the requirement for neutrality) and Kolmogorov complexity. We connect these quantified metrics with search trajectory visualisations, and find that more complex phenotypes are under-represented by fewer genotypes and are harder for evolution to discover. Less complex phenotypes, on the other hand, are over-represented by genotypes, are easier to find, and frequently serve as stepping-stones for evolution.

Keywords: Neutral networks · Genotype-to-phenotype mapping · Algorithm modeling · Algorithm analysis · Search trajectories · Complex networks · Visualisation · Kolmogorov complexity

1 Introduction

Neutral networks have been found to play an important role in natural and artificial evolution [2,17]. The notion of neutral networks derives from the idea that a search space can be explored by neutral moves that do not change fitness, as well as by moves improving fitness. Nodes of such a network are the genotypes being visited and edges between them are the variation steps taken by a searcher on that network. Each node, being a genotype also carries a fitness which can be used to determine whether a move from one node to another node is allowed or not. Some researchers have claimed that neutral moves are extremely important to allow evolutionary progress to proceed [13,19], and our long-standing interest and understanding of the role of neutrality in genetic programming (GP) systems [1] is deepened by the examination we report here.

© The Author(s), under exclusive license to Springer Nature Switzerland AG 2023
G. Pappa et al. (Eds.): EuroGP 2023, LNCS 13986, pp. 52–67, 2023.
https://doi.org/10.1007/978-3-031-29573-7_4

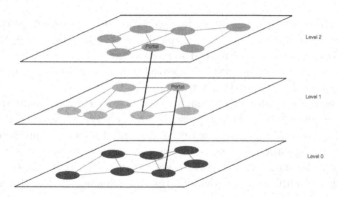

Fig. 1. Sketch of a network of neutral networks. Each level depicts one neutral network, with a discrete fitness value corresponding to its level. Nodes depict genotypes (genetic programs) which are connected within a level, reachable by neutral moves, with few nodes allowing jumps to a lower level (better fitness). The fitness of a node is measured by executing it and comparing the function it stands for with a target relation. The neutral networks are connected through what are called *portal* nodes to other neutral networks at a lower (better) fitness level.

Figure 1 shows a sketch of how to conceptualize the search space of a GP algorithm: As a network of (neutral) networks [9]. A single searcher in such a search process likely starts at a less-fit neutral network level (fitness level 2), and moves through the network by hopping from node to node via mutations or other variation operations that are mostly neutral. Occasionally, however, a *portal* node is found this way, which allows the searcher to enter another neutral network on a better fitness level. At that point, search again moves through the neutral network until it finds another portal.

Studying such search trajectories and connecting them with quantified metrics of genotypes and phenotypes allows to better understand the genotype-to-phenotype maps (G-P maps) and the search behaviour of evolutionary algorithms. In this research, we adapt a recent graph-based model, search trajectory networks (STNs) [15, 16] to analyse and visualise search trajectories of a simple linear GP system used to evolve Boolean functions. Search trajectory networks are a data-driven, graph-based model of search dynamics where nodes represent a given state of the search process and edges represent search progression between consecutive states. We connect this visualisation with an examination of the statistical behavior of those searchers navigating the corresponding genotype space. Following more recently formulated ideas about stiff G-P maps, we can tie the complexity of phenotypes to their potential for serving as stepping stones to a solution of the problem.

We define stiff G-P maps as those maps that have a strong correlation between the complexity of the genotype and the complexity of the corresponding phenotype. In nature, such maps can be found in the molecular world, and in computing they are found in the conventional maps of GP. However, there also exist other kinds of maps that allow the correlation between genotype complexity and phenotype complexity to relax. Typically, they are found in developmental systems allowing the complexity of the phenotype to grow over time under the influence of a genotype. The results reported here might not apply to such kind of maps, though they are certainly realizable in GP.

This research provides novel insights into how G-P maps result in the heterogeneity of phenotypes being represented by genotypes. Namely, it has been observed that some phenotypes are over-represented by a large number of genotypes (high redundancy) whereas others are under-represented by few (low redundancy). We measure the correlation between the redundancy of a phenotype and its complexity and demonstrate its influence on search trajectories toward a goal defined by fitness.

2   The LGP System

2.1   Boolean LGP Algorithm

The GP algorithm used in our research is a linear genetic programming (LGP) system where a sequential representation of computer programs is employed to encode an evolutionary individual [4]. Such a linear genetic program often consists of a set of imperative instructions to be executed sequentially. Registers are used to either read input variables (input registers) or to enable computational capacity (calculation register). One or more registers can be designated as the output register(s) such that the final stored value(s) after the program is executed will be the program's output.

In this study, we use an LGP algorithm for a three-input, one-output Boolean function search application, similar to our previously examined LGP system [10–12]. Each instruction has one return, two operands and one Boolean operator. The operator set has four Boolean functions {AND, OR, NAND, NOR}, any of which can be selected as the operator for an instruction. Three registers R_1, R_2, and R_3 receive the three Boolean inputs, and are write-protected in a linear genetic program. That is, they can only be used as an operand in an instruction. Registers R_0 and R_4 are calculation registers, and can be used as either a return or an operand. Register R_0 is also the designated output register, and the Boolean value stored in R_0 after a linear genetic program's execution will be the final output of the program. All calculation registers are initialized to FALSE before execution of a program. An example linear genetic program with three instructions is given as follows:

$$I_1 : R_4 = R_2 \text{ AND } R_3$$
$$I_2 : R_0 = R_1 \text{ OR } R_4$$
$$I_3 : R_0 = R_3 \text{ AND } R_0$$

A linear genetic program can have any number of instructions, however, for the ease of sampling in this study, we use linear genetic programs that have a fixed length of six instructions.

2.2   Genotype, Phenotype, and Fitness

The *genotype* in our GP algorithm is a unique linear genetic program. Since we have a finite set of registers and operators, as well as a fixed length for all programs, the genotype space is finite and we can calculate its size. For each instruction, two registers can be chosen as return registers and any of the five registers can be used as one of two operands. Finally, an operator can be picked from the set of four possible Boolean functions. Thus, there are $2 \times 5 \times 5 \times 4 = 200$ unique instructions. Given the fixed length of six instructions for all linear genetic programs, we have a total number of $200^6 = 6.4 \times 10^{13}$ possible different programs.

The *phenotype* in our GP algorithm is a Boolean relationship that maps three inputs to one output, represented by a linear genetic program, i.e., $f : \mathbf{B}^3 \rightarrow \mathbf{B}$, where $\mathbf{B} = \{\texttt{TRUE}, \texttt{FALSE}\}$. There are thus a total of $2^{2^3} = 256$ possible Boolean relationships. Having 6.4×10^{13} genotypes to encode 256 phenotypes, our LGP algorithm must have a highly redundant genotype-phenotype mapping. We define the *redundancy* of a phenotype as the total number of genotypes that map to it.

We choose the *fitness* of a linear genetic program as the deviation of the phenotype's behavior from a target Boolean function and want to minimize that deviation in the search process. Given three inputs, there are $2^3 = 8$ combinations of Boolean inputs. The Boolean relationship encoded by a linear genetic program can be seen as an 8-bit string representing the outputs that correspond to all 8 possible combinations of inputs. Formally, we define fitness as the Hamming distance of this 8-bit output and the target output. For instance, if the target relationship is $f(\mathrm{R}_1, \mathrm{R}_2, \mathrm{R}_3) = \mathrm{R}_1$ AND R_2 AND R_3, represented by the 8-bit output string of 00000001, the fitness of a program encoding the FALSE relationship, i.e., 00000000, is 1. Fitness is to be minimized and falls into the range between 0 and 8, where 0 is the perfect fitness and 8 is the worst.

3   Kolmogorov Complexity

Dingle et al. [5] report a very general result on complexity limited discrete input-output maps. Based on algorithmic information theory they state that the probability of finding certain outputs depends on their Kolmogorov complexity. In particular, the probability to find an output $x \in O$ can be bounded by a quantity that depends exponentially on its Kolmogorov complexity:

$$P(x) \leq 2^{-(K(x|f,n)+\mathcal{O}(1))}, \tag{1}$$

where $K(x|f,n)$ is the shortest program that produces x, given f and n, where f is the computable input-output map $f : I \rightarrow O$ and n characterizes the size of the input space. For binary inputs their number would be 2^n. While this gives only an upper bound, it is (negatively) exponentially dependent on complexity, and if one compares two outputs, this fact can be used to predict the prevalence of one output over the other. Even more astonishing, this estimate becomes independent of the particulars of the map, in the above mentioned asymptotic case of a limited complexity map where $K(f) + K(n) << K(x) + \mathcal{O}(1)$:

$$K(x|f, n) \approx K(x) + \mathcal{O}(1). \tag{2}$$

Later, Dingle et al. [7] apply these findings to a variety of systems, among them the RNA G-P map (from linear sequence to 2D structure) and to others.

Here we shall use these ideas to explain and predict the phenotypic trajectories of adaptive walkers in the fitness landscape of Boolean functions. In the context of our LGP algorithm, we define the Kolmogorov complexity (K-complexity) of a phenotype (Boolean relationship) as the minimal *effective* length of its underlying linear genetic programs. The effective length of a linear genetic program is the number of its effective instructions. An instruction of a program is effective when its execution influences the final result of the output, here the content of register R_0. We can then conceptualize the search process as an adaptive walk in the network of solutions (phenotypes), and, by repeating the process with a number of runs, we can visualise the prevalence of certain transitions (hops of searchers in the network).

4   Sampling and Metrics Estimation

Although finite, the genotype space of our LGP algorithm is enormous with a size of 6.4×10^{13} and can be challenging for exhaustive enumeration. Therefore, we randomly sample one billion linear genetic programs ($\approx 0.00156\%$ of the total possible programs) to approximate the genotype space.

These one billion programs are then mapped to the Boolean relationships (phenotypes) they represent, allowing us to estimate the redundancy of each phenotype as the total number of sampled genotypes that map to each phenotype. 239 out of the 256 phenotypes are represented by our sampled genotypes, among which phenotype FALSE has the greatest redundancy of almost 109 million genotypes, i.e., $> 1\%$ of the total number of sampled genotypes.

We first investigate the phenotypic effects of point mutations in our LGP system by sampling one million genotypes and their one-step mutants. Given the high redundancy in the G-P map, we observe that about 73.8% of the sampled point mutations are neutral. For the 26.2% non-neutral mutations, we compute the phenotypes of the genotype pairs for each mutation and measure the Hamming distance of these phenotypes. Figure 2A shows the distribution of such pairwise phenotypic distances. We can see that the majority of non-neutral point mutations results in small phenotypic changes but also that there is a substantial number of mutations with larger step sizes (4 or even 8 bits).

Next, we would like to examine the relation between redundancy of a phenotype and its complexity. Recall that the K-complexity of a phenotype is the minimal effective length of its underlying programs. Thus, the goal is to search for the shortest effective program that can encode a given Boolean function (phenotype). Again we randomly sample one billion linear genetic programs with varying lengths drawn from the range between 5 and 20. We then perform

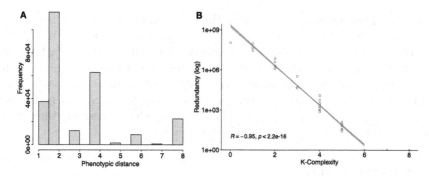

Fig. 2. Estimated metrics and characteristics. (A) Distribution of phenotypic distance of one-step genotype neighbours. Neutral mutations (73.8% of all mutations) are excluded from the graph. Spikes at even distances are caused by the fitness function. (B) Correlation of redundancy and K-complexity in log-linear scale. Phenotypes are represented as circles. Note the log scale; the straight line is the best fit to an exponential decay, and the shadow depicts the 95% confidence interval of the linear regression. Spearman's rank correlation coefficient R and p-value are also provided.

the structural intron removal algorithm [4] to identify the effective length of each program. We record for each phenotype the minimal effective length of its sampled underlying programs, and use it to estimate the K-complexity of that phenotype. Figure 2B shows the correlation of redundancy and K-complexity for all the phenotypes we sampled and measured. A strong and significant negative correlation is observed, which means that more complex phenotypes are represented by fewer genotypes, as suggested by [5].

To study the search trajectories for our LGP system, we perform adaptive walks where only neutral or improving point mutations are accepted. For a comparison, we set three target phenotypes with increasing difficulties, i.e., an easy target of phenotype 240 (redundancy 46 million, K-complexity 1), a medium target 20 (redundancy 3130, K-complexity 4), and a hard target 30 (redundancy 772, K-complexity 4). Two search scenarios are implemented, where first we always start with a randomly generated genotype of the most distant phenotype from the target, i.e., fitness of 8, and second we randomly generate a genotype without any consideration on its fitness. We call the first scenario *fixed start* search and the second *random start* search. We collect 100 runs for each scenario with each target phenotype, where in each run we initialize a linear genetic program and let it walk in the genotypic space for 2,000 steps. These results are used for the visualisation of the search trajectories.

5   Search Trajectory Networks

Search trajectory networks (STNs) [15, 16] are a graph-based tool to visualise and analyse the dynamics of any type of meta-heuristic: evolutionary, swarm-based or single-point, on both continuous and discrete search spaces. Originally, the model tracks the trajectories of search algorithms in genotypic space, where nodes represent visited genotypes. However, for very large search spaces, techniques have been proposed to cluster sets of genotypes into *locations* [16] which can even group genotypes with the same phenotype or behavior [18], in order to have coarser models that can be visualised and interpreted.

In order to define a graph-based model, we need to specify its nodes and edges. We start by giving these general definitions before describing three STN models we propose here to visualise GP search spaces.

5.1 General Definitions

Representative solution. A solution (genotype) to the optimization problem at a given time step that represents the status of the search algorithm (e.g. best in the population in a given iteration, incumbent solution for single point meta-heuristics).

Location. A non-empty subset of solutions that results from a predefined coarsening of the search space.

Trajectory. Given a sequence of representative solutions in the order in which they are encountered during the search process, a search trajectory is defined as a sequence of locations formed by replacing each solution with its corresponding location.

Nodes (N). The set of locations in a search trajectory of the search process being modeled.

Edges (E). Directed, connecting two consecutive nodes in the search trajectory. Edges are weighted with the number of times a transition between two given nodes occurred during the process of sampling and constructing the STN.

STN. Directed graph STN $= (N, E)$, with nodes N and edges E as defined above.

5.2 The Proposed STN Models

We propose three models with increasing coarsening, that is, with nodes grouping an increasing number of candidate solutions, in order to visualise the large and extremely neutral LGP search space under study.

1. **Genotype STN.** The locations (nodes) are unique genotypes in the search space, and edges represent transitions between genotypes.
2. **Genotype-Phenotype STN.** The locations (nodes) are phenotypes grouping connected components in the Genotype STN that share the same phenotype. Edges represent transitions between (compressed) nodes.
3. **Phenotype STN.** The locations (nodes) are unique phenotypes in the search space, and edges represent consecutive transitions between phenotypes.

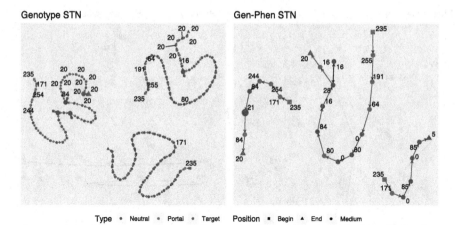

Fig. 3. Visualisation of the Genotype and Genotype-Phenotype STN models for target 20, using a force-directed graph layout. (Color figure online)

For constructing the STN models, multiple runs of adaptive walks (described in Sect. 4) are performed, and the visited locations and their transitions are aggregated into a single graph model. Notice that some locations and transitions may appear multiple times during the sampling process. However, the graph model retains as nodes each unique location, and as edges each unique transition between visited locations. Counters are maintained as attributes of the graph, indicating the frequency of occurrence of each (unique) node and edge.

5.3 Network Visualisation

Visualisation is a powerful and aesthetically inspiring way of appreciating network structure, which can offer insights not easily captured by network metrics alone. Node-edge diagrams are the most familiar form of network visualisation, where nodes are assigned to points in the two-dimensional Euclidean space and edges connect adjacent nodes by lines or curves. Nodes and edges can be decorated with visual properties such as size, color and shape to highlight relevant characteristics.

To illustrate our proposed STN models, we conduct a preliminary experiment using phenotype target 20 (medium difficulty), with three runs and 50 steps for the adaptive walks. Each run starts from a randomly generated genotype that has phenotype 235. Figures 3 and 4 illustrate the STN models. Our STN visualisations use node colors to identify four types of nodes: (1) neutral nodes, whose adjacent outgoing node has the same fitness, (2) portals, which link to a node with improved fitness, (3) target nodes, which have the required phenotype, and (4) (for the phenotype STNs only), we differentiate portal nodes with a direct link to the target. The shape of nodes identifies three positions in the search trajectories: (1) begin of trajectories, (2) end of trajectories, (3) intermediate

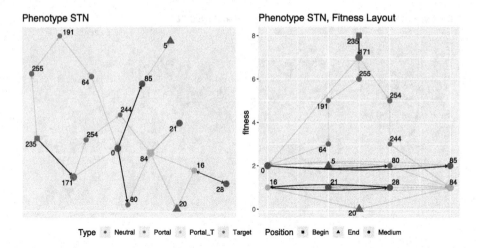

Fig. 4. Visualisation of the Phenotype STN model for target 20, using a force-directed graph layout (left) and a grid layout with fitness in the y coordinate. (Color figure online)

locations in the trajectories. Node labels indicate phenotype, while node sizes and edge darkness are proportional to their sampling frequency.

On the genotype STN (left plot) in Fig. 3 we can observe the three trajectories corresponding to the three adaptive walks conducted. The trajectories are long (remember walks have 50 steps in this experiment) and do not overlap, that is, they all visit different genotypes. Two of the trajectories reach the target (phenotype 20) while one of them ends in a different phenotype. To avoid a cluttered image, the genotype STN plot shows the node labels for portal and target nodes only. Notice the long chains of neutral nodes (dark gray) before finding a portal (blue nodes) to improving fitness, also several different genotypes in red correspond to the target phenotype. The genotype-phenotype STN (right plot) shows shorter trajectories as expected as nodes now represent subnetworks joining connected genotypes with the same phenotype. Still, the three trajectories do not have overlapping nodes, indicating that the three walks visit different regions of the search space. Interestingly, there are still long chains of neutral moves (dark gray nodes), especially visible in the middle trajectory; we can see how the trajectory enters in and out of phenotypes 0 and 80, before finding a portal to phenotype 84.

A key aspect of network visualisation is the graph-layout, which accounts for the positions of nodes in the 2D Euclidean space. Graphs are mathematical objects, they do not have a unique visual representation. Many graph-layout algorithms have been proposed. *Force-directed* layout algorithms [8], are based on physical analogies defining attracting and repelling forces among nodes. They strive to satisfy generally accepted aesthetic criteria such as an even distribution of nodes on the plane, minimizing edge crossings, and keeping a similar edge lengths. We use force-directed layouts for visualizing the STNs models in Figs. 3

and 4 (left plot). For the phenotype STNs, we also introduce a layout that takes advantage of the fitness values. The idea is to use the fitness values as the nodes' y coordinates, while the x coordinates are placed as a simple grid (Figs. 4 and 5), where nodes are centered according to the number of nodes per fitness level. These plots allow us to appreciate the progression of the search trajectories towards lower (better) fitness values, as well as the amount of neutrality present in the search space.

Our graph visualisations were produced using the igraph and ggraph packages of the R programming language. The phenotype STN model seen in Fig. 4 is more compact, having fewer nodes and edges as compared the the genotype and genotype-phenotype STN models. Most importantly, the phenotype STN model shows search overlaps across the different trajectories. That is, there are nodes that have more than one incoming edge, they are hubs, indicating locations that attract the search process. For the remainder of our analyses, we decided to use the phenotype STN model with the fitness-based graph layout. We argue that this combination has a greater potential to reveal interesting aspects of the search dynamic, as it allows the observation of locations of the search space where the process converges. The other models are however interesting to appreciate additional details.

5.4   Comparing Three Targets with Increasing Difficulty

As described in Sect. 4, for adaptive walks we set three targets with increasing difficulties (240: easy; 20: medium; 30: hard) and two search scenarios (fixed start and random start). Figure 5 shows the phenotype STNs for the six configurations. The nodes and edges are as defined in Sect. 5, the fitness-based graph layout is used, and the arrow heads as well as the node labels are omitted to keep the images less cluttered. Notice that the edges are either descending to lower fitness levels or neutral at the same fitness levels. The neutral edges are visualised as curves where the edges above point to the left and the edges below point to the right.

Search proceeds through hops, indicated by links of different darkness symbolizing how often they were traversed during the sampling process. The target node (red triangle at the bottom of each graph) is reached via different search pathways. The size of nodes is proportional to how many times it was visited during the adaptive walks, so large nodes represent locations that attract the search process. For the medium and hard targets (phenotypes 20 and 30), many search trajectories do not reach the target, they end at phenotypes with Hamming distance 1, i.e. close to the target (visualised by large pink triangles at fitness level 1).

It is interesting to observe that the varying size of nodes is more pronounced for more difficult targets, signalling that transitions have become more heterogeneous at those levels. Clearly, the landscape becomes more difficult to navigate closer to a difficult target and the number of one-step mutant neighbors to a target is smaller than for an easy target.

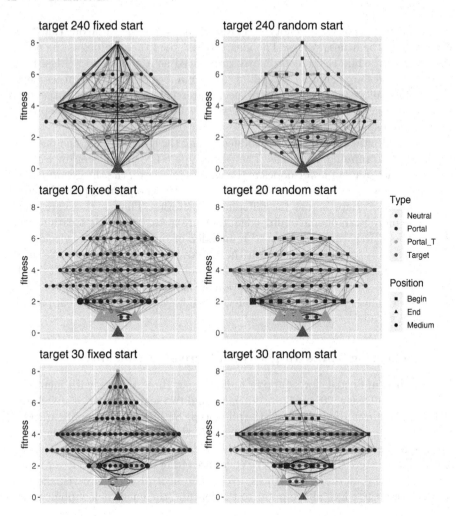

Fig. 5. Phenotype STNs when searching for target phenotypes of different difficulty (240: easy; 20: medium; 30: hard). The plots aggregate 100 trajectories, which start from either a fixed phenotype (left plots) or a random phenotype (right plots). The target node (red triangle at the bottom of each graph) is reached via different search pathways. The size of nodes and the darkness of edges indicate their sampling frequency. Arrow heads and node labels are omitted to simplify the images. (Color figure online)

We can see that most phenotypes are portals (blue nodes) offering the possibility of jumping to a lower level fitness, but clearly, many neutral moves happen on the way to the target, at each fitness level.

The graph layout reflects the structure of the search space - most phenotypes are located at around half Hamming distance to the target, that is at fitness levels 3, 4 and 5. The square node at the top of the left plots reflects the fact that these trajectories start with a fixed phenotype, while the bottom triangle in all

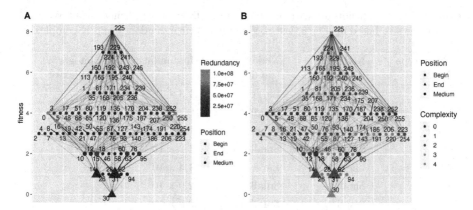

Fig. 6. Fitness changes for target 30 fixed start runs, with phenotype redundancy (A) and Kolmogorov complexity (B) marked by color. Only non-neutral mutations that improve fitness are shown as edges. Nodes are labeled with numbers representing their phenotypes. Shapes stand for different positions in a search. Size of a node indicates frequency of visit.

plots reflects that the target was found. Notice that the size of the red triangle is the largest for the easy target 240, and then gradually decreases in size for targets 20 and 30. This makes sense as the harder the target the less frequently it is reached by the search process within 2,000 steps.

6   Discussion

We now want to connect the observations from these visualisations with the complexity considerations mentioned in Sect. 3. We focus first on target 30, shown in the last row of Fig. 5.

We are interested in more details of the search, especially given the heterogeneity at the end of the search, close to the target. Table 1 shows the phenotypes closest to the target (one-bit mutants) found by the searchers. In Fig. 6, we look at the frequency of fitness-changing jumps from phenotype to phenotype in fixed start runs. We label each node with the phenotype it stands for, with a side by side comparison of nodes color-marked by redundancy (A) and complexity (B). In this figure we have removed the neutral edges to declutter the images.

We can see that their size strongly correlates with both their redundancy (positively) and with their complexity (negatively). Recall that larger node size indicates more frequent visits by searchers in the process of looking for the target. The exponential relationship indicated by Eq. 1 seems to bear out: Searchers are much more likely to pass through low complexity/high redundancy nodes – in this case phenotypes 14 and 31 – than through the other one-bit mutant neighbors found, 26, 28, 62 or 94.

We can extend this analysis to the mutants of the target with two-bit phenotypic distances as Fig. 6 shows all fitness changing moves of searchers for

Table 1. Mutant phenotypes with one-bit distance from target phenotype 30. We characterize the size based on Fig. 5, last row and list their redundancy and K-complexity.

Phenotype Number	Node Size (fixed start)	Node Size (random start)	Redundancy	Kolmogorov Complexity
14	Large	Large	1.3×10^6	2
22	N/A	Small	0	8
26	Small	Small	1.2×10^3	4
28	Small	Small	1.2×10^3	4
31	Large	Large	1.4×10^6	2
62	Small	Small	2.9×10^3	4
94	Small	Small	2.9×10^3	4

Table 2. Selected two-bit mutants of phenotype 30: One-bit mutants to the most frequent 1-bit neighbors 14 and 31 of the target node 30.

Phenotype (to 14)	Redundancy	Kolmogorov Complexity	Phenotype (to 31)	Redundancy	Kolmogorov Complexity
6	3.0×10^3	4	15	4.7×10^7	1
10	7.1×10^6	2	23	5.5×10^3	4
12	7.1×10^6	2	27	1.2×10^4	4
15	4.7×10^7	1	29	1.2×10^4	4
46	4.6×10^4	3	63	2.9×10^8	1
78	4.6×10^4	3	95	2.9×10^8	1
142	2.0×10^3	4	159	3.1×10^3	4

target 30. If we focus on two-bit mutants (fitness 2), we can see that most transitions happen from the highly redundant phenotypes, first 15, followed by transitions from 63 and 95. Most of them transition to the highly redundant phenotypes 14 and 31 on fitness level 1. We can examine in more detail the redundancy/complexity of two-bit mutants. Due to the quick combinatorial explosion, we have done that in Table 2 only for the two most representative nodes of fitness distance 1, phenotypes 14 and 31. Nodes 15, 63 and 95 stand out as the most redundant nodes. They thus provide most avenues to better fitness, with phenotypes 10 and 12 doing the same to a somewhat lesser extent.

Thus we can explain the dynamics of the search process post-facto by looking at the redundancy/complexity of phenotypes in the neighborhood of the target. We do not need to know many details of the search, except what constitutes the neighborhood of a node, to figure out where most searchers will come from.

Both, redundancy and complexity, require – of course – measurements to allow this explanation. While they are different (redundancy can be measured for all nodes in parallel), it might be argued that one has to have a clear picture of the fitness landscape for this analysis. This is correct for a measurement of

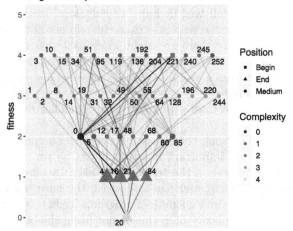

Fig. 7. Fitness changes for target 20 with fixed starting phenotype 221 (the square). Fitness-improving search trajectories for 100 runs with 2,000 steps, all going through low-complexity phenotypes to the target from the starting point. Phenotype complexity is marked by color.

redundancy, but the relationship with complexity is not based on anything other than the structure of the phenotypes themselves. Thus, in principle, it can be performed completely separate from the search. Evolution is doing here nothing else than seeking out the most probable pathways to the target. In other words, we can not only explain the search dynamics post-facto, but we can try to predict, at least approximately, a search dynamic before it happens. This is in line with what other research groups have found in their respective systems [3,6,14].

Suppose we start at phenotype 221 to reach target 20 (see Fig. 7). The one-bit neighbors of 221 are the set: $\{93, 157, 205, 213, 217, 220, 223, 253\}$. Some of those nodes are mutants pointing in the wrong direction of fitness, and can be removed from this list because selection would not allow them. That leaves us with $\{93, 157, 213, 217, 220\}$. However, a brief inspection of the redundancies of these phenotypes tells us that they are considerably less redundant (1.3×10^6 maximum for 213) than the phenotypes of the neutral network, which has nodes of redundancies of up to 4.7×10^7. As a result, the nodes $\{15, 51, 204, 240\}$ are the most likely nodes to be accessed from 221 on the neutral level and more likely than the one-bit mutations. In fact, we can see that only phenotype 220 of the one-bit mutants appears to be accessed. There is one interesting twist here: Phenotype 0, being the most likely phenotype in the whole network, is a two-bit mutation from 221 in the right direction. We can see that it is accessed more frequently than other nodes, both directly and indirectly from 221. Also, phenotype 48, with an redundancy of 7.1×10^6 is accessed, again a two-bit mutation from 221. The figure shows that both nodes have lower complexity than 221 which we know is correlated with their redundancy. If we recall Fig. 2A, statistics shows that two-bit mutations are actually more frequent than one-bit mutations in this system, followed by four-bit mutations. It seems that the step size is less of a concern for the searchers than the redundancy of phenotypes!

We note in passing that there are many more pathways to a better fitness solution when not only the rearrangement of instructions is possible (as would be the case in a transition from a program with six effective instructions) but when also an increase in the number of effective instructions were possible (as would be the case in a transition from a program with a smaller number of effective instructions).

Why is there such a strong correlation between phenotypic redundancy and K-complexity? This is an important question since - as we have seen - redundancy has such an influence on the trajectories taken by adaptive searchers in this fitness landscape. The answer has to do with the hard length limit in our system, which allows a maximum length of programs of six instructions. Suppose a phenotype has a K-complexity of 2, thus is not using the other 4 instructions theoretically available, they are rendered non-effective. A brief combinatorial consideration allows us to estimate that there are maximally $200^4 = 1.6 \times 10^9$ programs with four neutral instructions, assuming all calculation registers are used as a destination. This will be an upper limit, of course, as many of those might well not be neutral, either by virtue of their order or because of their internal composition. Nevertheless, it is a huge number of neutral variations of the same program. Compare that to an individual with five out of the six instructions being effective. There only is one instruction left that can be neutral, leaving a maximum of 200 neutral variations for this program.[1]

In summary, the reason why K-complexity is negatively correlated with redundancy of programs and thus phenotypes is the combinatorics in the neutral space! While this presupposes a hard limit on the total length of programs (effective plus non-effective code), a soft limit can allow similar effects to play out, like in RNA. It will probably not be as clearly visible, but should still be expected to emerge in such systems. These considerations are not restricted to the particular system examined here: The combinatorics of neutral spaces determines the redundancy of phenotypes and thus to a substantial degree the search trajectories in length-changing evolutionary systems in general.

References

1. Banzhaf, W.: Genotype-phenotype-mapping and neutral variation—a case study in genetic programming. In: Davidor, Y., Schwefel, H.-P., Männer, R. (eds.) PPSN 1994. LNCS, vol. 866, pp. 322–332. Springer, Heidelberg (1994). https://doi.org/10.1007/3-540-58484-6_276
2. Banzhaf, W., Leier, A.: Evolution on neutral networks in genetic programming. In: Yu, T., Riolo, R., Worzel, B. (eds.) Genetic Programming – Theory and Practice III, pp. 207–221. Kluwer (2006)

---

[1] Note that multiple programs can contribute to the same phenotype as specified by its behavior.

3. Barrick, J.E.: Limits to predicting evolution: insights from a long-term experiment with *Escherichia coli*. In: Evolution in Action: Past, Present and Future. GEC, pp. 63–76. Springer, Cham (2020). https://doi.org/10.1007/978-3-030-39831-6_7
4. Brameier, M., Banzhaf, W.: Linear Genetic Programming. Springer, Heidelberg (2007). https://doi.org/10.1007/978-0-387-31030-5
5. Dingle, K., Camargo, C., Louis, A.: Input-output maps are strongly biased towards simple outputs. Nat. Commun. **9**, 761 (2018)
6. Dingle, K., Novev, J., Ahnert, S., Louis, A.: Predicting phenotype transition probabilities via conditional algorithmic probability approximations. J. Roy. Soc. Interface (2023)
7. Dingle, K., Valle Perez, G., Louis, A.: Generic predictions of output probability based on complexities of inputs and outputs. Sci. Rep. **10**, 4415 (2020)
8. Fruchterman, T.M.J., Reingold, E.M.: Graph drawing by force-directed placement. Softw. Pract. Exp. **21**(11), 1129–1164 (1991)
9. Gao, J., Li, D., Havlin, S.: From a single network to a network of networks. Natl. Sci. Rev. **1**, 346–356 (2014)
10. Hu, T., Banzhaf, W.: Neutrality and variability: two sides of evolvability in linear genetic programming. In: Rothlauf, F., et al. (eds.) Proceedings of the 11th Annual Conference on Genetic and Evolutionary Computation, pp. 963–970 (2009)
11. Hu, T., Payne, J.L., Banzhaf, W., Moore, J.H.: Robustness, evolvability, and accessibility in linear genetic programming. In: Silva, S., Foster, J.A., Nicolau, M., Machado, P., Giacobini, M. (eds.) EuroGP 2011. LNCS, vol. 6621, pp. 13–24. Springer, Heidelberg (2011). https://doi.org/10.1007/978-3-642-20407-4_2
12. Hu, T., Payne, J.L., Banzhaf, W., Moore, J.H.: Evolutionary dynamics on multiple scales: a quantitative analysis of the interplay between genotype, phenotype, and fitness in linear genetic programming. Genet. Program. Evol. Mach. **13**, 305–337 (2012)
13. Kimura, M.: The Neutral Theory of Molecular Evolution. Cambridge University Press, Cambridge (1983)
14. Lobkovsky, A.E., Wolf, Y.I., Koonin, E.V.: Predictability of evolutionary trajectories in fitness landscapes. PLoS Comput. Biol. **7**(12), e1002302 (2011)
15. Ochoa, G., Malan, K.M., Blum, C.: Search trajectory networks of population-based algorithms in continuous spaces. In: Castillo, P.A., Jiménez Laredo, J.L., Fernández de Vega, F. (eds.) EvoApplications 2020. LNCS, vol. 12104, pp. 70–85. Springer, Cham (2020). https://doi.org/10.1007/978-3-030-43722-0_5
16. Ochoa, G., Malan, K.M., Blum, C.: Search trajectory networks: a tool for analysing and visualising the behaviour of metaheuristics. Appl. Soft Comput. **109**, 107492 (2021)
17. Reidys, C., Stadler, P., Schuster, P.: Generic properties of combinatory maps: neutral networks of RNA secondary structures. Bull. Math. Biol. **59**, 339–397 (1997)
18. Sarti, S., Adair, J., Ochoa, G.: Neuroevolution trajectory networks of the behaviour space. In: Jiménez Laredo, J.L., Hidalgo, J.I., Babaagba, K.O. (eds.) EvoApplications 2022. LNCS, vol. 13224, pp. 685–703. Springer, Cham (2022). https://doi.org/10.1007/978-3-031-02462-7_43
19. Wright, A.H., Laue, C.L.: Evolvability and complexity properties of the digital circuit genotype-phenotype map. In: Proceedings of the Annual Conference on Genetic and Evolutionary Computation, pp. 840–848 (2021)

GPAM: Genetic Programming
with Associative Memory

Tadeas Juza and Lukas Sekanina[(✉)][iD]

Faculty of Information Technology, Brno University of Technology, Bozetechova 2,
612 66 Brno, Czech Republic
xjuzat00@stud.fit.vutbr.cz, sekanina@fit.vutbr.cz

Abstract. We focus on the evolutionary design of programs capable of capturing more randomness and outliers in the input data set than the standard genetic programming (GP)-based methods typically allow. We propose Genetic Programming with Associative Memory (GPAM) – a GP-based system for symbolic regression which can utilize a small associative memory to store various data points to better approximate the original data set. The method is evaluated on five standard benchmarks in which a certain number of data points is replaced by randomly generated values. In another case study, GPAM is used as an on-chip generator capable of approximating the weights for a convolutional neural network (CNN) to reduce the access to an external weight memory. Using Cartesian genetic programming (CGP), we evolved expression-memory pairs that can generate weights of a single CNN layer. If the associative memory contains 10% of the original weights, the weight generator evolved for a convolutional layer can approximate the original weights such that the CNN utilizing the generated weights shows less than a 1% drop in the classification accuracy on the MNIST data set.

Keywords: Genetic programming · Associative memory · Neural network · Weight compression · Symbolic regression

1 Introduction

By *symbolic regression*, we mean a search in the space of mathematical expressions (or programs) to find the model that best fits a given data set [5]. When genetic programming (GP) is employed to solve a symbolic regression problem, the search is conducted by an evolutionary algorithm utilizing a specific problem encoding and genetic operators. GP often provides unique solutions because resulting programs can be composed of arbitrary elementary functions, under arbitrary constraints, and in a multi-objective optimization scenario allowing one to concurrently minimize the program's error and other parameters such as solution size.

In this paper, we focus on the evolutionary design of expressions capable of capturing more randomness and even outliers in the input data set than the

G. Pappa et al. (Eds.): EuroGP 2023, LNCS 13986, pp. 68–83, 2023.
https://doi.org/10.1007/978-3-031-29573-7_5

common GP-based methods typically allow. We propose *Genetic Programming with Associative Memory* (GPAM) – a GP-based system for symbolic regression which can utilize a small associative memory to store various data points to better approximate the original data set. Suppose the data set D is composed of pairs (x_i, y_i). During evolution, GPAM is building an expression $G(x)$ and updating an associative memory $AM(x)$ with the goal of minimizing the error w.r.t a chosen error metric. The resulting pair (G, AM) then produces the output y' such that if x is in AM then $y' = AM(x)$, otherwise $y' = G(x)$. In principle, the condition for reading from AM can be more complex; for example, if the input is somehow similar to x (but not exactly x) then AM can still be used. Another issue is the proper sizing of AM because AM represents an overhead that has to be considered during the evaluation.

To evaluate GPAM, we propose a new benchmark set. It consists of data points generated by the mathematical functions commonly used for benchmarking GP-based symbolic regression methods. However, some fraction of data points is replaced by randomly generated values. Note that this approach does not strictly follow the standard symbolic regression scenario because we assume no test data set.

The second part of this paper is devoted to the case study, which in fact, motivated the design of GPAM. It deals with the efficient processing of deep neural networks (DNN) on edge devices. When DNN inference engine is implemented as a specialized accelerator on a chip, the accelerator has to perform millions of accesses to the external weight memory (DRAM) to compute a single output (e.g., to determine the class in a classification task). The energy needed for one access to DRAM can be 173× higher than the single multiplication energy [15]. Hence, various techniques have been developed to reduce this cost, for example, bit width reduction and weight compression [2,4].

In this work, we propose a more radical approach. To avoid many expensive memory transfers, we try to approximate the weights using a suitable expression and a small memory directly on a chip. Hence, for a given set of weights (e.g., those belonging to one DNN layer), GPAM is employed to deliver a simple program that can access a small local memory. The program's execution will emulate a stream of weights that has to be read (under the standard setup) from external memory during one inference. As this is a completely new approach, we present a proof of concept of weight generation for a small convolutional neural network (CNN) trained as an MNIST classifier [9]. Using Cartesian genetic programming (CGP) [11], we evolved expression-AM pairs showing less than 1% drop in the accuracy of classification (with respect to the original trained CNN) if the AM size is 10% of the original weight memory size.

The rest of the paper is organized as follows. Section 2 briefly surveys relevant research. The proposed method (GPAM) is presented in Sect. 3. Results for five basic symbolic regression problems are reported in Sect. 4. The utilization of GPAM for weight generation in CNNs is evaluated in Sect. 5. Finally, discussion and conclusions are provided in Sect. 6.

2 Related Work

The related work summarizes genetic-based symbolic regression and techniques used to process DNNs efficiently, including weight compression techniques.

2.1 Symbolic Regression and Genetic Programming

The goal of *symbolic regression* is to find a mapping $y(x) = F(x)$ using a data set of paired examples $D = \{(x_i, y_i)\}_{i=1}^n$. GP excels in this task, especially if the problem size is reasonable and additional objectives (such as minimizing the model size) are defined. The pioneering work of John Koza was performed with tree GP [5]. This approach first specifies a set of elementary functions and a set of constants that can be used as building blocks of candidate expressions. Then, it defines operators enabling the construction of completely random expressions and modifying them (mutation, crossover, etc.). An evolutionary algorithm is employed to search in the space of such expressions. A candidate solution's quality (fitness) is obtained by executing it on a data set and measuring the error concerning desired objectives.

Later, other GP branches such as linear genetic programming and Cartesian genetic programming [11] were utilized for symbolic regression. Co-evolutionary GP [12], semantic GP [17], lexicase selection [7], and other methods have improved the quality of the original approaches. Detailed benchmarking of relevant GP-based methods is available in [6].

The concept of *memory* has been incorporated into GP by Koza [5] in the form of read and write functions working with a small fixed number of individually named storage elements. Teller [16] has extended this approach by introducing a linear array of indexed memory elements. Memory is typically required when an agent or a robot controller is evolved. Later, Langdon [8] showed that GP can automatically evolve simple abstract data structures such as stacks, queues, and lists. In another research direction, GP is employed to evolve the so-called *associative memories* – simple models developed to recall output patterns for certain input patterns using simple operations [18]. These models are usually understood as a subclass of artificial neural networks; thus, they differ from the ordinary content-addressable memory considered in this paper.

2.2 Efficient Processing of DNNs

As applications utilizing DNNs are now implemented using small, resource-constrained devices (e.g., mobile phones or nodes of the Internet of Things), efficient processing of DNNs has emerged as an essential topic for research and industry [1,15]. Hardware accelerators of DNN inference usually consist of the following major components: (1) an array of processing units – each of them implementing elementary arithmetic operations of DNNs such as "multiply and accumulate" (MAC); (2) memory subsystems including registers, a small but fast local memory (buffers), and large external memory for the weights; (3) a programmable controller.

An input (trained) DNN is mapped on the resources available in the accelerator to minimize the energy or latency of the inference process. Table 1 clearly
indicates that reducing the access to external memory (DRAM) has to be optimized with the highest priority. It has to be noted that one memory access can
require 128× more energy than reading the local memory (SRAM), and 173×
more energy than float multiplication in a MAC unit.

Table 1. Energy of operations on a chip (for 45 nm technology) [15].

Operation	Type	Width	Energy [pJ]
Add	Int	8	0.003
Add	Int	32	0.1
Add	Float	32	0.9
Multiply	Int	8	0.2
Multiply	Int	32	3.1
Multiply	Float	32	3.7
Read	SRAM	32	5.0
Read	DRAM	32	640

2.3 Weight Compression

A straightforward approach to reducing the energy associated with memory
accesses is minimizing the size of the weight memory. Various techniques in
this direction that are relevant to DNNs were proposed in [4].

In the *weight sharing* method (also known as the *weight compression*), a
group of similar weights is replaced by a single value. Instead of storing all the
DNN weights, only a limited number of shared values and a codebook, where
original weights are replaced by their corresponding indexes, are stored in the
external memory. Shared weight values are typically obtained with a clustering
algorithm like K-means. The weight-sharing principle can be applied at the entire
CNN level or separately for each layer. In [2], NSGA-II is adopted to find the
most suitable trade-off between the compression rate and the DNN accuracy
drop, which is unavoidable when a lossy compression scheme is applied. The
results obtained for recent CNN models, trained with the ImageNet dataset,
exhibit over 5× memory compression rate at an acceptable accuracy loss.

Another relevant work is an indirect encoding for neuroevolution called *compositional pattern-producing networks* (CPPNs) [13]. CPPNs are inspired by
developmental processes in organisms. They use simple functions and their com-
positions to generate complex behaviors. One of their applications is to generate
patterns of weights in neural networks. Suppose a two-dimensional field (layer)
in which the neuron's position is defined by coordinates (a, b). Similarly, a neuron
has coordinates (c, d) in another field. Then the weight of a connection between
these two neurons can be expressed as a function f (encoded as a CPPN), i.e.

$w_{(a,b)\to(c,d)} = f(a, b, c, d)$ [14]. We thus obtain a simple and scalable prescription (defined by CPPN) that can generate any weight between neurons in the two fields independent of their sizes. This mechanism can be seen as a weight generator, eliminating the need to store the weights between two layers. This approach has many interesting applications exploiting regularity, symmetry, and self-repetition provided by the used encoding.

3 Proposed Method

In this section, we first introduce GPAM in general. Then, its particular application to weight compression will be presented.

3.1 The GPAM Approach

Many outliers and even completely random values can exist in data sets collected in real-world applications. Expressions (programs) typically evolved with GP using elementary functions (composed of building blocks) are not usually able to capture them perfectly. However, the exact values of some of these outliers can be crucial when the evolved model is applied. Furthermore, there are use cases (Sect. 3.2) in which we need to model a given data set, and *no generalization is needed* concerning additional data points.

One of the possible solutions is GP utilizing associative memory. GP then evolves an expression capable of modeling the vast majority of data points but can employ a small associative memory to memorize some data points directly. Associative memory (also called content-addressable memory) is a special digital memory used in various high-speed searching applications. It compares input data (a key) against a table of stored data (keys), and returns the address of matching data or an associated data record directly.

Let us assume a data set D consisting of n pairs (x_i, y_i). However, this one-dimensional case can easily be generalized to other dimensions. In GPAM, GP is used to concurrently evolve expression G and fill in a small associative memory (AM) with a subset D_{AM} ($k_{AM} = |D_{AM}|$) of data points from D, where $k_{AM} \ll n$. The resulting value y' is computed using Eq. 1:

$$y'(x) = \begin{cases} AM(x) & \text{if } x \text{ is in } AM \\ G(x) & \text{otherwise} \end{cases} \tag{1}$$

It is assumed that the condition "x is in AM" can be evaluated quickly; this is possible with specialized hardware. The condition can be modified; for example, one can test whether a value *similar* to x_i is in AM. The similarity test can be based on rounding, truncation, XOR-ing across the input dimensions, or other transformations of x_i. In this paper, we deal with the exact equality test solely.

We propose to use CGP to evolve G because we seek a hardware-friendly implementation of the arithmetic expression, which is a typical output of

CGP [11]. CGP evolves directed acyclic graphs mapped onto a two-dimensional $n_c \times n_r$ (columns $\times$ rows) array of programmable nodes [11]. Our experiments will be performed with nodes implementing floating-point arithmetic operations; the use of reduced bit fix-point number representation is planned for future work. The fitness function is based on minimizing an error metric; in particular, we used the following function:

$$Fitness(G, AM, D) = \sum_{i=1}^{n}(y_i - y'(x_i))^2. \tag{2}$$

In CGP [11], a candidate solution with n_i inputs and n_o outputs is represented by a string of integers. CGP usually employs only a mutation operator which modifies c_m integers of the chromosome. During the CGP initialization phase, the initial expressions are randomly generated, and AM is seeded with k_{AM} randomly selected pairs from D. The content of AM can be modified by another mutation operator, which replaces a pair randomly picked in AM with another pair randomly taken from D, with the probability p_{mutmem}.

3.2   GPAM for Weight Generation

Convolutional neural networks are complex computational models that must be designed and then trained using suitable data from a given application domain [3]. To reduce memory access to external memory (and thus energy) in hardware accelerators of CNN inference, we utilize GPAM to evolve an internal on-chip configurable generator of CNN weights. The goal of this paper is to evaluate whether such a generator can, in principle, be constructed. Its hardware implementation will be the subject of another study.

Suppose that a CNN-based image classifier is fully trained, its classification accuracy on a test set is A_c, and its set of weights is denoted S, where S_j is a subset of weights ($S_j \subset S$) belonging to layer j. The objective is to approximate S_j by an evolved generator (consisting of the expression G_j and associative memory AM_j) in such a way that the (G_j, AM_j) generates the approximate weights sequentially, one weight in one step. The approximate weights are then consumed by subsequent processing units in CNN accelerator to determine the resulting class. Because the weights are not exactly the original ones, the objective is to keep the drop in accuracy acceptable. The size of external memory is reduced because it has to contain only the content of AM and some auxiliary constants (such as pre-computed averages of kernel values) for each layer.

Figure 1 specifies the instantiation of CGP when it generates the weights of a convolutional layer. CGP operates with the following inputs:

1. input channel index
2. output channel index
3. weight window coordinate w_x
4. weight window coordinate w_y
5. the average of the weights in the kernel (w_{avg})

The first four inputs are scaled to interval $[-0.5, +0.5]$. It is expected that these inputs can cheaply be generated by on-chip counters. All five inputs are aggregated (Agg in Fig. 1) to create a single input to AM. In this work, we concatenate these values; however, other approaches (such as XOR-ing) are possible to reduce the bit width of the AM input.

The output (i.e., y' produced by (G_j, AM_j)) is then added to the average w_{avg} and interpreted as the weight at position (w_x, w_y) in the corresponding channels. Our preliminary experiments revealed that generating these differences w.r.t w_{avg} gives better results than an approach generating the weights directly.

In the fitness function, which has the form of Eq. 2, the outputs produced by the evolved solution are compared against the original weights with the aim of minimizing the error. The resulting accuracy drop of the image classification is checked at the end of evolution using the test image data and for the best-evolved solution only. This approach is computationally less expensive than direct computing the classification accuracy for all candidate solutions on test image data.

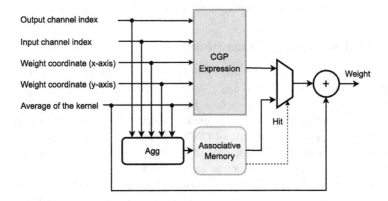

Fig. 1. GPAM as a generator of weights for a single layer of a CNN.

In summary, the configurable on-chip weight generator is configured for each layer. Its configuration includes a program represented by G_j, memory content AM_j, and average values of kernels.

4   Results for Symbolic Regression Benchmarks

This section presents the proposed benchmark problems and the results of GPAM on these benchmarks for various settings of the method.

4.1   Benchmarks

Table 2 gives five well-known benchmark functions taken from the literature [10] to evaluate GPAM. In each function, we replaced some percentage of data points (controlled by the parameter τ) with randomly generated values from interval $[-10, 10]$[1]. The modified data points remain unchanged across all the experiments reported in this paper.

Figure 2 shows the Nguyen-7 benchmark problem (green color) with the randomly generated points for $\tau = 5\%$, 10%, 20%, 25%, and 50% (cumulatively added as indicated by different colors).

Table 2. Baseline functions used to generate benchmark problems for GPAM.

Name	Function	Size	Interval	Step
Koza-1	$f(x) = x^4 + x^3 + x^2 + x$	40	$[-1, 1]$	0.05
Nguyen-7	$f(x) = ln(x + 1) + ln(x^2 + 1)$	20	$[0, 2]$	random
Nguyen-10	$f(x_0, x_1) = 2 * sin(x_0) * cos(x_1)$	100	$[-1, 1]$	random
Korns-1	$f(x_0, x_1, x_2, x_3, x_4) = 1.57 + (24.3 * x_3)$	10 000	$[-50, 50]$	random
Korns-4	$f(x_0, x_1, x_2, x_3, x_4) = -2.3 + 0.13 * sin(x_2)$	10 000	$[-50, 50]$	random

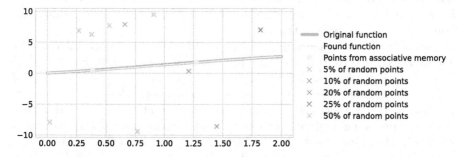

Fig. 2. The Nguyen-7 benchmark problem (green color) with the randomly generated points for various τ (cumulatively added). One of evolved solutions (for $\tau = 50\%$ (10 values) and $k_{AM} = 60\%$) is depicted in yellow.

4.2   Setup

Based on some trial runs, the CGP parameters were fixed as summarized in Table 3. According to [11], CGP employs the $(1 + \lambda)$ search strategy, where λ offspring are generated from one parent. The fitness function is as defined by Eq. 2.

4.3   Memory Sizing

Two key parameters significantly influence the quality of results: the size of associative memory (k_{AM}) and the number of randomly generated points in the data set (τ). We performed 30 independent runs of CGP for various settings

---

[1] Data sets are available at https://doi.org/10.5281/zenodo.7583555.

Table 3. Parameters of CGP for solving the benchmark functions.

Parameter	Value
n_i	function dimension
n_o	1
$n_c \times n_r$	20×2
L-back	maximum
λ	4
Generations	5000
c_m	2 (the integers to mutate in CGP)
p_{mutmem}	0.2 (the AM mutation probability)
Function set	$+, -, *, \%, sin(x), cos(x), e^x, log(x)$
(extension for Korns)	$x^2, x^3, sqrt(x), tan(x), tanh(x)$

of k_{AM} and τ, and with CGP parameters according to Table 3. Resulting box plots, showing the obtained fitness values, are presented in Fig. 3. Note that the fitness values are normalized with respect to the number of data points in all figures showing box plots in this paper.

Except for Korns-1 benchmark, CGP provides very good and stable results (the fitness is close to 0) in the reference experiment ($k_{AM} = 0$, $\tau = 0$). If the number of randomly modified points in a data set increases when the size of AM is constant, the fitness (i.e., the error) also increases. However, the error is decreased if the AM capacity is expanding for constant τ. The GPAM is thus capable of providing a suitable expression and selecting data points accommodated in AM to compensate for the higher error intensity (τ) to some extent. This trend is visible for the four test functions, i.e., except Korns-1 (for which we did not find a suitable CGP setup for successful evolution).

In Fig. 2, one of solutions ($ln(abs(e^x + x * x * x))$) evolved with $\tau = 50\%$ (10 values) and $k_{AM} = 60\%$ is shown in yellow. This solution perfectly fits the original function on the considered interval.

The experiments were performed on the Intel Xeon CPUs E5-2630 v4 running at 2.20 GHz. For the most demanding setup (Korns-1 with $k_{AM} = 30\%$), the median execution time is 280 s on a single core.

4.4 Role of Constants in GPAM

We hypothesize that the values stored in AM can be utilized by CGP as good constants to accelerate the search process. Hence, we allowed CGP to use constants in such a way that for each node input, a constant value is assigned with the probability $p_{const} = 0.1$ during random initialization as well as mutation. Four constant handling strategies are compared:

1. No constants allowed.
2. Random constants (from interval $[-1\,000, 1\,000]$ or $[-10\,000, 10\,000]$ for Korns benchmarks).

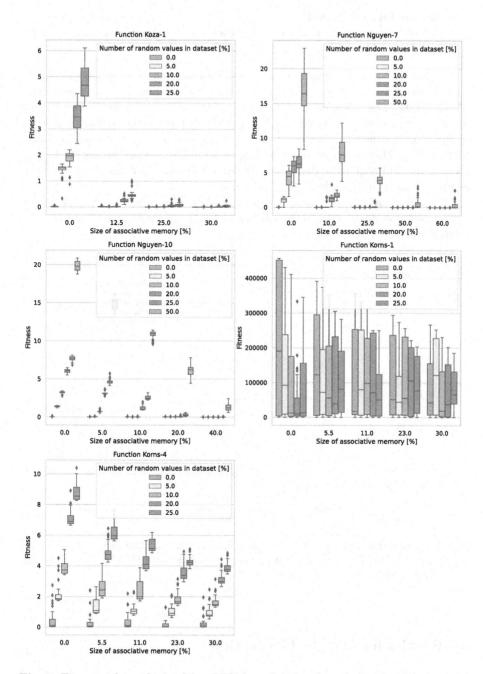

Fig. 3. Fitness values obtained by GPAM on five benchmark problems for various associative memory sizes (k_{AM}) and numbers of randomly modified data points in data sets (τ).

3. Constants taken from AM.
4. Combined constants (random constants with probability $0.5.p_{consts}$ and constants from AM with probability $0.5.p_{consts}$).

For setups 1–4, we repeated all experiments from the previous section. Figure 5 provides detailed results for the data set containing 25% randomly generated points. The remaining are omitted because of the space limit. In addition to the box plots, we performed a one-way ANOVA test (with a significance level of 95%) and verified that all constant handling methods are statistically indistinguishable. This observation is also supported by the median fitness (including the first and third quartile distance) computed in each generation from the best results of the 30 independent runs, see Fig. 4.

To summarize, the constants taken from AM do not help to obtain better results but are used more frequently by GPAM than other constants (again, verified statistically).

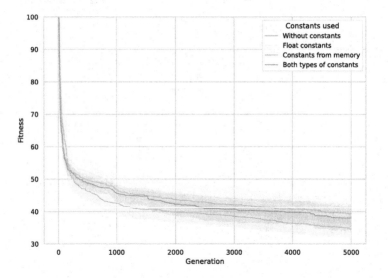

Fig. 4. Median (including the first and third quartile distance) of fitness values computed in each generation from the best results of 30 independent GPAM runs conducted for Nguyen-7 with $k_{AM} = 10\%$ and $\tau = 25\%$.

5   Results for Weight Generation

We evaluated the weight generation method (introduced in Sect. 3.2) on a simple CNN consisting of six layers; see Fig. 6. This CNN was trained using PyTorch to classify the MNIST data set (a digit classification task from a 28×28-pixel

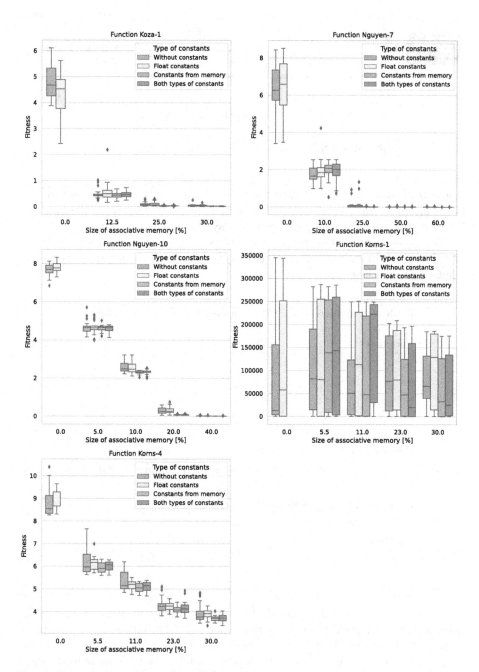

Fig. 5. Fitness values obtained by GPAM on five benchmark problems for various constant handling methods, associative memory sizes (k_{AM}), and 25% randomly modified data points in data sets.

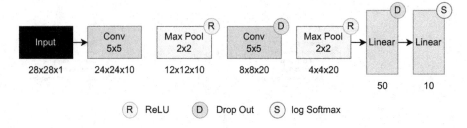

Fig. 6. CNN used as a benchmark problem for GPAM.

image [9]) and achieved 97.4% accuracy on test images. As we focus on convolutional layers in this paper, GPAM has to generate 250 weights for the first and 5000 weights for the second convolutional layer.

Table 4 summarizes the setup of CGP parameters for the evolutionary design of weight generators. The setup is based on the results of the previous study and a few additional test runs.

Table 4. Parameters of CGP when evolving the weight generator.

Parameter	Value
n_i	5
n_o	1
$n_c \times n_r$	20×10
L-back	maximum
λ	4
Generations	5000
c_m	2 (the integers to mutate in CGP)
p_{mutmem}	0.2 (the AM mutation probability)
Float constant probability	0.05
AM constant probability	0.05
Function set	$+, -, *, \%, sin(x), cos(x), e^x, log(x),$ $x^2, x^3, sqrt(x), tan(x), tanh(x)$

Figure 7 (left) shows box plots with the fitness values obtained from 30 independent runs of GPAM utilizing various sizing of AM. Three sets of box plots correspond with the (i) weight generator for the first convolution (left), (ii) weight generator for the second convolution (middle), and (iii) combined generator composed of the best solutions obtained from (i) and (ii). We observe the same trend as in the previous experiments; the error is reduced if a bigger AM is available. The lowest errors are obtained for the combined approach because we report the normalized values (w.r.t. the size of the data set).

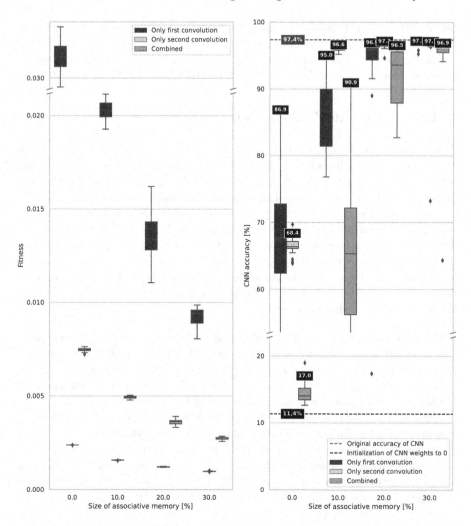

Fig. 7. Fitness and Accuracy (on the MNIST test data) obtained from 30 runs of GPAM utilizing various sizing of the associative memory. The numbers in rectangles represent the best obtained values (the outliers not considered)

Figure 7 (right) then shows the accuracy box plots on the MNIST test data for the three considered weight generators. If the associative memory contains 10% of the original weights, the evolved weight generator for the second convolutional layer can approximate the original weights such that the CNN utilizing the generated weights shows less than a 1% drop in the classification accuracy. Other trade-offs are seen in Fig. 7.

We also analyzed how many expressions generated by CGP in the experiments are primitive constants (after simplifying the evolved expressions using a math solver). The number of constant-producing expressions is 11% and 15% for the first and second convolutional layers, respectively.

Finally, for the most demanding setup ($k_{AM} = 30\%$), the median single-core execution time is 850 s on the Intel Xeon CPU E5-2630 v4 running at 2.20 GHz.

6   Discussion and Conclusions

We focused on the automated design of programs (in the form of expression with associative memory) for solving the symbolic regression problem in which the data set contains some randomly generated values (outliers). We proposed a new approach – GPAM – and showed on several benchmark problems that GPAM can solve some instances. Then we applied the same concept to the design of an on-chip generator capable of producing (selected) weights of a small CNN. Through these experiments, we proved that GPAM can work for smaller problem instances. However, there are many open problems. Among them, the scalability issues must be addressed with the highest priority. The following paragraphs address the limitations of the current approach and discuss the subproblems that future research should target.

- To evolve an expression, a basic version of CGP was employed. More sophisticated GP-based symbolic regression methods could provide better results.
- The evaluation has to be performed on more challenging data sets (including data sets in which we do not know the rates for random points/outliers) and difficult classification problems in the case of the weight generator.
- Similarly to Dupuis et al. [2], all operations were conducted in the floating point arithmetic. In one case study, our compression rate is around 10 and a small AM is needed. On more complex problems, Dupuis et al. [2] reports a compression rate of around 5 with a similar accuracy drop but without specialized memory. However, the literature tells us that an 8-bit weight quantization can provide the same accuracy as the original 32-bit float representation of the weights, i.e., the compression rate is $32/8 = 4$ [15]. We plan to redesign our implementation to support the quantized weights to improve the compression rate.
- Other fitness functions should be tested, e.g., those reflecting the importance of weights (with respect to the classification accuracy), because DNN weights do not contribute to resulting accuracy uniformly.
- The AM can be employed with various setups to improve performance and reduce the AM cost in hardware. The input key matching scheme can be relaxed to allow for approximate key matching (only the exact key matching is considered now).
- For each layer, an independent expression and AM are currently evolved. To reduce the area on a chip, various co-design approaches (including co-evolutionary techniques) can be adapted to allow sharing of sub-expressions and AM contents among the layers.

Acknowledgements. This work was supported by the Czech science foundation project 21-13001S, and it was partly carried out under the COST Action CA19135 (CERCIRAS).

References

1. Capra, M., Bussolino, B., Marchisio, A., Shafique, M., Masera, G., Martina, M.: An updated survey of efficient hardware architectures for accelerating deep convolutional neural networks. Future Internet **12**(7), 113 (2020)
2. Dupuis, E., Novo, D., O'Connor, I., Bosio, A.: A heuristic exploration of retraining-free weight-sharing for CNN compression. In: 27th Asia and South Pacific Design Automation Conference, ASP-DAC, pp. 134–139. IEEE (2022)
3. Goodfellow, I., Bengio, Y., Courville, A.: Deep Learning. MIT Press, Cambridge (2016)
4. Han, S., Mao, H., Dally, W.J.: Deep compression: compressing deep neural network with pruning, trained quantization and Huffman coding. In: 4th International Conference on Learning Representations, ICLR (2016)
5. Koza, J.R.: Genetic Programming: On the Programming of Computers by Means of Natural Selection. MIT Press, Cambridge (1992)
6. La Cava, W., et al.: Contemporary symbolic regression methods and their relative performance. In: Vanschoren, J., Yeung, S. (eds.) Proceedings of the Neural Information Processing Systems Track on Datasets and Benchmarks. vol. 1 (2021)
7. La Cava, W.G., Helmuth, T., Spector, L., Moore, J.H.: A probabilistic and multi-objective analysis of lexicase selection and ϵ-lexicase selection. Evol. Comput. **27**(3), 377–402 (2019)
8. Langdon, W.B.: Genetic Programming and Data Structures: Genetic Programming + Data Structures = Automatic Programming! Springer, Cham (1998)
9. LeCun, Y., Cortes, C., Burges, C.: MNIST handwritten digit database. ATT Labs. https://yann.lecun.com/exdb/mnist (2010)
10. McDermott, J., et al.: Genetic programming needs better benchmarks. In: Proceedings of the 14th International Conference on Genetic and Evolutionary Computation, pp. 791–798. ACM (2012)
11. Miller, J.F.: Cartesian Genetic Programming. Springer, Berlin (2011)
12. Schmidt, M.D., Lipson, H.: Coevolution of fitness predictors. IEEE Trans. Evol. Comput. **12**(6), 736–749 (2008)
13. Stanley, K.O.: Compositional pattern producing networks: a novel abstraction of development. Genet. Program Evolvable Mach. **8**(2), 131–162 (2007)
14. Stanley, K.O., D'Ambrosio, D.B., Gauci, J.: A hypercube-based encoding for evolving large-scale neural networks. Artif. Life **15**(2), 185–212 (2009)
15. Sze, V., Chen, Y., Yang, T., Emer, J.S.: Efficient processing of deep neural networks. Synth. Lect. Comput. Archit. **15**(2), 1–341 (2020)
16. Teller, A.: The evolution of mental models. In: Kinnear, K.E., Jr. (ed.) Advances in Genetic Programming, pp. 199–219. MIT Press, Cambridge (1994)
17. Vanneschi, L., Castelli, M., Silva, S.: A survey of semantic methods in genetic programming. Genet. Program Evolvable Mach. **15**(2), 195–214 (2014). https://doi.org/10.1007/s10710-013-9210-0
18. Villegas-Cortez, J., Olague, G., Aviles, C., Sossa, H., Ferreyra, A.: Automatic synthesis of associative memories through genetic programming: a first co-evolutionary approach. In: Di Chio, C., et al. (eds.) EvoApplications 2010. LNCS, vol. 6024, pp. 344–351. Springer, Heidelberg (2010). https://doi.org/10.1007/978-3-642-12239-2_36

MAP-Elites with Cosine-Similarity for Evolutionary Ensemble Learning

Hengzhe Zhang[1]([✉]), Qi Chen[1], Alberto Tonda[2], Bing Xue[1],
Wolfgang Banzhaf[3], and Mengjie Zhang[1]

[1] Victoria University of Wellington, Wellington, New Zealand
{hengzhe.zhang,qi.chen,bing.xue,mengjie.zhang}@ecs.vuw.ac.nz
[2] UMR 518 MIA-Paris, INRAE, Paris, France
alberto.tonda@inrae.fr
[3] Michigan State University, East Lansing, MI, USA
banzhafw@msu.edu

Abstract. Evolutionary ensemble learning methods with Genetic Programming have achieved remarkable results on regression and classification tasks by employing quality-diversity optimization techniques like MAP-Elites and Neuro-MAP-Elites. The MAP-Elites algorithm uses dimensionality reduction methods, such as variational auto-encoders, to reduce the high-dimensional semantic space of genetic programming to a two-dimensional behavioral space. Then, it constructs a grid of high-quality and diverse models to form an ensemble model. In MAP-Elites, however, variational auto-encoders rely on Euclidean space topology, which is not effective at preserving high-quality individuals. To solve this problem, this paper proposes a principal component analysis method based on a cosine-kernel for dimensionality reduction. In order to deal with unbalanced distributions of good individuals, we propose a zero-cost reference points synthesizing method. Experimental results on 108 datasets show that combining principal component analysis using a cosine kernel with reference points significantly improves the performance of the MAP-Elites evolutionary ensemble learning algorithm.

Keywords: Evolutionary ensemble learning · Quality diversity optimization · Multi-dimensional Archive of Phenotypic Elites

1 Introduction

Ensemble learning methods have gained popularity in recent years due to their ability to reduce the variance of unstable machine learning algorithms without increasing bias. Typically, the generalisation loss of an ensemble model $\mathbb{E}_F$ for a given dataset $\{X, Y\}$ can be decomposed into two terms, as shown in Eq. (1):

$$\mathbb{E}_F = \underbrace{\mathbb{E}_{f \in F}\left[(f(X) - Y)^2\right]}_{\text{average loss}} - \underbrace{\mathbb{E}_{f \in F}\left[(f(X) - \mathbb{E}_{f' \in F}\left[f'(X)\right])^2\right]}_{\text{ambiguity}} \quad (1)$$

© The Author(s), under exclusive license to Springer Nature Switzerland AG 2023
G. Pappa et al. (Eds.): EuroGP 2023, LNCS 13986, pp. 84–100, 2023.
https://doi.org/10.1007/978-3-031-29573-7_6

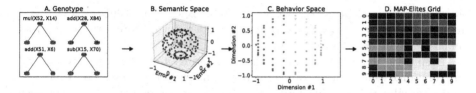

Fig. 1. The workflow of MAP-Elites

On the right side of this equation, the first term represents the average loss between the prediction of each model $f(X)$ and the target Y, and the second term ambiguity represents the difference between the prediction of each model $f(X)$ and the average prediction among models in the ensemble $\mathbb{E}_{f'}[f'(X)]$. For an evolutionary ensemble learning method, having two groups of base learners with the same average fitness values often means they have the same average loss. However, they may have different predictive accuracy, due to the difference in their ambiguity. Typically, a more diverse group of base learners has larger ambiguity and thus provides a more accurate prediction. In theory, we can optimize E_f with an evolutionary algorithm. However, evaluating the fitness value of an ensemble model may be computationally expensive in practice, and a more practical way is to implicitly optimize E_f by maintaining a set of high-quality and diverse individuals during the evolutionary process.

In this work, we focus on using genetic programming (GP) [2] to evolve a set of high-quality and diverse regressors for ensemble learning. GP has been widely used in regression tasks due to its flexible representation. However, the traditional GP framework mainly focuses on finding regressors minimizing the training error during the evolutionary process, making it ineffective at obtaining a diverse set of regressors in a single run. In order to obtain a set of complementary regressors, one idea is to take the semantics of regressors into consideration. The semantics of each GP individual represents the predictions for a set of samples. The target semantics is a point in the semantic space representing the target labels $\{y_1, \ldots, y_n\}$. In semantic GP for ensemble learning for regression, a desired ensemble model is a set of regressors with complementary semantics, thus the combined prediction of this kind of regressor can be approximately equal to the target semantics. To generate a desired ensemble model, it is important to develop novel selection operators that highlight both quality and diversity.

In the field of evolutionary computation (EC), there are a variety of techniques for finding diverse individuals with high quality. The Multi-dimensional Archive of Phenotypic Elites (MAP-Elites) [22] is a representative example. As shown in Fig. 1, MAP-Elites defines a behavioral space for a given problem that describes the desired property of high-quality solutions. In this example, a cosine-kernel-based principal component analysis (KPCA) method that only considers the angle distance between individuals is used to define the behavioral space. The general concept behind MAP-Elites is to divide the behavioral space into multiple cells and retain the best individual in each cell to maintain population diversity. Based on this idea, the MAP-Elites algorithm has been used to evolve an ensemble of classifiers [24], where the MAP-Elites algorithm employs a grid to record a diverse set of well-performing classifiers with different semantics from an ensemble model in a single run.

Despite the many benefits MAP-Elites can bring, it is still not widely adopted in evolutionary ensemble learning due to the difficulty in defining the behavioral space. Initially, defining behavioral descriptors requires domain knowledge, such as a handcrafted descriptor named the entropy of instructions in linear genetic programming (LGP) [11], which is a GP variant with a sequence of instructions to represent GP programs. Recent research demonstrates that behavioral descriptors for each GP individual can be automatically obtained based on its semantic vector [25]. For a regression problem with n training samples, its semantic vector is n-dimensional. When n is large, the curse of dimensionality causes exponential growth of the number of cells in a MAP-Elites grid. Recently, autoencoders (AE) have been used to automatically discover behavioral descriptors on robot control tasks [7] and classification tasks (Neuro-MAP-Elites) [25]. AE is a deep-learning-based dimensionality reduction method that uses a bottleneck architecture to compress high-dimensional data into low-dimensional representations. For evolutionary machine learning tasks, the optimal behavioral space should be able to describe the distribution of high-quality individuals. This means that AE for generating the behavioral space should be trained on high-quality individuals. To achieve this goal, Neuro-MAP-Elites trains a variational auto-encoder (VAE) [14] on good individuals from the final population of a GP run. Then, the pre-trained VAE can define a good MAP-Elites grid for evolving diverse and high-performing individuals in another GP run.

There are two potential limitations with Neuro-MAP-Elites. First, the behavioral descriptor generated by VAE may not be effective to find complementary learners. Considering a case where letting the semantic vectors of three individuals A, B, C be $A = \{y_1 - 100, \ldots, y_n - 100\}, B = \{y_1 - 500, \ldots, y_n - 500\}, C = \{y_1 + 100, \ldots, y_n + 100\}$. If selecting two individuals with the largest Euclidean distance to form an ensemble model, they will be $\{B, C\}$. However, the optimal set is $\{A, C\}$ because the average prediction results of these two individuals match the semantic target $\{y_1 \ldots y_n\}$. Unfortunately, Euclidean-space-based VAE may prefer $\{B, C\}$ and thus does not perform well for evolutionary ensemble learning. The second issue is that training a VAE on good individuals obtained from the final population of a GP run is inefficient and may misguide the evolutionary process. Compared with using good individuals in a single GP run, it is more efficient to use the target semantics in supervised learning tasks to generate reference points to train a dimensionality reduction model. Moreover, due to the mismatched distributions of the initial and the final populations in GP, a VAE trained on well-performing individuals in the previous GP run might not be helpful to the initial population.

In this paper, we propose a new ensemble learning method based on MAP-Elites and GP, named MEGP, with the following objectives:

- Considering that it is difficult for Euclidean-space-based VAE to find complementary individuals, we propose using cosine-kernel-based PCA for dimensionality reduction in MAP-Elites to better find complementary base learners.

KPCA with cosine-kernel focuses on the relative angle to the target semantics, thereby encouraging GP to find diverse and complementary regressors to create an ensemble model.

– We propose a zero-cost method for generating reference points representing good solutions in the semantic space for training a dimensionality reduction model. A dimensionality reduction model trained on reference points can be viewed as a good behavioral descriptor and can be used in MAP-Elites.

2 Related Work

2.1 Semantic GP

In recent years, semantic GP has attracted considerable attention. The key idea of semantic GP is to use semantic information in genetic operators or selection operators to generate offspring with high behavioral correlation with their parents. In terms of genetic operators, a considerable number of semantic-based crossover and mutation operators have been developed to fulfill the semantics for the new generation [21,30]. As for selection operators, there are some works that consider selecting parent individuals based on semantic vectors instead of fitness values [8,16], which improves population diversity and thus results in better performance.

2.2 GP-Based Ensemble Learning

The idea of using multiple GP models to form an ensemble model can be traced back to BagGP [15], where multiple runs of GP are performed within the bagging framework. However, it is possible to maintain a diverse set of models in a single GP run since it is a population-based method. In spatial structure with bootstrapped elitism (SS+BE) [10], the niching method [13] and a bootstrapping strategy are used to form an ensemble of GP models in a single GP run. A similar idea of using niching in GP to form an ensemble has been applied to vehicle routing problems [35]. Recently, an algorithm named 2SEGP [34] shows that purely relying on the bootstrapping strategy can also yield satisfactory results. In a GP-based feature construction scenario, it is also possible to rely on the randomness of base learners to produce an ensemble model that outperforms XGBoost [37]. When the base learner is not random enough and the bootstrapping strategy is not allowed, a diverse set of base learners can still be produced by using the quality-diversity optimization framework [25].

2.3 Quality Diversity Optimization

In recent years, quality diversity (QD) optimization has been widely used to tackle the problem of deceptive landscapes [36] and produce diverse solutions [9]. QD algorithms can be classified into grid-based and archive-based methods, based on whether they rely on a discretized behavioral space to maintain population diversity or not. Grid-based QD optimization methods discretize the behavioral space to preserve diversity, with MAP-Elites being a typical example.

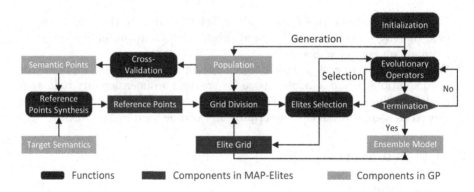

Fig. 2. All components in MEGP.

As shown in Fig. 1, MAP-Elites first maps an individual from a high-dimensional semantic space to a low-dimensional behavioral space. Then, it divides the behavioral space into multiple grids and retains only the best individual in each cell, where all individuals in the MAP-Elites grid can be used to form an ensemble model. MAP-Elites was initially developed for robot design [22], but it has been applied to a variety of other problems, including agent control [27], airfoil optimization [12], workforce scheduling [32], and the traveling thief problem [26]. In the GP domain, MAP-Elites was initially applied to program synthesis and symbolic regression tasks [4,11]. Subsequently, it has been extended to classification tasks for evolving an ensemble of classifiers [24].

As for archive-based methods, a typical example is Novelty Search with Local Competition (NSLC) [18]. The key idea of NSLC is to use an external archive to keep diverse individuals and use a multi-objective optimization algorithm to breed individuals based on diversity and local ranking. There are a lot of differences between MAP-Elites and NSLC. One key difference is that the MAP-Elites algorithm uses a grid to explicitly keep the structure, while NSLC implicitly keeps the structure based on a distance measure. In the evolutionary ensemble learning domain, both grid-based and archive-based QD methods have been studied [3,25], but a comparison between them is still lacking.

3   The Proposed Ensemble Learning Algorithm

This work presents a MAP-Elites-based ensemble GP method, named MEGP. First, we introduce the algorithmic framework. Then we describe dimensionality reduction methods in MEGP and a method for generating reference semantic points that can be used in training a dimensionality reduction model.

3.1   The Overall Framework

MEGP introduces MAP-Elites into the GP-based ensemble learning scenario. The pseudocode for MEGP is presented in Algorithm 1, and all components of MEGP are shown in Fig. 2. MEGP follows the conventional framework of GP, but differs from it in the following ways:

Algorithm 1. MEGP

Input: Population Size N, Number of Generations max_gen, Dimensionality of the
    MAP-Elites Grid G, Training Data $\{(x_1, y_1), \ldots, (x_n, y_n)\}$
Output: MAP-Elites Grid E
 1: Randomly initialize a population of GP individuals $P = \{\Phi_1 \ldots \Phi_N\}$
 2: $E \leftarrow$ MAP-Elites grid initialization with P ▷ MAP-Elites Grid
 3: $gen \leftarrow 0$
 4: **while** $gen \leq max\_gen$ **do**   ▷ Main loop
 5:     $P \leftarrow$ mutation and crossover(P)
 6:     **for** $\Phi \in P$ **do**   ▷ Evaluation
 7:         $\{\hat{y}_1, \ldots, \hat{y}_n\} \leftarrow$ cross-validation$(\Phi, \{\hat{x_1}, \ldots, \hat{x_n}\})$
 8:         $Y_\Phi \leftarrow \{\hat{y}_1, \ldots, \hat{y}_n\}$
 9:     $PE \leftarrow$ selecting top-50% individuals from $P \cup E$
10:     $\{R_1, \ldots, R_{2|PE|}\} \leftarrow$ reference point synthesis $(\{y_1, \ldots, y_n\}, \{\hat{Y}_1, \ldots, \hat{Y}_{|PE|}\})$
11:     $\{Z_1, \ldots, Z_{|PE|}\} \leftarrow$ dimensionality reduction $(PE, \{R_1, \ldots, R_{2|PE|}\})$
12:     $E \leftarrow$ grid division $(PE, \{Z_1, \ldots, Z_{|PE|}\}, G)$
13:     $E \leftarrow$ elites selection(E)
14:     $P \leftarrow$ random selection(E) ▷ Selection
15:     $gen \leftarrow gen + 1$
    return E

- Multi-tree Representation: MEGP uses multiple GP trees to represent a single individual, and a linear model is used to combine these GP trees to make a prediction. The multi-tree GP is used due to its more expressive and flexible representation ability [17].
- Cross-validation Loss: MEGP uses an efficient leave-one-out cross-validation method [6] in the fitness function to evaluate each GP individual Φ based on a ridge regressor, which allows mitigating the over-fitting issue.
- Ensemble Learning: MEGP uses all individuals $e \in E$ in the final MAP-Elites grid to form an ensemble model. For an unseen data point x', the prediction result is the average of all prediction results, i.e., $\frac{\sum_{e \in E} e(x')}{|E|}$.

3.2 Angle-Based Dimensionality Reduction

The mapping of individuals from a high-dimensional semantic space to a low-dimensional behavior space is a key step in MAP-Elites. We propose to employ cosine-kernel principal component analysis (KPCA) for dimensionality reduction in MEGP. A dimensionality reduction algorithm maps the semantics $\{\hat{y}_1^i, \ldots, \hat{y}_n^i\}$ of an individual i in the semantic space to a low-dimensional point $\{z_1^i, z_2^i\}$ in the behavior space, where n is the number of data points/instances in the training dataset. PCA is a simple and efficient algorithm for dimensionality reduction tasks. However, the standard PCA algorithm focuses on capturing the variance in Euclidean space. Sometimes, individuals with bad fitness values can have large Euclidean distances from others, but they are not good candidates for an ensemble model and should not be included in the behavior space. In order to

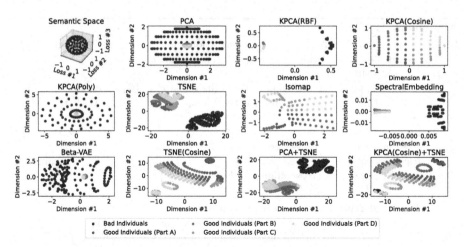

Fig. 3. An illustrative example of different dimensionality reduction techniques for inducing a behavioral space.

solve this problem, a cosine kernel function defined as $cos(i,j) = \frac{i \cdot j^{\top}}{\|i\|\|j\|}$ for any two points i, j in the semantic space is used to transform points from the n-dimensional semantic space to another n-dimensional implicit feature space. In the implicit feature space, each dimension represents the cosine similarity between a data point and the others. Next, PCA is applied to the implicit feature space to generate a two-dimensional behavior space. Because cosine similarity ignores scale, the implicit feature space only preserves the angle distance between points. Thus, individuals are only considered novel if they approach the target semantics from a different angle, i.e., from a different direction.

To illustrate why the cosine-kernel-based PCA method is suitable for MAP-Elites, Fig. 3 provides an example of dimensionality reduction results for a three-dimensional semantic space, where the central point represents the optimal predictive result. The purple data points in Fig. 3 represent a group of bad individuals located far from the target semantics. The remaining data points represent a group of good individuals. A perfect behavioral space should keep the best individual in each cell and remove all inferior individuals. However, Fig. 3 shows that many conventional dimensionality reduction methods fail to achieve this goal. For example, with PCA, large parts of behavioral space are filled with purple data points, indicating that several bad individuals will be retained due to their excellent diversity. In this example, only KPCA with a cosine kernel and Isomap place good individuals in the entire space. However, Isomap focuses to preserve the local structure in a low-dimensional space. Thus, it may fail to perform well if good and bad points are connected and distributed on a single manifold, such as the "Swiss roll" data [1]. In contrast, KPCA with a cosine kernel only considers the angles between points during the dimensionality reduction procedure. Thus, individuals with different fitness values will fall within the same region if the cosine similarity between predicted values is high.

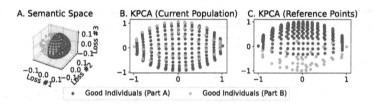

Fig. 4. An illustrative example to show the effect of constructing a dimensionality reduction model on the current population and symmetrical reference semantic points.

3.3 Reference Semantic Points

To train a dimensionality reduction model to generate a behavior space capturing the distribution of high-quality individuals only, previous research [24] used the semantics of good individuals in the final population of a GP run. These semantic points are used to construct a behavior space and are referred to as reference points. However, the target semantics $\{y_1, y_2, \ldots, y_n\}$ is available for supervised learning tasks. Consequently, for each semantic point $\{\hat{y_1}, \hat{y_2}, \ldots, \hat{y_n}\}$ in the current population P and the current MAP-Elites grid E, a reference point can be generated with $\{(1-\alpha)*y_1 + \alpha*\hat{y_1}, (1-\alpha)*y_2 + \alpha*\hat{y_2}, \ldots, (1-\alpha)*y_n + \alpha*\hat{y_n}\}$, where α is a hyperparameter indicating how close a synthetic reference point is to the target semantics. α is empirically set as 0.1 in this paper. Notably, for each individual, we not only synthesize a reference point based on α but also generate a symmetric reference point with $-\alpha$. Generating a symmetric reference point guarantees that the average of all reference points equals the target semantics. After obtaining reference points, we can train a dimensionality reduction model using these points and then apply the trained model to the semantic points of the current population to construct a MAP-Elites grid.

Figure 4 shows the behavioral space of two imbalanced sets of data points using reference points or not. As shown in Fig. 4B, if constructing a KPCA model on the current population, the blue points representing individuals over-estimating the value of the first sample and the green points representing individuals under-estimating the value of the first sample will be mixed. This mix makes it very hard to obtain complementary base learners by selecting the best one from each cell. Conversely, if KPCA is constructed with symmetrical reference points, complementary points will be dispersed across distinct regions of a behavioral space. MAP-Elites can easily obtain a collection of complementary base learners.

Nevertheless, it is important to note that pre-training a dimensionality reduction model based on reference points is risky. Figure 5 provides an example of dimensionality reduction results based on online and offline modes. The online mode means training a dimensionality reduction model on the current population, whereas offline means training a model on reference points. Both kinds of models will be applied to the current population for dimensionality reduction. In Fig. 5, the colored points on the outer circle represent models in the current population, while the red points on the inner circle represent synthetic reference

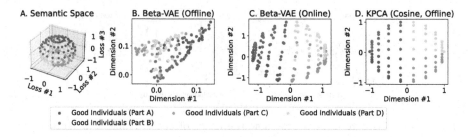

Fig. 5. Online versus offline dimensionality reduction paradigms.

points. Figure 5B provides dimension reduction results in the offline mode, while Fig. 5C provides a comparative example of dimension reduction results in the online mode. Comparing these two plots reveals that VAE fails to generate an appropriate behavioral space for mismatched population distribution, as colored points are highly concentrated in the center. In contrast, if we train a VAE on the current population, the behavioral space can maintain the current population's structure, proving that the problem is the mismatched distribution, not using a VAE. Furthermore, Fig. 5D shows that KPCA with a cosine kernel is not significantly affected by this issue, illustrating another advantage of using KPCA. To sum up, synthesizing reference points is helpful to generate a good behavioral space, but it should be paired with an appropriate dimensionality reduction method to alleviate the negative impact of the mismatched distribution.

4   Experiment Settings

In this section, several experiments are conducted to answer the following questions:

- Does the cosine kernel PCA-based dimensionality reduction method result in a better ensemble model in MEGP when compared to the commonly used dimensionality methods?
- Does a behavioral space generated by reference points improve the predictive performance of the ensemble model?

4.1   Datasets

In this paper, we conduct experiments on the Penn Machine Learning Benchmark (PMLB) [28], a curated list of datasets derived from OpenML datasets [33]. For comparison of dimensionality reduction algorithms and reference points, experiments are carried out on 108 datasets in PMLB with less than 10000 instances due to limited computational resources. For comparison with other algorithms, experiments are performed on standard PMLB with 122 datasets. Among these datasets, 63 were synthesized by the Friedman function and are synthetic datasets, while the remaining 59 are real-world datasets.

Table 1. Parameter settings for MEGP.

Parameter	Value
Population Size	1000
Maximal Number of Generations	50
Crossover and Mutation Rates	0.9 and 0.1
Maximum Tree Depth	8
Maximum Initial Tree Depth	2
Number of Trees in An Individual	10
Dimensionality of the MAP-Elites Grid	10
Functions	+, -, *, AQ, Sin, Cos, Abs, Max, Min, Negative

4.2   Experimental Protocol

For the following experiments, we follow a conventional experimental protocol in the evolutionary computation domain, i.e., each algorithm is tested on each dataset with 30 independent runs. In each run, 80% of the data is used as the training data, and the remaining is used as the test data. After runs are finished, a Wilcoxon rank sum test with a significance level of 0.05 is used to verify the effectiveness of the proposed method. As for the comparison with other machine learning algorithms, we follow the convention of SRBench [5], i.e., each algorithm is tested on each dataset with 10 independent runs. The hyper-parameters of benchmark algorithms are tuned using the halving-grid search method [19] to ensure that the prediction performance of benchmark algorithms is fully exploited.

4.3   Parameter Settings

Table 1 presents the parameter settings of MEGP. The population size and crossover rate are conventional settings for GP [8]. Analytical quotient (AQ) is used in MEGP to replace the division operator in order to avoid division by zero. AQ is defined as $AQ(a, b) = \frac{a}{\sqrt{1+b^2}}$, where a and b represent two input variables.

4.4   Benchmark Dimensionality Reduction Methods

Here, we select 9 popular dimensionality reduction methods for comparisons because these methods are widely used in the machine learning field [29]. A brief introduction of investigated methods is as follows:

- Principal Component Analysis (PCA) [31]: PCA is a linear dimensionality reduction method that finds new dimensions to maximize variance in the data.

Table 2. Experimental results of nine dimensionality reduction methods in MEGP ("+", "∼" or "-" mean that a method in a row is significantly better than, similar to, and worse than the method in the column).

	t-SNE(COSINE)	PCA	KPCA(RBF)
KPCA(COSINE)	12(+)/96(∼)/0(−)	45(+)/63(∼)/0(−)	62(+)/45(∼)/1(−)
TSNE(Cosine)	—	30(+)/78(∼)/0(−)	54(+)/54(∼)/0(−)
PCA	—	—	46(+)/60(∼)/2(−)
	KPCA(POLY)	**TSNE**	**Beta-VAE**
KPCA(COSINE)	74(+)/34(∼)/0(−)	58(+)/50(∼)/0(−)	71(+)/37(∼)/0(−)
TSNE(Cosine)	75(+)/33(∼)/0(−)	53(+)/55(∼)/0(−)	67(+)/41(∼)/0(−)
PCA	70(+)/38(∼)/0(−)	12(+)/94(∼)/2(−)	47(+)/61(∼)/0(−)
KPCA(RBF)	70(+)/37(∼)/1(−)	0(+)/69(∼)/39(−)	8(+)/81(∼)/19(−)
KPCA(POLY)	—	0(+)/39(∼)/69(−)	5(+)/39(∼)/64(−)
TSNE	—	—	32(+)/76(∼)/0(−)
	Isomap	**SpectralEmbedding**	
KPCA(COSINE)	63(+)/45(∼)/0(−)	64(+)/44(∼)/0(−)	
TSNE(Cosine)	58(+)/50(∼)/0(−)	55(+)/53(∼)/0(−)	
PCA	31(+)/76(∼)/1(−)	30(+)/77(∼)/1(−)	
KPCA(RBF)	7(+)/64(∼)/37(−)	1(+)/82(∼)/25(−)	
KPCA(POLY)	0(+)/41(∼)/67(−)	0(+)/42(∼)/66(−)	
TSNE	20(+)/87(∼)/1(−)	17(+)/91(∼)/0(−)	
Beta-VAE	2(+)/87(∼)/19(−)	1(+)/94(∼)/13(−)	
Isomap	—	8(+)/97(∼)/3(−)	

- Kernel PCA with RBF/Polynomial/Cosine Kernel [31]: These three methods are based on PCA, with the difference of using RBF/Polynomial/Cosine kernels to calculate the similarity between points instead of the covariance.
- T-distributed Stochastic Neighbor Embedding (t-SNE-Euclidean/Cosine) [20]: t-SNE is a non-linear dimensionality reduction method that keeps both the local and the global structure. The key idea is to minimize the Kullback-Leibler divergence between high-dimensionality representation and low-dimensionality representation through gradient descent.
- Beta-VAE [14]: Beta-VAE is a deep-learning-based dimensionality reduction method. It maps input variables into a multivariate latent distribution. Unlike AE, it optimizes the reconstruction error and Kullback-Leibler divergence simultaneously to make the latent distribution approximate the expected distribution. A hyperparameter β is used to control the tradeoff between minimizing the reconstruction error and Kullback-Leibler divergence.
- Isomap [1]: Isomap is a manifold learning method to keep the local structure. It tries to keep the geodesic distance between points the same in the high-dimension and the low-dimension space.
- SpectralEmbedding [23]: Spectral embedding is similar to KPCA, but with the difference in that the eigen-decomposition is performed on a Laplacian matrix rather than on a kernel-matrix.

5   Experimental Results

5.1   Comparisons of MAP-Elites Using Different Dimensionality Reduction Methods

In this section, we present the experimental results of using 9 dimensionality reduction methods in MAP-Elites. MAP-Elites with cosine-kernel-based PCA significantly outperform beta-VAE on 71 out of the 108 datasets, see Table 2. On the other 37 datasets, the two methods have comparable performance. To examine the results in more detail, we plot curves of the test score of the ensemble model, average fitness of individuals in the MAP-Elites grid, and mean negative cosine similarity of individuals in the MAP-Elites grid against the number of generations in Fig. 6, Fig. 7, and Fig. 8, respectively. Figure 6 demonstrates that using KPCA with a cosine kernel is superior to using other methods in terms of the test R^2 score. To find out the reasons, Fig. 7 shows the average fitness of all base learners in the MAP-Elites grid. It indicates that some dimensionality reduction methods such as KPCA (POLY) and KPCA (RBF), make MAP-Elites select individuals with an average fitness lower than 0.8 and this may impair the accuracy of the ensemble model. Moreover, the superior performance of MAP-Elites with KPCA not only comes from selecting good fitness individuals but also from selecting individuals with a high level of diversity. To validate whether cosine-kernel-based KPCA is useful for keeping archive diversity, the average negative cosine similarity of base learners is presented in Fig. 8. Here, we use cosine similarity as opposed to Euclidean distance because a large negative cosine similarity indicates good complementarity between base learners, whereas a large Euclidean distance may be caused by base learners with very low accuracy. As shown in Fig. 8, the negative cosine similarity of base learners consistently decreases as evolution goes on when using PCA as the dimensionality reduction method. In contrast, the average negative cosine similarity of cosine-kernel-based KPCA stays at a stable level after 30 iterations, and it is higher than the results of PCA, providing evidence that using cosine-kernel PCA as a dimensionality reduction method is beneficial. It is worth noting that cosine KPCA does not have the best negative cosine similarity because half of the individuals with poor performance will be filtered out as shown in Algorithm 1, and such an elimination process may reduce the negative cosine similarity. Other methods, like KPCA (POLY) and KPCA (RBF), are less affected by this process because they select a large number of bad individuals.

5.2   Impact of Using Reference Points

In this section, we investigate whether inducing a behavioral space from reference points is beneficial. We compare the prediction performance on the test set with and without reference points. Several dimensionality reduction techniques, such as t-SNE and spectral embedding, are omitted because they cannot predict unseen data points. For the remaining methods, the results with and without reference points are shown in Fig. 9. As shown, reference points improve the

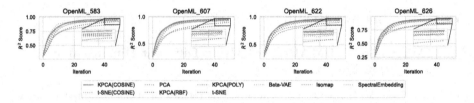

Fig. 6. Test R^2 score with respect to the number of generations

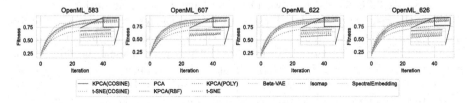

Fig. 7. Average fitness of individuals in an archive with respect to the number of generations

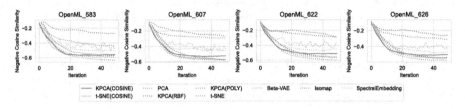

Fig. 8. Average negative cosine similarity of semantic vectors in the archive with respect to the number of generations

predictive performance of cosine-kernel-based KPCA on 39 datasets and do not degrade it on any other dataset. However, reference points do not work well with PCA and Beta-VAE, and even worsen performance on 46 and 7 datasets, respectively. These results validate our assumptions in Sect. 3. Consequently, we can conclude that using reference points to develop a dimensionality reduction model is useful, but it should be paired with suitable dimensionality reduction techniques.

5.3 Comparison with Other Machine Learning and Symbolic Regression Methods

To validate the efficacy of the proposed method, we compare MEGP to 14 symbolic regression methods and 8 machine learning methods on 122 datasets from SRBench. Figure 10 demonstrates the distribution of test R^2 scores for various algorithms. The red dot denotes the mean values of the median R^2 scores for all datasets. This figure depicts that MEGP outperforms other SR and ML methods on average for both synthetic and real-world datasets. For example, on real-world

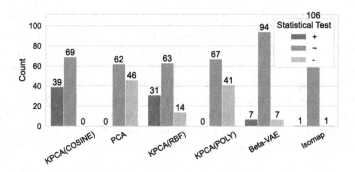

Fig. 9. Impact of reference semantic points on different dimensionality reduction techniques ("+", "∼" or "-" mean that using reference points is significantly better than, similar to, and worse than not using reference points on the specific dimensionality reduction method).

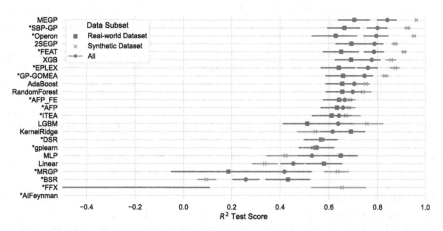

Fig. 10. Experimental results on 122 PMLB datasets (Results for AIFeynman are out of bounds and are therefore not shown).

datasets, MEGP has an average R^2 score of 0.704, outperforming a representative GP-based ensemble learning method 2SEGP [34], which has an average R^2 score of only 0.692. The advantage is significant with a p-value of $5 * 10^{-5}$.

6 Conclusions

In this paper, a new GP-based ensemble learning method named MEGP is proposed. First, MEGP uses an angle-based dimensionality reduction method in MAP-Elites to preserve good and complementary individuals. Meanwhile, MEGP synthesizes reference points to deal with an imbalanced distribution of good individuals. Experimental results show that MEGP with cosine-kernel KPCA outperforms MEGP with PCA on 45 datasets and is comparable to PCA on 63 datasets. Also, reference points improve its performance on 39 datasets and do not hurt it on others. Experimental results on SRBench demonstrate that MEGP outperforms 22 ML and SR algorithms across 122 datasets. This paper only examines the performance of MEGP in the regression scenario, it would be intriguing to study MEGP in the classification scenario in the future. Furthermore, while this paper focuses on improving MAP-Elites, the findings may also be applicable to archive-based quality-diversity optimization methods, which merit further investigation in the future. Last, finding new ways to aggregate GP models is also a promising direction to investigate in the future.

References

1. Balasubramanian, M., Schwartz, E.L.: The isomap algorithm and topological stability. Science **295**(5552), 7–7 (2002)
2. Banzhaf, W., Nordin, P., Keller, R.E., Francone, F.D.: Genetic Programming: An Introduction: On the Automatic Evolution of Computer Programs and its Applications. Morgan Kaufmann Publishers Inc., Burlington (1998)
3. Boisvert, S., Sheppard, J.W.: Quality diversity genetic programming for learning decision tree ensembles. In: Hu, T., Lourenço, N., Medvet, E. (eds.) EuroGP 2021. LNCS, vol. 12691, pp. 3–18. Springer, Cham (2021). https://doi.org/10.1007/978-3-030-72812-0_1
4. Bruneton, J.P., Cazenille, L., Douin, A., Reverdy, V.: Exploration and exploitation in symbolic regression using quality-diversity and evolutionary strategies algorithms. arXiv preprint arXiv:1906.03959 (2019)
5. Cava, W.L., et al.: Contemporary symbolic regression methods and their relative performance. In: 35th Conference on Neural Information Processing Systems Datasets and Benchmarks Track (Round 1) (2021)
6. Cawley, G.C., Talbot, N.L.: Fast exact leave-one-out cross-validation of sparse least-squares support vector machines. Neural Netw. **17**(10), 1467–1475 (2004)
7. Cazenille, L.: Ensemble feature extraction for multi-container quality-diversity algorithms. In: Proceedings of the Genetic and Evolutionary Computation Conference, pp. 75–83 (2021)
8. Chen, Q., Xue, B., Zhang, M.: Preserving population diversity based on transformed semantics in genetic programming for symbolic regression. IEEE Trans. Evol. Comput. **25**(3), 433–447 (2020)
9. Cully, A., Clune, J., Tarapore, D., Mouret, J.B.: Robots that can adapt like animals. Nature **521**(7553), 503–507 (2015)
10. Dick, G., Owen, C.A., Whigham, P.A.: Evolving bagging ensembles using a spatially-structured niching method. In: Proceedings of the Genetic and Evolutionary Computation Conference, pp. 418–425 (2018)
11. Dolson, E., Lalejini, A., Ofria, C.: Exploring genetic programming systems with MAP-Elites. In: Banzhaf, W., Spector, L., Sheneman, L. (eds.) Genetic Programming Theory and Practice XVI. GEC, pp. 1–16. Springer, Cham (2019). https://doi.org/10.1007/978-3-030-04735-1_1
12. Gaier, A., Asteroth, A., Mouret, J.B.: Aerodynamic design exploration through surrogate-assisted illumination. In: 18th AIAA/ISSMO Multidisciplinary Analysis and Optimization Conference, p. 3330 (2017)

13. Goldberg, D.E., Richardson, J., et al.: Genetic algorithms with sharing for multimodal function optimization. In: Genetic Algorithms and Their Applications: Proceedings of the 2nd International Conference on Genetic Algorithms, vol. 4149 (1987)

14. Higgins, I., et al.: beta-VAE: learning basic visual concepts with a constrained variational framework. In: International Conference on Learning Representations (2017)

15. Iba, H.: Bagging, boosting, and bloating in genetic programming. In: Proceedings of the 1st Annual Conference on Genetic and Evolutionary Computation-Volume 2, pp. 1053–1060 (1999)

16. La Cava, W., Helmuth, T., Spector, L., Moore, J.H.: A probabilistic and multiobjective analysis of lexicase selection and ε-lexicase selection. Evol. Comput. **27**(3), 377–402 (2019)

17. La Cava, W., Singh, T.R., Taggart, J., Suri, S., Moore, J.H.: Learning concise representations for regression by evolving networks of trees. In: International Conference on Learning Representations (2018)

18. Lehman, J., Stanley, K.O.: Evolving a diversity of virtual creatures through novelty search and local competition. In: Proceedings of the 13th Annual Conference on Genetic and Evolutionary Computation, pp. 211–218 (2011)

19. Li, L., Jamieson, K., DeSalvo, G., Rostamizadeh, A., Talwalkar, A.: Hyperband: a novel bandit-based approach to hyperparameter optimization. J. Mach. Learn. Res. **18**(1), 6765–6816 (2017)

20. Van der Maaten, L., Hinton, G.: Visualizing data using t-sne. J. Mach. Learn. Res. **9**(11) (2008)

21. Moraglio, A., Krawiec, K., Johnson, C.G.: Geometric semantic genetic programming. In: Coello, C.A.C., Cutello, V., Deb, K., Forrest, S., Nicosia, G., Pavone, M. (eds.) PPSN 2012. LNCS, vol. 7491, pp. 21–31. Springer, Heidelberg (2012). https://doi.org/10.1007/978-3-642-32937-1_3

22. Mouret, J.B., Clune, J.: Illuminating search spaces by mapping elites. arXiv preprint arXiv:1504.04909 (2015)

23. Ng, A., Jordan, M., Weiss, Y.: On spectral clustering: analysis and an algorithm. In: Advances in Neural Information Processing Systems, vol. 14 (2001)

24. Nickerson, K., Hu, T.: Principled quality diversity for ensemble classifiers using map-elites. In: Proceedings of the Genetic and Evolutionary Computation Conference Companion, pp. 259–260 (2021)

25. Nickerson, K., Kolokolova, A., Hu, T.: Creating diverse ensembles for classification with genetic programming and neuro-map-elites. In: Medvet, E., Pappa, G., Xue, B. (eds.) European Conference on Genetic Programming (Part of EvoStar), pp. 212–227. Springer, Cham (2022). https://doi.org/10.1007/978-3-031-02056-8_14

26. Nikfarjam, A., Neumann, A., Neumann, F.: On the use of quality diversity algorithms for the traveling thief problem. In: Proceedings of the Genetic and Evolutionary Computation Conference, pp. 260–268 (2022)

27. Nilsson, O., Cully, A.: Policy gradient assisted map-elites. In: Proceedings of the Genetic and Evolutionary Computation Conference, pp. 866–875 (2021)

28. Olson, R.S., La Cava, W., Orzechowski, P., Urbanowicz, R.J., Moore, J.H.: Pmlb: a large benchmark suite for machine learning evaluation and comparison. BioData Min. **10**(1), 1–13 (2017)

29. Pedregosa, F., et al.: Scikit-learn: machine learning in python. J. Mach. Learn. Res. **12**, 2825–2830 (2011)

30. Pietropolli, G., Manzoni, L., Paoletti, A., Castelli, M.: Combining geometric semantic GP with gradient-descent optimization. In: Medvet, E., Pappa, G., Xue, B. (eds.) European Conference on Genetic Programming (Part of EvoStar), pp. 19–33. Springer, Cham (2022)
31. Schölkopf, B., Smola, A., Müller, K.R.: Kernel principal component analysis. In: Gerstner, W., Germond, A., Hasler, M., Nicoud, J.D. (eds.) International Conference on Artificial Neural Networks, pp. 583–588. Springer, Cham (1997). https://doi.org/10.1007/BFb0020217
32. Urquhart, N., Hart, E.: Optimisation and illumination of a real-world workforce scheduling and routing application (WSRP) via Map-Elites. In: Auger, A., Fonseca, C.M., Lourenço, N., Machado, P., Paquete, L., Whitley, D. (eds.) PPSN 2018. LNCS, vol. 11101, pp. 488–499. Springer, Cham (2018). https://doi.org/10.1007/978-3-319-99253-2_39
33. Vanschoren, J., Van Rijn, J.N., Bischl, B., Torgo, L.: Openml: networked science in machine learning. ACM SIGKDD Explor. Newsl. **15**(2), 49–60 (2014)
34. Virgolin, M.: Genetic programming is naturally suited to evolve bagging ensembles. In: Proceedings of the Genetic and Evolutionary Computation Conference, pp. 830–839 (2021)
35. Wang, S., Mei, Y., Zhang, M.: Novel ensemble genetic programming hyper-heuristics for uncertain capacitated arc routing problem. In: Proceedings of the Genetic and Evolutionary Computation Conference, pp. 1093–1101 (2019)
36. Wang, Y., Xue, K., Qian, C.: Evolutionary diversity optimization with clustering-based selection for reinforcement learning. In: International Conference on Learning Representations (2021)
37. Zhang, H., Zhou, A., Zhang, H.: An evolutionary forest for regression. IEEE Trans. Evol. Comput. **26**(4), 735–749 (2022)

Small Solutions for Real-World Symbolic Regression Using Denoising Autoencoder Genetic Programming

David Wittenberg[✉][ID] and Franz Rothlauf[ID]

Johannes Gutenberg University, Mainz, Germany
{wittenberg,rothlauf}@uni-mainz.com

Abstract. Denoising Autoencoder Genetic Programming (DAE-GP) is a model-based evolutionary algorithm that uses denoising autoencoder long short-term memory networks as probabilistic model to replace the standard recombination and mutation operators of genetic programming (GP). In this paper, we use the DAE-GP to solve a set of nine standard real-world symbolic regression tasks. We compare the prediction quality of the DAE-GP to standard GP, geometric semantic GP (GSGP), and the gene-pool optimal mixing evolutionary algorithm for GP (GOMEA-GP), and find that the DAE-GP shows similar prediction quality using a much lower number of fitness evaluations than GSGP or GOMEA-GP. In addition, the DAE-GP consistently finds small solutions. The best candidate solutions of the DAE-GP are 69% smaller (median number of nodes) than the best candidate solutions found by standard GP. An analysis of the bias of the selection and variation step for both the DAE-GP and standard GP gives insight into why differences in solution size exist: the strong increase in solution size for standard GP is a result of both selection and variation bias. The results highlight that learning and sampling from a probabilistic model is a promising alternative to classic GP variation operators where the DAE-GP is able to generate small solutions for real-world symbolic regression tasks.

Keywords: Genetic Programming · Estimation of Distribution Algorithms · Denoising Autoencoders · Symbolic Regression

1 Introduction

Symbolic regression (SR) is the task of finding a mathematical expression that best fits a given dataset. In advance, it is unknown how many functions or terminals are used in an expression. Therefore, the encoding has to flexibly handle different solution sizes for which the variable-length parse trees used in genetic programming (GP) are particularly well suited [13].

Many GP variants have been proposed for solving SR tasks where some have proven to perform well compared to state-of-the-art machine learning algorithms [19]. Variants include multi-objective optimization approaches [24] or

© The Author(s), under exclusive license to Springer Nature Switzerland AG 2023
G. Pappa et al. (Eds.): EuroGP 2023, LNCS 13986, pp. 101–116, 2023.
https://doi.org/10.1007/978-3-031-29573-7_7

improved selection methods [14]. Recently, geometric semantic genetic programming (GSGP) has gained much attention where the idea is to apply recombination and mutation operators to the behavioral vector space, i.e., the output vector of a candidate solution when applied to a training set [17]. GSGP returns high-quality solutions for SR but has the disadvantage that the variation step stacks candidate solutions together and generates, by construction, offspring candidate solutions where the size of the candidate solutions (the number of nodes in a GP parse tree) grows exponentially with the number of generations [15], hindering the interpretability of the candidate solutions which is considered as one of the key advantages of GP compared to other regression methods [15,30].

To tackle this problem, Virgolin et al. [30] recently demonstrated that model-based evolutionary algorithms (MBEAs) may be an alternative to GSGP and GP. MBEAs replace the classic recombination and mutation operators by sampling from a learned probabilistic model. The authors study the gene-pool optimal mixing evolutionary algorithm for GP (GP-GOMEA). GP-GOMEA first learns linkages in a tree template (similar to the probabilistic prototype tree (PPT) [23]) and then uses gene-pool optimal mixing (GOM) as variation operator to transfer relevant sub-structures to the offspring. On a set of real-world regression problems, Virgolin et al. found that a model-based EA like GP-GOMEA outperforms GP. However, in their experiments they explicitly constrain the size of the solution space by setting a strict solution size limit of 15, 31, or 63 nodes (maximum depth of three, four, or five), which strongly reduces the prediction quality of GP. Moreover, the authors note that the fixed tree template to learn linkages may be a limitation of GP-GOMEA [30].

This work uses DAE-GP [33], a more flexible and neural network-based MBEA approach, to solve real-world SR tasks, where the idea is to use denoising autoencoder long short-term memory networks (DAE-LSTM) as probabilistic model. Compared to previous model-based approaches, the DAE-GP does not impose any assumptions about the relationships between problem variables which enables the model to flexibly identify hidden relationships in training data. In contrast to GP-GOMEA [30], DAE-LSTM do also not require any limitations on the maximum tree size. During model building, the DAE-GP first learns the properties of the parent population by learning to reconstruct candidate solutions that we propagate through the DAE-LSTM. During model sampling, new offspring solutions are generated by propagating small variations of these candidate solutions through the learned DAE-LSTM. Previous work on DAE-GP focused mainly on simple toy problems, such as the generalization of the royal tree problem [31,33]. Recently, Wittenberg & Rothlauf [32] presented the first results for a real-world SR problem. The authors demonstrated that the DAE-GP outperforms standard GP in the number of fitness evaluations while introducing significantly smaller best candidate solutions. The results are promising but limited to the Airfoil problem. Moreover, it is not clear why differences in solution size exist [32].

This paper extends and builds upon the results of Wittenberg & Rothlauf [32]. On a set of nine real-world regression datasets, we compare the prediction quality of the DAE-GP to standard GP and benchmark the results to both GSGP and GP-GOMEA. We find that the DAE-GP finds solutions of similar fitness, but uses much less fitness evaluations than GSGP and GP-GOMEA. Moreover, the DAE-GP consistently generates small solutions. Our results demonstrate that the best candidate solutions of the DAE-GP are 69% smaller than the best solutions generated by standard GP. An analysis of the bias of the selection and variation step for both the DAE-GP and standard GP gives insight into why differences in solution size exist: the strong increase in solution size for standard GP is a result of both selection and variation bias. The results highlight that learning and sampling from a probabilistic model is a promising alternative to classic recombination and mutation that can help to generate small solutions for real-world SR.

We organize the remainder of the paper as follows. In Sect. 2, we briefly summarize related work. Section 3 focuses on DAE-LSTM. In Sect. 4, we introduce the considered SR problems and compare the prediction quality of the DAE-GP to standard GP, GSGP, and GP-GOMEA. We proceed by comparing the solution sizes and analyze the learning behavior of both the DAE-GP and standard GP. Section 5 draws conclusions.

2   Related Work

The DAE-GP is a MBEA that belongs to the class of estimation of distribution genetic programming (EDA-GP) algorithms [33]. Research on EDA-GP has moved from simple univariate to more complex multivariate models.

We can roughly categorize research on EDA-GP into two research streams [11]: the first one focuses on models relying on a probabilistic prototype tree (PPT) [23]. Given that a is equal to the maximum arity of the functions in the function set, a PPT is an a-ary tree of maximum depth d_{max}. At each node of the tree, the idea is to learn a multinomial distribution over the set of allowed functions and terminals. In 1997, Salustowicz and Schmidhuber [23] introduced PPT as probabilistic model in the first EDA-GP called probabilistic incremental program evolution (PIPE). Based on the univariate PIPE, e.g., the bivariate estimation of distribution programming (EDP) [36] or the multivariate program optimization with linkage estimation (POLE) [8,10] followed. For the MAX, the deceptive MAX, and the royal tree problem, Hasegawa & Iba [10] report that POLE outperforms standard GP in the number of fitness evaluations [10]. Recently, Virgolin et al. [30] suggested GP-GOMEA, which combines a PPT-like tree template (the linkage model) with gene-pool optimal mixing (GOM) as variation operator. Since GOM is essentially a hill-climber that iteratively improves parent solutions given the linkage model, GP-GOMEA belongs to the class MBEA but not to the class of EDA-GP. For a set of real-world regression problems, the authors found that GP-GOMEA generates better solutions than GP. However, the size of the solution space was restricted (15, 31, or 63 nodes), making a direct comparison to GP difficult [30].

The second research stream focuses on EDA-GP models that rely on a grammar [11]. A grammar is a set of production rules, where Ratle and Sebag [21] presented stochastic grammar-based genetic programming (SG-GP) as the first grammar-based EDA-GP in 2001. SG-GP uses stochastic context-free grammar (SCFG) as probabilistic model [21]. Relevant extensions are, e.g., program with annotated grammar estimation (PAGE) [9] or grammar-based genetic programming with a Bayesian network (BGBGP) [34,35]. For the asymmetric royal tree problem and the deceptive MAX, Wong et al. [35] report that BGBGP outperforms POLE, PAGE, and grammar-based GP in the number of fitness evaluations [35].

Besides the two main research streams, some authors also introduced EDA-GP variants that do not rely on a PPT or a grammar. One such variant is the DAE-GP by Wittenberg et al. [33] that we use in this work and that relies on DAE-LSTM as probabilistic model. DAE-LSTM are artificial neural networks that flexibly learn dependencies between problem variables by reconstructing candidate solutions through a latent space. For a generalization of the royal tree problem, the authors found that the DAE-GP needs less fitness evaluations than standard GP to find the optimal solution. The results indicate that the DAE-GP can better identify promising areas of the solution space compared to standard GP resulting in a more efficient search in the number of fitness evaluations, especially in large solution spaces [33]. As a follow-up work and based on previous work by Probst & Rothlauf [20], Wittenberg [31] studied how denoising affects the exploration and exploitation behavior of the DAE-GP. The results indicate that denoising can be used to control the diversity behavior in search [31]. However, results in both works [31,33] are limited to the generalization of the royal tree problem, which is a problem of artificial nature. Wittenberg & Rothlauf [32] presented the first results for real-world SR and demonstrated that their DAE-GP significantly outperforms standard GP in the number of fitness evaluations and generates significantly smaller best candidate solutions. However, results are limited to the Airfoil problem and do not give any insight into why differences in solution size exist [32]. This paper extends these results to nine real-world SR tasks and analyzes the bias of the selection and variation step of both standard GP and DAE-GP to answer the question why differences in solution size exist.

3   Denoising Autoencoder LSTM

DAE-LSTM are autoencoders that consist of an encoding and a decoding LSTM (AE-LSTM) [25]: the encoding LSTM maps a candidate solution represented as a linear sequence in prefix expression to the latent space; the decoding LSTM maps the latent space back to a candidate solution. To prevent the model from learning the simple identity function, candidate solutions given as input to the model are denoised, which transforms the AE-LSTM into a DAE-LSTM. DAE-LSTM are used as probabilistic model in the DAE-GP, where the following two steps are repeated at each generation: first, the DAE-LSTM is trained to learn the properties of selected candidate solutions (model building). Then, new candidate solutions are propagated through the trained model to generate a new offspring population (model sampling).

This paper uses the architecture presented in [31][1]. Section 3.1 gives a brief overview on model building and sampling. Section 3.2 presents Levenshtein tree edit, a new denoising strategy that we introduce in this paper.

3.1 Model Building and Sampling

During model building, the aim is to capture relevant properties of the parent population in the latent space by learning to reconstruct candidate solutions given as input. We first split our parent population into a training and a validation set and then iterate over the candidate solutions x present in our training set X: we first denoise (mutate) the input candidate solution x according to our denoising strategy (we can adjust the type and strength of denoising), which transforms x into $\tilde{x}$. Then, we use the encoding and decoding LSTM to propagate $\tilde{x}$ through the DAE-LSTM, which generates the output candidate solution o. Finally, we compute the reconstruction error between the output o and the original (not the corrupted) input candidate solution x and update our trainable parameters θ using gradient descent. We apply early stopping on a hold-out validation set U to assure that the DAE-LSTM does not overfit to the training data in X, where we stop training as soon as the validation error does not improve over 200 epochs.

During model sampling, we use those parameters θ that minimize the validation error. We generate new candidate solutions o forming the offspring population P_{g+1} by propagating random candidate solutions from the training set X through the trained model. We first randomly pick a candidate solution $x \in X$ and denoise x using the same denoising strategy as during model building. Then, we propagate $\tilde{x}$ through the DAE-LSTM and add the resulting output candidate solution o to the offspring. Similar to [31], we assure that the DAE-LSTM only generates syntactically valid candidate solutions where we use syntax control during model sampling.

3.2 A New Denoising Strategy: Levenshtein Tree Edit

Reconstructing the input during model building can be a trivial task if the hidden layer of the AE-LSTM is sufficiently large: we learn the simple identity function which means that we replicate the candidate solutions given as input to the model. To avoid this and to learn a more useful representation in the latent space, we apply denoising on input candidate solutions, which transforms the AE-LSTM into a DAE-LSTM. The idea is to slightly corrupt input candidate solutions, making the model robust to noise that is present in the parent population [29,33]. We can formally describe the process as follows: at each generation g, we first select promising candidate solutions x from the population P_g, forming the training set X (of size N), and denoise x using the corruption function $c(x)$ by

$$\tilde{x}^i = c(x^i) \ \forall i \in \{1, .., N\}, \tag{1}$$

---

[1] For a detailed description on DAE-LSTM, refer to Wittenberg [31].

where $\tilde{x}^i$ represents the corrupted version of the i-th candidate solution of X, which we use as input to the DAE-LSTM [33].

The stronger we denoise x (the more parts of x we corrupt), the more we force the DAE-LSTM to focus on general properties of the parent population. Recently, Wittenberg [31] analyzed the influence of corruption strength on the exploration and exploitation behavior of the DAE-LSTM. Exploration means that many new candidate solutions are introduced into search, whereas exploitation describes that an algorithm focuses stronger on previously seen solution spaces. In general, we have to find a good balance between exploration and exploitation in search [22]. For the generalization of the royal tree problem, Wittenberg found that a strong corruption of input candidate solutions results in a strong exploration of the solution space. In contrast, lowering the corruption strength increases the exploitation of the solution space. Therefore, the author recommends to carefully choose the appropriate denoising strategy, where the strength of corruption can help to balance the level of exploration or exploitation needed for a given problem.

Wittenberg [31] introduced Levenshtein edit as a new corruption function. Levenshtein edit operates on the string representation of a candidate solution (the linear representation in prefix expression) and uses insertion (add one node), deletion (remove one node), and substitution (replace one node by another node) to transform x into $\tilde{x}$. One advantage of Levenshtein edit is that we can easily control the corruption strength by increasing or decreasing the edit percentage p $(0 < p < 1)$: given a candidate solution x of size m, with $x_j, j \in \{1, 2, .., m\}$, each node x_j has a probability p of being corrupted. Moreover, the use of Levenshtein edit allows variations of solution sizes (number of nodes) [31].

However, the approach suggested by Wittenberg [31], which operates on the string representation of a candidate solution, has the disadvantage that it easily generates syntactically invalid GP solutions and ignores the semantics of a candidate solution. For example, when dealing with GP parse trees, we usually assume that replacing a node closer to the root of a tree has a stronger impact on fitness than replacing a node that is closer to a leaf node.

Consequently, we want to suggest Levenshtein tree edit, which is similar to Levenshtein edit [31] in that we use insertion, deletion, and substitution as edit operators. However, we apply these edit operators on the tree representation of a candidate solution. Given a function set F and a terminal set T, we proceed as follows to apply the edit operators on x: insertion adds a new branch (of depth 1) to a random leaf node of x. Here, we first randomly choose a leaf node x_j and insert a function $f \in F$ of arity r at the position of x_j. Then, we define x_j as the child node of f (the position of x_j is randomly chosen) and sample $r - 1$ new terminals $t \in T$, forming the arguments of f. Deletion shrinks x by randomly choosing a branch (of depth 1) and replacing that branch with one of the branch's arguments. Here, we only consider branches where the arguments of the root of the branch are terminals to prevent the deletion operator from removing larger parts of x. Substitution uses simple point mutation, where we replace a randomly chosen node x_j by a new node of same arity r.

To use Levenshtein tree edit as corruption function in the DAE-LSTM, we first generate a sequence s, with s_j, $j \in \{1, 2, .., m\}$ of random edit operations where each s_j is, with uniform probability, either insertion, deletion, or substitution. Given an edit percentage p ($0 < p < 1$), we then iterate over s and apply s_j on x with probability p. Thus, we can control corruption strength by adjusting the edit percentage p which leads to a corruption of x that is relative to its solution size m.

Using Levenshtein tree edit as denoising strategy has several advantages: (1) we assure that x keeps being syntactically valid, (2) we introduce variance in solution size, and (3) we can easily control the corruption strength by adjusting p, allowing a fine-grained corruption (also semantically) of x. The larger p, the stronger the corruption.

4   Experiments

We apply the DAE-GP to real-world symbolic regression tasks and compare its performance to standard GP, GSGP, and GP-GOMEA. We first describe the experimental setup, where we introduce the datasets and explain which parameters we use for both the DAE-GP and standard GP. Then, we present and discuss the results. We focus our analysis on prediction quality and candidate solution size.

4.1   Experimental Setup

Table 1 shows the set of nine real-world regression datasets used in the experiments and provides an overview of the number of features and observations used in the datasets. The datasets are commonly used in the GP literature [15,16,30] and available in the UCI Machine Learning Repository [4].

Table 1. Real-world regression datasets.

Dataset	#Features	#Observations	Source
Airfoil	5	1,503	[1]
Boston Housing	13	506	[7]
Concrete	8	1,030	[37]
Energy Cooling	8	768	[27]
Energy Heating	8	768	[27]
Parkinsons	20	5,875	[26]
Wine Red	11	1,599	[3]
Wine White	11	4,898	[3]
Yacht	6	308	[6]

For our experiments, we use the evolutionary framework DEAP [5] and the neural network framework Keras [2]. We initialize the population with ramped half-and-half (RHH), where we set the minimum and maximum initialization depth to $d_{min} = 2$ and $d_{max} = 6$, respectively, and allow the algorithms to search for solutions up to a maximum tree depth of $d_{max} = 17$. Similar to [15,30], we use the function set $F = \{+, -, *, \div_{AQ}\}$, where addition, subtraction, multiplication, and the analytic quotient (AQ) are binary functions in the function set F. For a number of regression tasks, Ni et al. [18] report that using AQ as an alternative to protected division helps GP to improve prediction quality [18]. We use the terminal set $T = Z \cup ERC$, which consists of problem features $Z = \{z_1, z_2, \ldots, z_n\}$ and ephemeral random integer constants $ERC \in [-5, \ldots, 5]$. We set the population size to 500, use a tournament selection of size 7, and stop the experiments as soon as they reach 10,000 fitness evaluations. We conduct 30 runs per dataset and algorithm (540 runs in total) and split the dataset at each run into 50% training set and 50% test set, where we use the root mean squared error (RMSE) to evaluate the prediction quality on both training and test set.

As variation operator, GP uses standard subtree crossover (90%) with internal node bias (90% functions, 10% terminals) and subtree mutation (10%) [13]. Note that subtree mutation uses full initialization with $d_{min} = 0$ and $d_{max} = 2$ to generate a new subtree. The DAE-GP uses model building and sampling as variation operator, where some hyperparameters need to be set in advance. We did not conduct a hyperparameter optimization but used standard settings: similar to [31], we set the number of hidden layers to one and the hidden dimension equal to the maximum size l of the candidate solutions that we use as input to the model, allowing the model to flexibly adjust its complexity depending on the size of the candidate solutions presented to the model. At each generation, we split the population into 50% training set and 50% validation set to track the validation error for early stopping. We train the DAE-LSTM with batches of 25 (10% of training set X) solutions, use adaptive moment estimation (Adam) [12] for gradient descent, and set the learning rate α equal to 0.001. As denoising strategy, we use Levenshtein tree edit and set the edit percentage to $p = 0.05$ for all experiments.

4.2   Prediction Quality

Table 2 shows the median RMSE of the best solution found of the DAE-GP and standard GP after 10,000 fitness evaluations, aggregated over a total of 540 runs, and benchmarks the results to GSGP [15] and GP-GOMEA [30]. Note that the results of GSGP and GP-GOMEA are results from the literature that were generated with different parameter settings: Virgolin et al. [30] restrict the solution size throughout a run to a maximum depth of $d_{max} = 4$ (they also tested $d_{max} = 3, 5$ but recommend $d_{max} = 4$), choose a population size of 2,000 and allow GP-GOMEA to run for 20 generations, resulting in a maximum number of 2,400,000 fitness evaluations (at each generation, each candidate solution is evaluated up to $2l$-2 times, with $l = 31$). The results reported in this paper

Table 2. Median Best RMSE of best candidate solution of standard GP, DAE-GP, GSGP, and GP-GOMEA with approximated number of fitness evaluations (FE) in brackets. Note that the results for GSGP and GP-GOMEA are taken from the literature. A bold test RMSE indicates better prediction quality. The asterisk (*) denotes significant differences on the test RMSE when comparing standard GP to DAE-GP.

Dataset	Set	standard GP (10,000 FE)	DAE-GP (10,000 FE)	GSGP [15] (250,000 FE)	GP-GOMEA [30] (2,400,000 FE)
Airfoil	Train	16.914	6.313	11.783	3.771
	Test	17.787	6.177*	11.280	**3.889**
Boston Housing	Train	4.090	4.037	-	3.606
	Test	**4.214**	4.376	-	4.501
Concrete	Train	8.947	9.864	8.510	6.985
	Test	9.016	10.089	8.886	**7.221**
Energy Cooling	Train	2.618	2.564	3.114	2.759
	Test	2.650	**2.581**	3.129	2.881
Energy Heating	Train	3.249	2.774	2.677	57.199
	Test	18.787	18.687	**2.739**	58.379
Parkinsons	Train	6.457	6.575	9.812	-
	Test	**6.469**	6.605	9.868	-
Wine Red	Train	0.567	0.546	0.632	0.627
	Test	0.561	**0.548**	0.636	0.638
Wine White	Train	0.630	0.621	0.729	0.731
	Test	0.632	**0.628**	0.735	0.736
Yacht	Train	1.938	3.049	6.529	0.883
	Test	2.083*	2.968	6.437	**1.153**

correspond to the results of the GP-GOMEA with improved linkage trees (LT-MI$_{\tilde{6}}$) that we transformed from the variance normalized mean squared error (NMSE) to the RMSE for better comparison. In contrast, Martins et al. [15] use a population size of 1,000 and choose 250 generations for GSGP, resulting in a maximum number of 250,000 fitness evaluations. Although a fair comparison between DAE-GP and GSGP and GP-GOMEA is difficult (DAE-GP as well as standard GP use only 10,000 fitness evaluations, but obtain similar prediction quality), the results are helpful for assessing the performance of DAE-GP.

We report the prediction quality on both the training and the test set. To determine the test RMSE, we first identify the best candidate solution on the training set and then evaluate that candidate solution on the test set. A bold test RMSE indicates a better prediction quality (lower RMSE) compared to all the other considered algorithms on a given dataset. The asterisk (*) denotes significant differences in prediction quality between standard GP and the DAE-GP on a given dataset. Here, we apply (pairwise) Mann-Whitney U tests, also known as Wilcoxon rank-sum tests, to test the hypothesis that the best fitness distributions are from the same population. We set the significance level to 0.05.

To counteract the problem of statistical errors when conducting mutliple pairwise comparisons, we use Bonferroni correction, where we divide the significance level (here 0.05) by the number of pairwise comparisons (here 9).

Although the results for GSGP and GP-GOMEA from the literature use a higher number of fitness evaluations, we find that there is not a clear winner algorithm that dominates the others in prediction quality: on the test set, both the DAE-GP and GP-GOMEA yield best results on 3 datasets, followed by standard GP (best on 2 datasets) and GSGP (best on one dataset). Differences in prediction quality are in many cases not significant.

When comparing standard GP to the DAE-GP (we know the best fitness distributions only for these algorithms), we find that the DAE-GP significantly outperforms standard GP only on the Airfoil dataset, whereas standard GP shows significantly better results than the DAE-GP only on the Yacht dataset. For 7 datasets, no significant differences can be found. However, when considering the number of fitness evaluations available to each algorithm, we notice that GP-GOMEA uses up to 240 times (GSGP up to 25 times) more fitness evaluations than both the DAE-GP and standard GP, without yielding considerably better prediction results. This demonstrates that standard GP performs quite well (when allowing the solutions to grow). Likewise, it indicates that the DAE-GP is able to well exploit relevant areas of the solution space and is able to transfer this knowledge to the offspring, yielding an efficient search in the number of fitness evaluations.

4.3 Analyzing the Search Behavior

We analyze the search behavior of both standard GP and the DAE-GP and focus on solution size (number of nodes). Analyzing the solution size is relevant since smaller candidate solutions tend to be more interpretable and can help practitioners to better understand the identified solution. Especially when we compare GP solutions to other successful regression methods, such as gradient boosting, we notice that the interpretability of the final candidate solution is one of the key advantages of GP. Therefore, some recent studies focus on finding small solutions for GP [15,30].

Table 3 shows the median solution size of the best candidate solutions found by the DAE-GP and standard GP for the considered datasets. Similar to Table 2, bold numbers indicate a smaller solution size distribution of the DAE-GP to standard GP where the asterisk denotes significant differences using the Mann-Whitney U test.

The results demonstrate that the best candidate solutions generated by the DAE-GP are significantly smaller than the candidate solutions generated by standard GP. The DAE-GP generates best candidate solutions with a median size of 23 (in the range of interpretable solutions [30]), while standard GP generates best candidate solutions with a median size of 75. The DAE-GP thus consistently generates best candidate solutions that are, in median, 69% smaller than the candidate solutions generated by standard GP, while yielding a similar prediction quality.

Table 3. Median solution size (number of nodes) of best candidate solution, where a bold number indicates a smaller solution size. The asterisk (*) denotes significant differences in solution size.

Dataset	standard GP	DAE-GP
Airfoil	98	**24***
Boston Housing	85	**23***
Concrete	74	**24***
Energy Cooling	69	**23***
Energy Heating	59	**23***
Parkinsons	94	**23***
Wine Red	69	**23***
Wine White	50	**23***
Yacht	92	**25***

Figure 1 plots the median solution size (left) and the median fitness on the test set (right) of the best candidate solution with 25th and 75th percentiles over the number of fitness evaluations, where results are aggregated over all nine datasets. It reflects how both best solution size and best fitness develop throughout search. We observe that the size of the best solutions (left) increases for both the DAE-GP and standard GP over the number of fitness evaluations. However, the slope of the increase is different when comparing the DAE-GP to standard GP: while the DAE-GP only slightly increases its solution size to a median of 23 nodes, standard GP consistently introduces larger solutions until a median solution size of 75, where the 25th and 75th percentiles indicate that also the variance in the solution size increases over the number of fitness evaluations. Meanwhile, we can only hardly observe any differences in solution quality when comparing the slope of the median best fitness of standard GP to DAE-GP

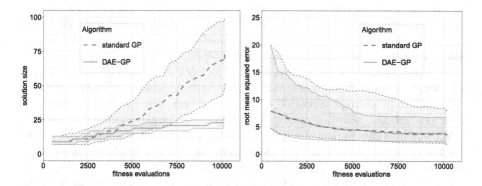

Fig. 1. Median solution size (left) and median fitness on test set (right) of best candidate solution over fitness evaluations across all datasets, with 25th and 75th percentiles.

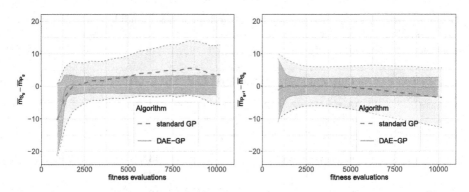

Fig. 2. Bias of selection step (left) and variation step (right) on solution size measured by average differences in solution size m between populations P over fitness evaluations across all datasets, with standard deviations.

(right). This is in line with previous findings from Table 3, where no clear winner in prediction quality could be found. However, variance in fitness is slightly higher for standard GP.

To better understand why the size of the best solution grows for both standard GP and DAE-GP, we investigate how both the selection and variation step bias the population towards smaller or larger solutions. Recall that the DAE-GP only replaces the variation step of standard GP. Therefore, differences in solution size must be a result of both the variation step that introduces new offspring solutions and the selection step that builds a subset of promising offspring solutions. Figure 2 plots the bias of the selection and variation step for both standard GP and the DAE-GP by measuring averaged solution size differences (differences in the number of nodes) between populations. Assuming that a population P_g is selected into population S_g at generation g, and that variation generates the offspring population P_{g+1}, we can calculate differences between average solution sizes $\overline{m}$ of populations P: on the left, we measure the bias of the selection step by $\overline{m}_{S_g} - \overline{m}_{P_g}$; on the right, we calculate the bias of the variation step by $\overline{m}_{P_g+1} - \overline{m}_{S_g}$. Positive values indicate that the selection or variation step biases search towards larger candidate solutions, whereas negative values indicate a bias towards smaller solutions.

The results are surprising. Interestingly, for both standard GP and the DAE-GP, selection (left) first biases search towards very small candidate solutions, with minimum average values of -10.3 in the very first generation. Then, in later generations, selection pushes search towards larger candidate solutions. While the bias in solution size is only slightly positive for the DAE-GP (up to maximum average values of 0.48) and remains on that low level, we observe that the bias towards larger solutions increases for standard GP (up to maximum average values of 5.57), with much larger standard deviations compared to the DAE-GP. Considering the variation step (right), we expect that the average bias towards larger or smaller solutions for both standard GP and the DAE-

GP is close to zero: standard GP mainly uses crossover, which only recombines (in a pairwise manner) sub-trees of previously selected candidate solutions. In contrast, the DAE-GP uses model building and sampling and transfers learned properties (including the solution size) to the offspring. For the DAE-GP, we find a bias towards smaller solutions only in the very first generations, which is probably because the complexity of the population presented to the model is highest in these generations. Later, average differences follow our expectations and remain close to zero (with average values of around −0.25). The results indicate that the DAE-GP well learns the properties of the parent population and is able to transfer these properties to the offspring. However, in very early generations, a slightly higher model complexity could help to further improve model quality. For standard GP, the average bias remains close to zero only until around 5,000 fitness evaluations. Then, the slope decreases to a minimum value of −3.5. We believe that the reason for the increasing bias towards smaller solutions is due to sub-tree mutation. Here, the idea is to replace a random sub-tree of a candidate solution by a new sub-tree, with the size of the new sub-tree being fixed (see Sect. 4.1). Since the solution size increases for standard GP over generations, the probability of replacing a larger sub-tree by sub-tree mutation increases over generations, which results in a bias of the standard GP variation operator towards smaller solutions. However, we also notice that the variance in solution size is very high compared to the DAE-GP and that this variance also increases with growing solution size.

The strong increase of best candidate solution sizes for standard GP compared to the DAE-GP is a result of both selection and variation: while the variation step of standard GP introduces offspring candidate solutions that vary much stronger in solution size compared to the DAE-GP, it is the selection step that identifies large solutions as promising solutions and pushes search towards these large solutions. In contrast, the DAE-GP introduces high-quality candidate solutions that vary much less in solution size, highlighting that model building and sampling may be a promising alternative compared to classic recombination and mutation to generate small solutions.

5   Conclusions and Future Work

This paper studied the DAE-GP, a novel neural network-based MBEA approach that uses denoising autoencoder long short-term memory networks as probabilistic model. For a set of nine standard real-world regression datasets, we find that the DAE-GP performs similar to GSGP and GP-GOMEA using a much lower number of fitness evaluations. In addition, it consistently generates significantly smaller solutions compared to standard GP, without the need of additional restrictions. We analyzed the bias of the selection and variation step for both the DAE-GP and standard GP to answer the question why differences in solution size exist and found that the strong increase in solution size for standard GP is a result of both selection and variation bias: while standard GP introduces offspring solutions with much larger variance in solution size compared to the

DAE-GP, it is the selection step that pushes search towards large solutions. The results highlight that sampling from a probabilistic model building is a promising alternative to classic recombination and mutation where the DAE-GP can be used to generate small solutions for real-world symbolic regression tasks.

In future work, we study how we can improve the interpretability of best candidate solutions for both the DAE-GP and GP. One option is to reduce solution complexity by algebraic simplifications on the solutions, as proposed by [15]. Furthermore, we want to analyze other model architectures, such as the transformer architecture [28], optimize the hyperparameters at each generation, and extend the analysis to more fitness evaluations and more benchmark problems.

Acknowledgements. We thank our group in Mainz for previous work and insightful discussions on this topic. Parts of this research were conducted using the supercomputer Mogon offered by Johannes Gutenberg University Mainz (hpc.uni-mainz.de), which is a member of the AHRP (Alliance for High Performance Computing in Rhineland Palatinate, www.ahrp.info) and the Gauss Alliance e.V. The authors gratefully acknowledge the computing time granted on the supercomputer Mogon at Johannes Gutenberg University Mainz (hpc.uni-mainz.de).

References

1. Brooks, T.F., Pope, D.S., Marcolini, M.A.: Airfoil self-noise and prediction, vol. 1218. In: National Aeronautics and Space Administration, Office of Management, Scientific and Technical Information Division (1989)
2. Chollet, F.: keras (2015). https://github.com/fchollet/keras
3. Cortez, P., Cerdeira, A., Almeida, F., Matos, T., Reis, J.: Modeling wine preferences by data mining from physicochemical properties. Decis. Support Syst. **47**(4), 547–553 (2009)
4. Dua, D., Graff, C.: UCI machine learning repository (2017). http://archive.ics.uci.edu/ml
5. Fortin, F.A., De Rainville, F.M., Gardner, M.A., Parizeau, M., Gagńe, C.: DEAP: evolutionary algorithms made easy. J. Mach. Learn. Res. **13**(1), 2171–2175 (2012)
6. Gerritsma, J., Onnink, R., Versluis, A.: Geometry, resistance and stability of the delft systematic yacht hull series. Int. Shipbuild. Prog. **28**(328), 276–297 (1981)
7. Harrison, D., Jr., Rubinfeld, D.L.: Hedonic housing prices and the demand for clean air. J. Environ. Econ. Manag. **5**(1), 81–102 (1978)
8. Hasegawa, Y., Iba, H.: Estimation of Bayesian network for program generation. In: Proceedings of the Third Asian-Pacific Workshop on Genetic Programming, pp. 35–46. Hanoi, Vietnam (2006)
9. Hasegawa, Y., Iba, H.: Estimation of distribution algorithm based on probabilistic grammar with latent annotations. In: Proceedings of the IEEE Congress on Evolutionary Computation (CEC 2007), pp. 1043–1050. IEEE (2007). https://doi.org/10.1109/CEC.2007.4424585
10. Hasegawa, Y., Iba, H.: A Bayesian network approach to program generation. IEEE Trans. Evol. Comput. **12**(6), 750–764 (2008). https://doi.org/10.1109/tevc.2008.915999
11. Kim, K., Shan, Y., Nguyen, X.H., McKay, R.I.: Probabilistic model building in genetic programming: a critical review. Genet. Program Evolvable Mach. **15**(2), 115–167 (2014). https://doi.org/10.1007/s10710-013-9205-x

12. Kingma, D.P., Ba, J.: Adam: a method for stochastic optimization. In: International Conference on Learning Representations. San Diego, CA, USA (2015)
13. Koza, J.R.: Genetic Programming: On the Programming of Computers by Means of Natural Selection. MIT Press, Cambridge, London (1992)
14. La Cava, W., Spector, L., Danai, K.: Epsilon-lexicase selection for regression. In: Proceedings of the Genetic and Evolutionary Computation Conference 2016 (GECCO 2016), pp. 741–748. Association for Computing Machinery, New York, NY, USA (2016). https://doi.org/10.1145/2908812.2908898
15. Martins, J.F.B.S., Oliveira, L.O.V.B., Miranda, L.F., Casadei, F., Pappa, G.L.: Solving the exponential growth of symbolic regression trees in geometric semantic genetic programming. In: Proceedings of the Genetic and Evolutionary Computation Conference (GECCO 2018), pp. 1151–1158. Association for Computing Machinery, New York, NY, USA (2018). https://doi.org/10.1145/3205455.3205593
16. de Melo, V.V., Vargas, D.V., Banzhaf, W.: Batch tournament selection for genetic programming: the quality of lexicase, the speed of tournament. In: Proceedings of the Genetic and Evolutionary Computation Conference (GECCO 2019), pp. 994–1002. Association for Computing Machinery, New York, NY, USA (2019). https://doi.org/10.1145/3321707.3321793
17. Moraglio, A., Krawiec, K., Johnson, C.G.: Geometric semantic genetic programming. In: Coello, C.A.C., Cutello, V., Deb, K., Forrest, S., Nicosia, G., Pavone, M. (eds.) PPSN 2012. LNCS, vol. 7491, pp. 21–31. Springer, Heidelberg (2012). https://doi.org/10.1007/978-3-642-32937-1_3
18. Ni, J., Drieberg, R.H., Rockett, P.I.: The use of an analytic quotient operator in genetic programming. IEEE Trans. Evol. Comput. **17**(1), 146–152 (2013). https://doi.org/10.1109/TEVC.2012.2195319
19. Orzechowski, P., La Cava, W., Moore, J.H.: Where are we now? A large benchmark study of recent symbolic regression methods. In: Proceedings of the Genetic and Evolutionary Computation Conference (GECCO 2018), pp. 1183–1190. Association for Computing Machinery, New York, NY, USA (2018). https://doi.org/10.1145/3205455.3205539
20. Probst, M., Rothlauf, F.: Harmless overfitting: using denoising autoencoders in estimation of distribution algorithms. J. Mach. Learn. Res. **21**(78), 1–31 (2020). http://jmlr.org/papers/v21/16-543.html
21. Ratle, A., Sebag, M.: Avoiding the bloat with stochastic grammar-based genetic programming. In: Collet, P., Fonlupt, C., Hao, J.-K., Lutton, E., Schoenauer, M. (eds.) EA 2001. LNCS, vol. 2310, pp. 255–266. Springer, Heidelberg (2002). https://doi.org/10.1007/3-540-46033-0_21
22. Rothlauf, F.: Design of Modern Heuristics: Principles and Application, 1st edn. Springer, Heidelberg (2011). https://doi.org/10.1007/978-3-540-72962-4
23. Salustowicz, R., Schmidhuber, J.: Probabilistic incremental program evolution. Evol. Comput. **5**(2), 123–141 (1997). https://doi.org/10.1162/evco.1997.5.2.123
24. Schmidt, M., Lipson, H.: Age-fitness pareto optimization. In: Riolo, R., McConaghy, T., Vladislavleva, E. (eds.) Genetic Programming Theory and Practice VIII, pp. 129–146. Springer, New York (2011). https://doi.org/10.1007/978-1-4419-7747-2_8
25. Srivastava, N., Mansimov, E., Salakhutdinov, R.: Unsupervised learning of video representations using LSTMs. In: Proceedings of the 32nd International Conference on Machine Learning (ICML 2015), pp. 843–852. ACM, Lille, France (2015). https://doi.org/10.5555/3045118.3045209

26. Tsanas, A., Little, M.A., McSharry, P.E., Ramig, L.O.: Accurate telemonitoring of Parkinson's disease progression by noninvasive speech tests. IEEE Trans. Biomed. Eng. **57**(4), 884–893 (2009)

27. Tsanas, A., Xifara, A.: Accurate quantitative estimation of energy performance of residential buildings using statistical machine learning tools. Energy Build. **49**, 560–567 (2012)

28. Vaswani, A., et al.: Attention is all you need. Adv. Neural. Inf. Process. Syst. **30**, 5998–6008 (2017)

29. Vincent, P., Larochelle, H., Bengio, Y., Manzagol, P.A.: Extracting and composing robust features with denoising autoencoders. In: Proceedings of the 25th International Conference on Machine Learning (ICML 2008), pp. 1096–1103. ACM, Helsinki, Finland (2008). https://doi.org/10.1145/1390156.1390294

30. Virgolin, M., Alderliesten, T., Witteveen, C., Bosman, P.A.N.: Improving model-based genetic programming for symbolic regression of small expressions. Evol. Comput. **29**(2), 211–237 (2021). https://doi.org/10.1162/evco_a_00278

31. Wittenberg, D.: Using denoising autoencoder genetic programming to control exploration and exploitation in search. In: Medvet, E., Pappa, G., Xue, B. (eds.) Genetic Programming (EuroGP 2022). LNCS, vol. 13223, pp. 102–117. Springer, Cham (2022). https://doi.org/10.1007/978-3-031-02056-8_7

32. Wittenberg, D., Rothlauf, F.: Denoising autoencoder genetic programming for real-world symbolic regression. In: Proceedings of the Genetic and Evolutionary Computation Conference Companion (GECCO 2022), pp. 612–614. Association for Computing Machinery, New York, NY, USA (2022). https://doi.org/10.1145/3520304.3528921

33. Wittenberg, D., Rothlauf, F., Schweim, D.: DAE-GP: denoising autoencoder LSTM networks as probabilistic models in estimation of distribution genetic programming. In: Proceedings of the 2020 Genetic and Evolutionary Computation Conference (GECCO 2020), pp. 1037–1045. ACM, New York, NY, USA (2020). https://doi.org/10.1145/3377930.3390180

34. Wong, P.K., Lo, L.Y., Wong, M.L., Leung, K.S.: Grammar-based genetic programming with Bayesian network. In: IEEE Congress on Evolutionary Computation (CEC 2014), pp. 739–746. IEEE, Beijing, China (2014)

35. Wong, P.K., Lo, L.Y., Wong, M.L., Leung, K.S.: Grammar-based genetic programming with dependence learning and Bayesian network classifier. In: Proceedings of the Genetic and Evolutionary Computation Conference (GECCO 2014), pp. 959–966. ACM, Vancouver, Canada (2014). https://doi.org/10.1145/2576768.2598256

36. Yanai, K., Iba, H.: Estimation of distribution programming based on Bayesian network. In: IEEE Congress on Evolutionary Computation (CEC 2003), pp. 1618–1625. IEEE, Canberra, Australia (2003). https://doi.org/10.1109/CEC.2003.1299866

37. Yeh, I.C.: Modeling of strength of high-performance concrete using artificial neural networks. Cem. Concr. Res. **28**(12), 1797–1808 (1998)

Context Matters: Adaptive Mutation for Grammars

Pedro Carvalho$^{(\boxtimes)}$ ⓘ, Jessica Mégane ⓘ, Nuno Lourenço ⓘ, and Penousal Machado ⓘ

Centre for Informatics and Systems of the University of Coimbra, Department of Informatics Engineering, University of Coimbra, Coimbra, Portugal
{pfcarvalho,jessicac,naml,machado}@dei.uc.pt

Abstract. This work proposes Adaptive Facilitated Mutation, a self-adaptive mutation method for Structured Grammatical Evolution (SGE), biologically inspired by the theory of facilitated variation. In SGE, the genotype of individuals contains a list for each non-terminal of the grammar that defines the search space. In our proposed mutation, each individual contains an array with a different, self-adaptive mutation rate for each non-terminal. We also propose Function Grouped Grammars, a grammar design procedure to enhance the benefits of the propose mutation. Experiments were conducted on three symbolic regression benchmarks using Probabilistic Structured Grammatical Evolution (PSGE), a variant of SGE. Results show our approach is similar or better when compared with the standard grammar and mutation.

Keywords: Adaptive Mutation · Grammar-design · Grammar-based Genetic Programming

1 Introduction

Grammar-based Genetic Programming (GP) algorithms have been an important tool for the evolution of computer programs since their inception [1–3]. The most popular approach is Grammatical Evolution (GE) which is notable for decoupling the genotype and the phenotype, using a grammar to translate a data structure into an executable program. The representation and variation operators used by GE present some known issues, such as low locality and high redundancy. The first means that small changes in the genotype can cause significant changes in the phenotype, and the second means that most modifications do not affect the phenotype. These characteristics result in a bad trade between exploration and exploitation, which makes the algorithm perform similarly to random search [4].

Structured Grammatical Evolution (SGE) [5] is a variant of GE that uses a different representation for the individuals. The genotype comprises several lists, one for each non-terminal in the grammar, and each list contains the indexes of the rules to be expanded. SGE shows better performance when compared to GE, and other grammar-based approaches [6], but also improved locality and lower

G. Pappa et al. (Eds.): EuroGP 2023, LNCS 13986, pp. 117–132, 2023.
https://doi.org/10.1007/978-3-031-29573-7_8

redundancy when compared to standard GE [7,8], in part due to its operators. This representation allows the recombination operator to be grammar-aware, preserving the list of each non-terminal. On the other hand, the grammar does not inform the mutation operator. Mutation in SGE changes the production rule of the non-terminal selected to mutate. This operator affects all genes with the same frequency, regardless of grammatical context. Using a static and equal value for all non-terminals fails to consider that not all mutations are equally destructive.

Specific genes can play an essential role in the solution and, when mutated, may completely ruin the phenotype behavior. On the other hand, some genes may have a tuning role; in this case, a mutation will only result in a minor adjustment to the solution. Despite these differences, both types of genes are equal in the eyes of mutation. Biological processes have evolved to prevent this phenomenon. Gerhart et al. [9] propose that there are core components vital to the individual which remain unchanged for long periods and regulatory genes that combine existing core components and change frequently. The result is a system that can quickly adapt to new environments through regulatory changes while preserving the core components that ensure individuals are functional. We can replicate this behavior in grammar-based algorithms using different adaptive mutation probabilities for each non-terminal. This approach enables the system to autonomously regulate mutation rates to match the impact of changes to that non-terminal. Note that this solution is only as effective as the correlation between non-terminals and mutation impact. It follows that the effectiveness of this mutation is related to grammar-design [10–13], as grammars with more rules enable finer tuning of mutation probabilities. The grouping of productions within each non-terminal is also relevant; separating low and high impact changes into separate rules should improve performance.

In this work, we propose Adaptive Facilitated Mutation, a biologically inspired grammar-aware self-adaptive mutation operator for SGE and its variants. Furthermore, we propose "Function Grouped Grammars", a method for grammar design that empirically outperforms grammars commonly used for regression in GE. We compare our approach to standard grammar and mutation and find that, when combined, Function Grouped Grammars and Adaptive Facilitated Mutation are statistically superior or similar to the baseline in three relevant GP benchmarks.

The remainder of this work is structured as follows: First, Sect. 2 presents the background necessary to understand the work presented. Section 3 presents the proposed mutation and grammar-design method. Section 4 details the experimentation setup used, and Sect. 5 the experimental results regarding performance and analysis of probabilities. Section 6 gathers the main conclusions and provides insights regarding future work.

2   Background

Evolutionary Algorithms (EAs) are optimization algorithms inspired by the biological processes of natural evolution. A population of individuals (candidate solutions) evolves over several generations, guided by a fitness function. Similar to nature, these individuals are subject to selection, reproduction, and genetic variation. These algorithms face known issues such as parameter tuning, premature convergence, and lack of diversity. Researchers propose novel representations [5,14,15], selection methods, genetic operators, and parameter selection approaches to address these issues.

2.1 Grammar-Based Genetic Programming

GP is an EA that evolves solutions as programs. Over the years, researchers have proposed many variants of GP, and grammar-based approaches gained more popularity as grammars are helpful to set restrictions to the search space [3].

GE [14] is the most popular grammar-based GP methods. The individuals' genotype is a string/vector of integers that is translated into a phenotype (an executable function) through a grammar. The individuals are subject to selection mechanisms, mutation, and crossover in each generation.

This approach is relevant but suffers from high redundancy [16], and poor locality [17,18], damaging the efficiency of evolution. Redundancy is measured by analyzing the proportion of effective mutations, and locality studies how well genotypic neighbors correspond to phenotypic neighbors. Standard GE showed performance similar to random search [4], which motivated researchers to propose different initialization methods [19], representations [5,15], genetic operators, but also to investigate grammar design [10].

Position Independent Grammatical Evolution (πGE) [20] uses a different representation and mapping mechanism that removes the positional dependency that exists in GE. Each codon of the genotype contains two values, *nont* that consists of the non-terminal and *rule* that states the rule index to expand. This method improves performance compared with GE [21]. However, another study found that this method also suffers from poor locality [7].

SGE [5,8] proposed a new representation for the genotype and variation operators, which resulted in better performance and fewer issues regarding locality and redundancy [6,7]. The genotype comprises a list of integers (one per grammar non-terminal), with each integer corresponding to the index of a production rule. This structure allows mutation to occur inside the same non-terminal and crossover to exchange the list of derivation options. Another advantage of this proposal is that only valid solutions are allowed. SGE imposes a depth limit on solutions. Once the limit is surpassed, only non-recursive productions are chosen, forcing individuals to consolidate into a valid genotype.

Probabilistic Structured Grammatical Evolution (PSGE) [15] is a recent proposal to SGE that uses a probabilistic grammar, namely a Probabilistic Context-Free Grammar (PCFG), to bias the search, and where codons of the genotype are floats. Each grammar production rule has a probability of being selected. These probabilities change based on the frequency of expansion of that rule on the best individual. If the rule is not expanded, the probability decreases. This proposal performed better or similarly compared to SGE and outperformed GE in all problems. The evolved grammar also provides information about the features more relevant to the problem [22]. Another probabilistic grammar approach is Co-evolutionary Probabilistic Structured Grammatical Evolution (Co-PSGE) [23]. In this method, each individual has a PCFG, which may suffer mutation to the probabilities values. This approach also showed similar or better performance than SGE.

2.2 Adaptive Mutation Rate

Although most works in the literature use a static parameter for mutation and crossover rates, research shows that dynamic parameters may improve the search and introduce more diversity to the population. Adaptive mutations have been widely proposed in the literature to tackle some of the issues that EAs present.

Self-adaptive Gaussian mutation has been widely used by Evolution Strategies (ES) [24] and adapted into EAs [25]. During the evolutionary process, the mutation rate varies, suffering a Gaussian mutation. This approach achieves better results than standard mutation and performs similarly to ES. Teo [26] studied a self-adaptive Gaussian mutation operator for the Generalized Generation Gap (G3) algorithm, and it outperformed the standard algorithm in two of the four problems tested.

Other approaches consider individuals' fitness when adapting the mutation probability. Libelli et al. [27] and Lis [28] approaches showed better performance than classical Genetic Algorithm (GA).

Adaptive Mutation Probability Genetic Algorithm (APmGA) [29] dynamically adjusts the mutation probability during the evolutionary process based on the variations of the population entropy between the current and previous generations.

Salinas et al. [30] proposed an EA where operators are GP trees. In each generation, the probability of an operator increases or decreases based on the individual's performance after the operator.

Gomez proposed Hybrid Adaptive Evolutionary Algorithm (HAEA) [31] to adapt the operator probabilities during evolution. Each individual encodes its genetic rates. The probabilities by a random value can change according to the fitness of the offspring This algorithm inspired other proposals that showed that although for some problems there are no significant improvements in performance, the algorithm can obtain similar results without the new to pre-tuning, needing less computational time [32–34].

Coelho et al. [35] presented a new hybrid self-adaptive algorithm based on ES guided by neighborhood structures and tested for combinatorial contains mutation probabilities, and the second contains integer values that control the strength of the disturbance. The results were similar to the other approaches tested and showed that the adaptive mutation could escape local optima and balance exploration and exploitation.

To our knowledge, there is only one adaptive mutation rate parameter proposal in GE. Fagan et al. [36] propose Fitness Reactive Mutation (FRM), an adaptive mutation that increases the mutation rate in case a fitness plateau is reached to diversify the population and decreases when a new optimum is found, using increments/decrements of 0.01. The approach found similar results as the fixing mutation rate.

2.3 Grammar-Design

It is possible to design grammar to produce syntactically constrained solutions or to incorporate domain knowledge by biasing the grammar. The grammar's design can significantly impact the search of GE [11–13].

Miguel Nicolau [11] proposes a method to reduce the number of non-terminals of the grammar. The authors compare standard GE with a standard and a reduced grammar and showed an empirical increase in performance. This work motivated a study with different types of grammars. This study shows that recursion-balanced grammar could also improve performance [10].

A recent work by Dick et al. [12] showed that GE is more sensitive to grammar design than Context-Free Grammar Genetic Programming (CFG-GP). The results suggest that CFG-GP is more sensitive to parameter tuning than grammar design.

Hemberg et al. [13] compared GEs with a depth-first mapping mechanism that uses three grammars: infix (standard GE), prefix and postfix. The results showed that different grammars can improve performance, although the authors report no significant differences.

Grammatical Evolution by Grammatical Evolution $((GE)^2)$ uses two grammars, the universal and the solution grammar. The universal grammar describes the rules to construct the solution grammar. The rules are used to map the individuals and can evolve towards biasing the search space. Results showed that the evolved grammars presented some bias towards some non-terminal symbols.

Manzoni et al. [37] showed theoretically that different grammars of equal quality impact the performance of (1+1)-EA. The structure of the grammar is problem dependent but can favor the search. A mutation operator that modifies the probability of selecting the grammar rules was also proposed.

3   Adaptive Facilitated Mutation

In biology, organisms have evolved to canalize the rate and effect of mutations on the phenotype. Gerhart et al. [9] propose that organisms adapt to new environments through regulatory changes that enable or disable pre-existing conserved components. This variation increases the probability of viable genetic mutation since core components remain unaffected by these regulatory changes. In sum, the modularity, adaptability, and compartmentation of genetic material in organisms allow facilitated variation through regulatory change.

Facilitated Mutation [38] (FM) is a biologically inspired mutation mechanism that aims to replicate the benefits of facilitated variation. This mechanism leverages the grammar's inherent compartmentation to regulate the mutation's frequency and destructiveness. With this mutation, each non-terminal has a different mutation probability, rather than the single mutation probability traditionally used by grammar-based EA.

In this work, we propose Adaptive Facilitated Mutation (AFM), an extension of FM using an adaptive mutation array that removes the need to set a mutation probability for each non-terminal manually. Each individual carries a *mutation array* containing these mutation probabilities, illustrated in Fig. 1. For each individual in the initial population, the mutation array is initialized using a specified *starting mutation probability*.

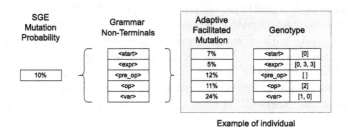

Fig. 1. In SGE, all genes are mutated based on a single probability (pictured left). Adaptive Facilitated Mutation uses a mutation array for each individual containing a probability for each non-terminal (pictured right).

In each generation, all individual's probabilities are adjusted using a random value sampled from a Gaussian distribution with $N(0, \sigma)$, where σ is a configurable parameter. Figure 2 illustrates the evolution of the mutation array.

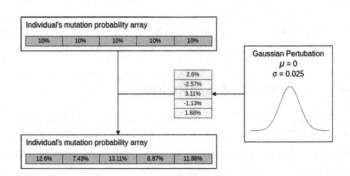

Fig. 2. Example of the first perturbation to the mutation probability array of an individual. This array is subsequently updated using a value sampled from a Gaussian distribution. This distribution is centered at 0 using a configurable standard deviation σ.

AFM only complements mutation operators by refining the frequency and impact of mutation. Once the mechanism determines which non-terminals to mutate, other operators should be used to alter the genotype within the defined scope. During crossover, the offspring individual inherits the mutation array from its fittest parent. More sophisticated inheritance mechanisms may improve this approach further, but we opted for a simple strategy to validate the approach. In this work, we use AFM for SGE and its variants, but this method is compatible with any grammar-based GP algorithms where it is possible to tie each codon to a corresponding non-terminal.

3.1   Grammar Design for Adaptive Facilitated Mutation

Since AFM leverages grammar structure, a purposefully designed grammar may enhance the method's performance. We hypothesize that AFM is more effective in grammars with multiple non-terminals containing related symbols. Additionally, a larger number of non-terminals may improve performance by enabling finer turning of mutation probabilities through a more detailed mutation array. Non-terminals commonly group symbols based on semantic similarity. For example, a grammar may use a non-terminal for all operators with a single expansion combining operators related to trigonometry (i.e., *sin*, *cos*) and the power function (i.e., *square*, *sqrt*). These same symbols can be grouped into several non-terminals based on function rather than semantics. Following this reasoning, trigonometric operations would be grouped in a specific non-terminal and the power function in a separate one. In Fig. 3, we illustrate how a grammar can be extended into a Function Grouped grammar.

<start> ::=<expr>
<expr> ::=<expr><op><expr> |
　　　　<pre_op>(<expr>) |
　　　　　　<var>
<op> ::=+|−|∗|/
<pre_op> ::=*sin|cos|sqrt|square*
<var> ::=1.0|x[n]

(a) Initial Grammar.

<start> ::=<expr_var>
<expr_var> ::=<expr>|<var>
<expr> ::=<expr_var><op><expr_var> |
　　　　<pre_op>(<expr_var>)
<op> ::=+|−|∗|/
<pre_op> ::=<trig_op>|<pow_op>
<trig_op> ::=*sin|cos*
<pow_op> ::=*sqrt|square*
<var> ::=1.0|x[n]

(b) Function Grouped Grammar.

Fig. 3. A grammar of semantically grouped non-terminals (3a) can be re-constructed based on functional groups (3b). This procedure extends the grammar and possibly improves performance when using Adaptive Facilitated Mutation.

4   Experimental Setup

We use PSGE [15] in all experiments as it has equal or superior performance to SGE in the selected tasks. We compare the algorithm using Standard Mutation (SM) and Facilitated Mutation (FM). Additionally, we compare the Standard Grammar (SG) and the Function Grouped Grammar (FG) that follows the principles outlined in Sect. 3 (complete grammar shown in Fig. 4).

We evaluate the performance of our method in three popular symbolic regression GP benchmarks, the Quartic polynomial, the Pagie polynomial, and Boston Housing [39,40]. The Quartic polynomial is defined by the mathematical expression shown in Eq. 1. The function is sampled in the interval $[-1, 1]$ with a step of 0.1.

$$x[0]^4 + x[0]^3 + x[0]^2 + x[0] \tag{1}$$

The Pagie polynomial is known to be a more difficult symbolic regression benchmark (Eq. 2). The outputs are computed in the interval: $-5 \leq x[0], x[1] \leq 5.4$, with step size 0.4. In Pagie, both features fall in the same range of values.

$$\frac{1}{1 + x[0]^{-4}} + \frac{1}{1 + x[1]^{-4}} \tag{2}$$

<start> ::=<expr_var>

<expr_var> ::=<expr>|<var>

<expr> ::=<expr_op>|

<pre_op>(<expr_var>)

<expr_op> ::=<expr><op><expr>|

(<expr><op><expr>)

<op> ::=+|−|*|/

<pre_op> ::=<trig_op>

|<exp_log_op>|*inv*

<trig_op> ::=*sin*|*cos*

<exp_log_op> ::=*exp*|*log*

<var> ::=1.0|x[n]

Fig. 4. Function Grouped Grammar used in the experiments. Number $x[n]$ terminals change to match the problem: 1 for Quartic, 2 for Pagie, 12 for Boston Housing

The third benchmark is Boston Housing [41]. This is a predictive modeling problem, where one needs to build a model to predict the price of Boston houses based on 13 features. There are 506 instances split into 90% for training and 10% for test. The 13 features that compose the dataset are heterogeneous regarding their intervals, with ranges varying from $0 \leq x[3] \leq 1$ to $0.32 \leq x[11] \leq 396.9$. This diversity likely reduces the effectiveness of our approach as all features are grouped in the $< var >$ non-terminal, making it difficult for the algorithm to distinguish between different types of features with different mutation rates. It is possible to adjust the grammar using expert knowledge, separating the variables into non-terminals based on orders of magnitude or function. This type of grammar design, while possibly effective, is outside the scope of this work.

The fitness functions used to evaluate the individuals consider the minimization of the Root Relative Squared Error (RRSE) between the individual's solution and the target on a data set.

Table 1 summarizes the parameters used in the experiments. All problems use the same population size, mutation and crossover rates, tournament size, max depth, and number of generations. In preliminary experimentation, we trialed four parameters for the AFM's Gaussian perturbation: $\sigma = [0.001, 0.0025, 0.005, 0.01]$ We found that facilitated mutations' σ parameter benefited from tuning when moved to different tasks. Consequently, each experiment uses the best σ value for the corresponding problem. We repeat all experiments 100 times to investigate statistically meaningful differences between the approaches.

Table 1. Parameters used in experiments for Quartic, Pagie, and Boston Housing.

Parameters	Quartic	Pagie	Boston Housing
Population Size	1000		
Generations	100		
Elitism	10%		
Mutation	Gaussian N(0,0.5)		
Mutation Probability	10%		
Adaptive Facilitated Mutation σ	0.0025		0.001
Crossover Probability	90%		
Tournament Size	3		
Max Depth	10		

5   Results

This section presents the results obtained for each problem in terms of the Mean Best Fitness (MBF) of 100 repetitions. We statistically compare the different approaches using the Mann-Whitney test with Bonferroni correction with a significance level $\alpha = 0.05$.

Figure 5 shows the results for the Quartic polynomial. At the end of evolution, all approaches achieve a similar result in this problem, except for SM+SG, which performs significantly worse than the others (see Table 2 for statistical analysis). Quartic likely has a local optimum that evolution cannot escape without our

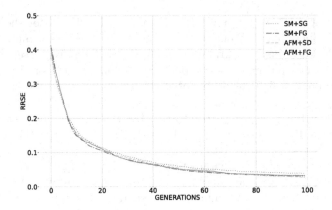

Fig. 5. Plot shows the mean best fitness of 100 runs for the Quartic polynomial.

proposed mechanism. While SM+FG and AFM+SG statistically outperform the baseline, AFM+FG's advantages are only empirical. We hypothesize that AFM+FG would join the other two methods with additional repetitions, creating two tiers of solution quality in Quartic.

Table 2. P-value for Mann-Whitney Statistical Tests using Bonferroni correction with significance level $\alpha = 0.05$ for Quartic polynomial. Bold indicates that the method in the corresponding row is statistically superior.

Quartic polynomial	SM+SG	SM+FG	AFM+SG	AFM+FG
SM+SG				
SM+FG	**0.048**			
AFM+SG	**0.004**	0.356		
AFM+FG	0.072	0.760	0.332	

Figure 6 shows the MBF for the Pagie polynomial across 100 generations. Looking at the results, one can see that the advantages of the proposed approaches are clear for this problem, particularly the FG grammar. From generation 25 until termination, both AFM solutions have a better MBF than their SM counterparts. While AFM is empirically superior to SM, FG amplifies the benefits of this approach.

The statistical analysis results (shown in Table 3) reveal that AFM+FG outperforms all SG approaches while SM+FG is similar. These results suggest that our proposed grammar may be marginally superior, but considerable benefits come from combining it with the appropriate mutation.

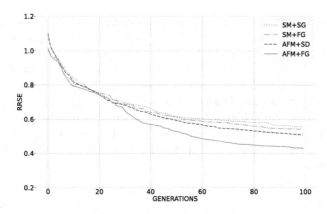

Fig. 6. Plot shows the mean best fitness of 100 runs for the Pagie polynomial.

Table 3. P-value for Mann-Whitney Statistical Tests using Bonferroni correction with significance level $\alpha = 0.05$ for Pagie polynomial. Bold indicates that the method in the corresponding row is statistically superior.

Pagie polynomial	SM+SG	SM+FG	AFM+SG	AFM+FG
SM+SG				
SM+FG	0.965			
AFM+SG	0.426	0.400		
AFM+FG	**0.002**	**0.001**	**0.001**	

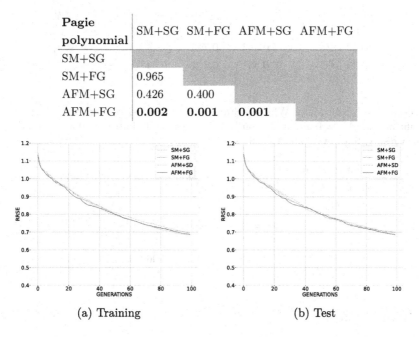

(a) Training   (b) Test

Fig. 7. Plot shows the mean best fitness of 100 runs for the Boston Housing dataset.

Finally, in Fig. 7 we show the MBF for the Boston Housing Training and Test. A brief perusal of the results indicates that AFM+FG achieves the best MBF in Test. When comparing the mutations, AFM generalizes better as it maintains a similar performance between train and test.

Note that SM+SG appears to be worst at generalizing to the test data. This is most evident when comparing SM+SG with both AFM approaches, as the differences between these methods are noticeably larger when moved to test data. Despite the highlighted differences, statistical tests for Boston Housing Training and Test reveal no statistical differences (Tables 4 and 5). Given that FG cannot account for the feature variety of Boston Housing, it is remarkable that the approach still achieves competitive results, especially in the test data. It is possible that a specifically designed FG that uses expert knowledge to group the features based on function could achieve even better and statistically significant results.

Table 4. P-value for Mann-Whitney Statistical Tests using Bonferroni correction with significance level $\alpha = 0.05$ for Boston Housing Training. Bold indicates that the method in the corresponding row is statistically superior.

Boston Housing Training	SM+SG	SM+FG	AFM+SG	AFM+FG
SM+SG				
SM+FG	0.530			
AFM+SG	0.263	0.763		
AFM+FG	0.531	0.361	0.548	

Table 5. P-value for Mann-Whitney Statistical Tests using Bonferroni correction with significance level $\alpha = 0.05$ for Boston Housing Test. Bold indicates that the method in the corresponding row is statistically superior.

Boston Housing Test	SM+SG	SM+FG	AFM+SG	AFM+FG
SM+SG				
SM+FG	0.929			
AFM+SG	0.377	0.333		
AFM+FG	0.364	0.403	0.963	

6 Conclusion

In this paper, we propose AFM, a mutation method that leverages grammar-based GP's properties to replicate natural evolutionary phenomena. This approach divides the single mutation probability commonly found in such approaches into a mutation array, where each grammar non-terminal has a corresponding mutation rate. Each individual has a mutation array that co-evolves with the genetic code. A randomly sampled value from a Gaussian distribution adjusts the mutation rates of all individuals in each generation.

We also propose a grammar-design approach, Function Grouped Grammars, to enhance the effectiveness of the mutation proposed. Function Grouped Grammars organize non-terminals based on functional similarity rather than the semantic similarity common in the field. We compare our proposals with a baseline (standard mutation and standard grammar) and find that, when combined, our approaches are statistically superior or similar to the baseline in three relevant GP benchmarks.

This approach still requires parameter tuning, one of the problems tackled by the literature by proposing adaptive mutations. However, few of these approaches consider different values for different symbols [35]. This work shows that an adaptive mutation rate can be beneficial for search and grammar design amplifies these benefits.

6.1   Future Work

The results of our experiments are promising, but additional tests are essential to validate our approach further. The benchmarks addressed are relevant in GP, but a more extensive (and varied) set of benchmarks could bring meaningful insights into the general applicability of the FG+FM.

Regarding AFM, we use a fixed starting mutation rate in all experiments. Our method leverages adaptability as a tool for improved evolution, but we did not investigate the potential of AFM as a replacement for mutation rate tuning. Further development of AFM may also lead to an implementation that does not rely on Gaussian distributions and *sigma* tuning for perturbations. In the future, we want to explore alternatives where AFM is a competitive, parameterless alternative to standard mutation. Another line of work is to experiment with different inheritance mechanisms during crossover. In this work, the offspring inherited the array of probabilities from the most fitted parent. It would be interesting to explore the random selection of the parent that passes the array or the application of the existing SGE crossover to the mutation arrays of the parents. Such approaches would better preserve the advantages of co-evolution, possibly improving results.

Function Grouped Grammars can also be further investigated. In this work, we still apply this idea conservatively. Considering the FG used in experiments, it is still possible to separate constants from variables and commutative operators from non-commutative operators. Barring small details, we use the same grammar for all tasks. Applying the same approach to problems requiring more complex grammar would be interesting, as such problems yield more opportunities for function grouping.

Finally, the applicability of these ideas to different, compatible grammar-based GP must also be investigated. While this works focuses on PSGE, the same approach is easily applicable in SGE [5], Co-PSGE [23], and even more different systems like πGE [20]. Any grammar-based approach where genes are tied to a non-terminal may benefit from FG+FM.

Acknowledgments. This work was funded by FEDER funds through the Operational Programme Competitiveness Factors - COMPETE and national funds by FCT - Foundation for Science and Technology (POCI-01-0145-FEDER-029297, CISUC - UID/CEC/00326/2020) and within the scope of the project A4A: Audiology for All (CENTRO-01-0247-FEDER-047083) financed by the Operational Program for Competitiveness and Internationalisation of PORTUGAL 2020 through the European Regional Development Fund.

The first author is funded by FCT, Portugal, under the grant UI/BD/151053/2021 and the second under the grant 2022.10174.BD.

References

1. Whigham, P.A., Science, D.O.C.: Grammatically-based genetic programming (1995)
2. Ryan, C., Collins, J.J., Neill, M.O.: Grammatical evolution: evolving programs for an arbitrary language. In: Banzhaf, W., Poli, R., Schoenauer, M., Fogarty, T.C. (eds.) EuroGP 1998. LNCS, vol. 1391, pp. 83–96. Springer, Heidelberg (1998). https://doi.org/10.1007/BFb0055930
3. McKay, R.I., Hoai, N.X., Whigham, P.A., Shan, Y., O'Neill, M.: Grammar-based genetic programming: a survey. Genet. Program. Evolvable Mach. **11**(3–4), 365–396 (2010). https://doi.org/10.1007/s10710-010-9109-y
4. Whigham, P.A., Dick, G., Maclaurin, J., Owen, C.A.: Examining the "best of both worlds" of grammatical evolution. In: Proceedings of the 2015 Annual Conference on Genetic and Evolutionary Computation, pp. 1111–1118. ACM (2015)
5. Lourenço, N., Assunção, F., Pereira, F.B., Costa, E., Machado, P.: Structured grammatical evolution: a dynamic approach. In: Ryan, C., O'Neill, M., Collins, J.J. (eds.) Handbook of Grammatical Evolution, pp. 137–161. Springer, Cham (2018). https://doi.org/10.1007/978-3-319-78717-6_6
6. Lourenço, N., Ferrer, J., Pereira, F.B., Costa, E.: A comparative study of different grammar-based genetic programming approaches. In: McDermott, J., Castelli, M., Sekanina, L., Haasdijk, E., García-Sánchez, P. (eds.) EuroGP 2017. LNCS, vol. 10196, pp. 311–325. Springer, Cham (2017). https://doi.org/10.1007/978-3-319-55696-3_20
7. Medvet, E.: A comparative analysis of dynamic locality and redundancy in grammatical evolution. In: McDermott, J., Castelli, M., Sekanina, L., Haasdijk, E., García-Sánchez, P. (eds.) EuroGP 2017. LNCS, vol. 10196, pp. 326–342. Springer, Cham (2017). https://doi.org/10.1007/978-3-319-55696-3_21
8. Lourenço, N., Pereira, F.B., Costa, E.: Unveiling the properties of structured grammatical evolution. Genet. Program. Evolvable Mach. **17**(3), 251–289 (2016). https://doi.org/10.1007/s10710-015-9262-4
9. Gerhart, J., Kirschner, M.: The theory of facilitated variation. Proc. Natl. Acad. Sci. **104**(1), 8582–8589 (2007). https://doi.org/10.1073/pnas.0701035104
10. Nicolau, M., Agapitos, A.: Understanding grammatical evolution: grammar design. In: Ryan, C., O'Neill, M., Collins, J.J. (eds.) Handbook of Grammatical Evolution, pp. 23–53. Springer, Cham (2018). https://doi.org/10.1007/978-3-319-78717-6_2
11. Nicolau, M.: Automatic grammar complexity reduction in grammatical evolution. In: The 3rd Grammatical Evolution Workshop: A Workshop of the 2004 Genetic and Evolutionary Computation Conference (GECCO 2004), Seattle, Washington, USA, 26–30 June 2004. Seattle, Washington, USA (2004)

12. Dick, G., Whigham, P.A.: Initialisation and grammar design in grammar-guided evolutionary computation. In: Proceedings of the Genetic and Evolutionary Computation Conference Companion (GECCO 2022), pp. 534–537. Association for Computing Machinery, New York, NY, USA (2022). https://doi.org/10.1145/3520304.3529051

13. Hemberg, E.: Pre-, in-and postfix grammars for symbolic regression in grammatical evolution (2008)

14. Ryan, C., O'Neill, M., Collins, J.J. (eds.): Handbook of Grammatical Evolution. Springer, Cham (2018). https://doi.org/10.1007/978-3-319-78717-6

15. Megane, J., Lourenco, N., Machado, P.: Probabilistic structured grammatical evolution. In: 2022 IEEE Congress on Evolutionary Computation (CEC), pp. 991–999. IEEE (2022). https://doi.org/10.1109/cec55065.2022.9870397

16. Thorhauer, A.: On the non-uniform redundancy in grammatical evolution. In: Handl, J., Hart, E., Lewis, P.R., López-Ibáñez, M., Ochoa, G., Paechter, B. (eds.) PPSN 2016. LNCS, vol. 9921, pp. 292–302. Springer, Cham (2016). https://doi.org/10.1007/978-3-319-45823-6_27

17. Thorhauer, A., Rothlauf, F.: On the locality of standard search operators in grammatical evolution. In: Bartz-Beielstein, T., Branke, J., Filipič, B., Smith, J. (eds.) PPSN 2014. LNCS, vol. 8672, pp. 465–475. Springer, Cham (2014). https://doi.org/10.1007/978-3-319-10762-2_46

18. Rothlauf, F., Oetzel, M.: On the locality of grammatical evolution. In: Collet, P., Tomassini, M., Ebner, M., Gustafson, S., Ekárt, A. (eds.) EuroGP 2006. LNCS, vol. 3905, pp. 320–330. Springer, Heidelberg (2006). https://doi.org/10.1007/11729976_29

19. Nicolau, M.: Understanding grammatical evolution: initialisation. Genet. Program. Evolvable Mach. 18(4), 467–507 (2017). https://doi.org/10.1007/s10710-017-9309-9

20. O'Neill, M., Brabazon, A., Nicolau, M., Garraghy, S.M., Keenan, P.: πgrammatical evolution. In: Deb, K. (ed.) GECCO 2004. LNCS, vol. 3103, pp. 617–629. Springer, Heidelberg (2004). https://doi.org/10.1007/978-3-540-24855-2_70

21. Fagan, D., O'Neill, M., Galván-López, E., Brabazon, A., McGarraghy, S.: An analysis of genotype-phenotype maps in grammatical evolution. In: Esparcia-Alcázar, A.I., Ekárt, A., Silva, S., Dignum, S., Uyar, A.Ş (eds.) EuroGP 2010. LNCS, vol. 6021, pp. 62–73. Springer, Heidelberg (2010). https://doi.org/10.1007/978-3-642-12148-7_6

22. Mégane, J., Lourenço, N., Machado, P.: Probabilistic grammatical evolution. In: Hu, T., Lourenço, N., Medvet, E. (eds.) EuroGP 2021. LNCS, vol. 12691, pp. 198–213. Springer, Cham (2021). https://doi.org/10.1007/978-3-030-72812-0_13

23. Mégane, J., Lourenço, N., Machado, P.: Co-evolutionary probabilistic structured grammatical evolution. In: Proceedings of the Genetic and Evolutionary Computation Conference, pp. 991–999. ACM (2022). https://doi.org/10.1145/3512290.3528833

24. Beyer, H., Schwefel, H.: Evolution strategies - a comprehensive introduction. Nat. Comput. 1, 3–52 (2004). https://doi.org/10.1023/A:1015059928466

25. Hinterding, R.: Gaussian mutation and self-adaption for numeric genetic algorithms. In: Proceedings of 1995 IEEE International Conference on Evolutionary Computation, vol. 1, p. 384. IEEE (1995)

26. Teo, J.: Self-adaptive mutation for enhancing evolutionary search in real-coded genetic algorithms. In: 2006 International Conference on Computing & Informatics, pp. 1–6. IEEE (2006)

27. Libelli, S.M., Alba, P.: Adaptive mutation in genetic algorithms. Soft Comput. 4(2), 76–80 (2000). https://doi.org/10.1007/s005000000042
28. Lis, J.: Genetic algorithm with the dynamic probability of mutation in the classification problem. Pattern Recogn. Lett. 16(12), 1311–1320 (1995)
29. Stark, N., Minetti, G.F., Salto, C.: A new strategy for adapting the mutation probability in genetic algorithms (2012)
30. Cruz-Salinas, A.F., Perdomo, J.G.: Self-adaptation of genetic operators through genetic programming techniques. In: Proceedings of the Genetic and Evolutionary Computation Conference, pp. 913–920. ACM (2017)
31. Gomez, J.: Self adaptation of operator rates in evolutionary algorithms. In: Deb, K. (ed.) GECCO 2004. LNCS, vol. 3102, pp. 1162–1173. Springer, Heidelberg (2004). https://doi.org/10.1007/978-3-540-24854-5_113
32. Gómez, J., León, E.: On the class of hybrid adaptive evolutionary algorithms (CHAVELA). Nat. Comput. 20(3), 377–394 (2021). https://doi.org/10.1007/s11047-021-09843-5
33. Montero, E., Riff, M.C.: Calibrating strategies for evolutionary algorithms. In: 2007 IEEE Congress on Evolutionary Computation. IEEE (2007)
34. Montero, E., Riff, M.-C.: Self-calibrating strategies for evolutionary approaches that solve constrained combinatorial problems. In: An, A., Matwin, S., Raś, Z.W., Ślęzak, D. (eds.) ISMIS 2008. LNCS (LNAI), vol. 4994, pp. 262–267. Springer, Heidelberg (2008). https://doi.org/10.1007/978-3-540-68123-6_29
35. Coelho, V.N., et al.: Hybrid self-adaptive evolution strategies guided by neighborhood structures for combinatorial optimization problems. Evol. Comput. 24(4), 637–666 (2016)
36. Fagan, D., Hemberg, E., Nicolau, M., O'Neill, M., McGarraghy, S.: Towards adaptive mutation in grammatical evolution. In: Proceedings of the Fourteenth International Conference on Genetic and Evolutionary Computation Conference Companion (GECCO Companion 2012). ACM Press (2012)
37. Manzoni, L., Bartoli, A., Castelli, M., Goncalves, I., Medvet, E.: Specializing context-free grammars with a (1 + 1)-EA. IEEE Trans. Evol. Comput. 24(5), 960–973 (2020)
38. Tiso, S., Carvalho, P., Lourenço, N., Machado, P.: Structured mutation inspired by evolutionary theory enriches population performance and diversity. arXiv preprint arXiv:2302.00559 (2023)
39. White, D.R., et al.: Better GP benchmarks: community survey results and proposals. Genet. Program. Evolvable Mach. 14(1), 3–29 (2012). https://doi.org/10.1007/s10710-012-9177-2
40. McDermott, J., et al.: Genetic programming needs better benchmarks. In: Proceedings of the Fourteenth International Conference on Genetic and Evolutionary Computation Conference (GECCO 2012), , pp. 791–798. ACM Press (2012)
41. Harrison, D., Jr., Rubinfeld, D.L.: Hedonic housing prices and the demand for clean air. J. Environ. Econ. Manag. 5(1), 81–102 (1978)

A Boosting Approach to Constructing an Ensemble Stack

Zhilei Zhou[1], Ziyu Qiu[1], Brad Niblett[2], Andrew Johnston[2],
Jeffrey Schwartzentruber[1], Nur Zincir-Heywood[1],
and Malcolm I. Heywood[1]

[1] Faculty of Computer Science, Dalhousie University, Nova Scotia, Canada
{ZhileiZhou,amousqiu,nzincirh,mheywood}@dal.ca,
jeffrey.schwartzentruber@gmail.com
[2] 2Keys Corporation - An Interac Company, Ottawa, Canada
{bniblett,ajohnston}@2keys.ca
http://www.2keys.ca

Abstract. An approach to evolutionary ensemble learning for classification is proposed using genetic programming in which boosting is used to construct a stack of programs. Each application of boosting identifies a single champion and a residual dataset, i.e. the training records that thus far were not correctly classified. The next program is only trained against the residual, with the process iterating until some maximum ensemble size or no further residual remains. Training against a residual dataset actively reduces the cost of training. Deploying the ensemble as a stack also means that only one classifier might be necessary to make a prediction, so improving interpretability. Benchmarking studies are conducted to illustrate competitiveness with the prediction accuracy of current state-of-the-art evolutionary ensemble learning algorithms, while providing solutions that are orders of magnitude simpler. Further benchmarking with a high cardinality dataset indicates that the proposed method is also more accurate and efficient than XGBoost.

Keywords: Boosting · Stacking · Genetic Programming

1 Introduction

Ensemble learning represents a widely employed meta-learning scheme for deploying multiple models under supervised learning tasks [1,7,23]. In general, there are three basic formulations: Bagging, Boosting or Stacking. We note, however, that most instances of ensemble learning adopt one scheme alone (Sect. 2). Moreover, Boosting and Bagging represent the most widely adopted approaches.

In this work, we investigate the utility of Boosting for constructing a Stacked ensemble (Sect. 3). Rather than members of an ensemble (i.e. programs) being deployed in parallel, our Stack assumes that programs are deployed sequentially in the order that they were originally discovered. Programs will be rewarded for

Supported by 2Keys Corporation - An Interac Company.

either suggesting a class label or declaring the prediction as 'ambiguous'. As we partner Stacking with Boosting, each program added to the Stack is explicitly focused on what previous programs could not classify. Thus, additional programs are only trained against exemplars associated with ambiguous predictions, i.e. the cardinality of the training partition actually decreases for each round of boosting. Post training, members of the Stack are sequentially visited until one makes a prediction (other than 'ambiguous'), implying that only a fraction of the programs comprising the Stack needs to be visited to suggest a label.

A benchmarking study demonstrates the effectiveness of the approach by first comparing with recent evolutionary ensemble learning algorithms on a suite of previously benchmarked low cardinality datasets (Sect. 4). Having established the competitiveness of the proposed approach from the perspective of classifier accuracy, we then benchmark using a high cardinality classification task consisting of $\approx 800,000$ records. The best of previous approaches is demonstrated not to scale under these conditions. Conversely, the efficiency of the proposed approach implies that parameter tuning is still feasible under the high cardinality setting.

2 Related Work

Ensemble learning appears in three basic forms, summarized as follows and assuming (without loss of generality) that the underlying goal is to produce solutions for a classification problem:

- **Bagging:** n classifiers are independently constructed and an averaging scheme adopted to aggregate the n independent predictions into a single ensemble recommendation.[1] The independence of the n classifiers is established by building each classifier on a *different* sample taken from the original training partition, or a 'bag'.
- **Boosting:** constructs n classifiers sequentially. Classifier performance is used to incrementally reweight the probability of sampling training data, $\mathcal{D}$, used to train the next classifier. Thus, each classifier encounters a sample of $\mathcal{D}/n$ training records. Post-training the ensemble again assumes an averaging scheme to aggregate the n labels into a single label.
- **Stacking:** assumes that a heterogeneous set of n classifiers is trained on a common training partition. The predictions from the n classifiers are then 'stacked' to define a $n \times |\mathcal{D}|$ dataset that trains a 'meta classifier' to produce the overall classification prediction.

Such a division of ensemble learning architectures reflects the view that the learning algorithm only constructs one model at a time. However, genetic programming (GP) develops multiple models simultaneously. Thus, one approach for GP ensemble learning might be to divide the population into islands and

---

[1] Classification tasks often assume the majority vote, although voting/weighting schemes might be evolved [3].

expose each island to different data subsets, where the data subset is constructed using a process synonymous with Bagging or Boosting (e.g. [9,12,13]). Some of the outcomes resulting from this research theme were that depending on the degree of isolation between populations, classifiers might result that were collectively strong, but individually weak [13], or that solution simplicity might be enabled by the use of ensemble methods [12]. Indeed, solution simplicity has been empirically reported for other GP ensemble formulations [15].

Another recurring theme is to assume that an individual takes the form of a 'multi-tree' [2,17,25]. In this case, an 'individual' is a team of n Tree-structured programs. Constraints are then enforced on the operation of variation operators in order to maintain context between programs in the multi-tree. This concept was developed further under the guise of 'levels of selection' in which selection can operate at the 'level' of a program or ensemble [26,29]. However, in order to do so, it was necessary to have different performance definitions for each level.

Virgolin revisits the theme of evolving ensembles under Bagging while concentrating on the role of selection [27]. Specifically, individuals are evaluated w.r.t. *all* bags. The cost of fitness evaluation over all bags is minimized by evaluating programs once and caching the corresponding fitness values. The motivation for evaluating performance over all the bags is to promote uniform development toward the performance goal.

Boosting and Bagging in particular have also motivated the development of mechanisms for decoupling GP from the cardinality of the training partition. In essence, GP performance evaluation is only ever conducted relative to the content of a data subset, DS, where $|DS| << |\mathcal{D}|$. However, the data subset is periodically resampled, with biases introduced to reflect the difficulty of labelling records correctly and the frequency of selecting data to appear in the data subset [11,24]. This relationship was then later explicitly formulated from the perspective of competitively coevolving ensembles against the data subset [14–16], i.e. data subset and ensemble experience different performance functions.

Several coevolutionary approaches have also been proposed in which programs are defined in one population and ensembles are defined by another, e.g. [14–16,21]. Programs are free to appear in multiple ensembles and the size of the ensemble is determined through evolution. The works of [14–16] have been extensively benchmarked over multi-class, high-dimensional and high-cardinality datasets. One limitation is the development of hitchhikers, i.e. programs that appear in the ensemble that never contribute to labelling data.

The concept of 'Stacking' has been less widely employed. However, the cascade-correlation approach to evolving neural networks might be viewed in this light [8]. Specifically, cascade-correlation begins with a single 'perceptron' and identifies the error residual. Assuming that the performance goal has not been reached, a new perceptron is added that receives as input all the previous perceptron outputs and the original data attributes. Each additional perceptron is trained to minimize the corresponding residual error. Potter and deJong demonstrated a coevolutionary approach to evolving neural networks using this architecture [19]. Curry et al. benchmarked such an approach for training layers of GP classifiers under the cascade-correlation architecture [6]. However, they discovered that the GP classifiers could frequently degenerate to doing no more than copying the input from the previous layer.

Finally, the concept of interpreting the output of a GP program as a dimension in a new feature space rather than a label is also of significance to several ensemble methods. Thus, programs comprising an ensemble might be rewarded for mapping to a (lower-dimensional) feature space that can be clustered into class-consistent regions [16,18]. Multiple programs appearing in an ensemble define the dimensions of the new feature space. Likewise, ensembles based on multi-trees have been rewarded for mapping to a feature space that maximizes the performance of a linear discriminant classifier [2].

3   Evolving an Ensemble Stack Using Boosting

The motivation of this work is to use a combination of boosting and stacking to address data cardinality while constructing the GP ensemble. We assume that classifiers are sequentially added to the ensemble. Our insight is that after adding each classifier, the data correctly classified is *removed* from the training partition. Thus, as classifiers are added to the ensemble the cardinality of the training partition decreases. Moreover, the next classifier evolved is explicitly directed to label what the ensemble cannot currently label. Post training, the ensemble is deployed as a 'stack' in which classifiers are applied in the order in which they were evolved. As will become apparent, we need not visit all members of the ensemble stack in order to make a prediction. In the following, we first present the evolutionary cycle adopted for developing the Boosted Ensemble Stack (Sect. 3.1) and then discuss how the Ensemble Stack is deployed (Sect. 3.2).

3.1   The Boosting Ensemble Stack Algorithm

Algorithm 1 summarizes the overall approach taken to boosting in this research. The training dataset, $\mathcal{D}$, is defined in terms of a matrix X^t of n (input) records (each comprising of d-attributes) and a vector Y^t of n labels (i.e. supervised learning for classification). We may sample records pairwise, $\langle \boldsymbol{x}_p \in X^t, y_p \in Y^t \rangle$, during training. The outer loop (Step 3) defines the number of 'boosting epochs', where this sets an upper limit on ensemble size. Step 5 initializes a new population, consisting of program decision stumps alone (single node programs). Step 6 performs fitness evaluation, ranking, parent pool selection, and variation for a limited number of generations (Max_GP_Epoch). Specifically, the following form is assumed:

1. Fitness evaluation (Step 7) assumes the Gini index in which the output of each program is modelled as a distribution (Algorithm 2). Fitter individuals are those with a higher Gini index.
2. Parent pool selection (PP) implies that the worst %Gap programs are deleted, leaving the parent pool as the survivors (Algorithm 1, Step 9).

3. Test for early stopping (Step 10) where this is defined in terms of fitness and 'bin purity' a concept defined below.
4. Variation operators (Step 21) are limited to 1) cloning %*Gap* parents, 2) adding a single new node to a clone where the parameters of the new node are chosen stochastically, and 3) mutating any of the parameters in the resulting offspring.

In assuming a performance function based on the Gini index, programs perform the following mapping: $\hat{y}_p = f(x_p)$ where x_p is input record p and $\hat{y}_p$ is the corresponding output from the program. There is no attempt to treat $\hat{y}_p$ as a predicted classification label for record p. Instead, $\hat{y}_p$ represents the mapping of the original input, x_p, to a scalar value on a 1-dimensional number line, $\hat{y}$. After mapping all p inputs to their corresponding $\hat{y}_p$ we quantize $\hat{y}$ into NumBin intervals (Step 11), as illustrated by Fig. 1. Each interval defines an equal non-overlapping region of the number line $\hat{y}$ and an associated bin. The bins act as a container for the labels, $y_p = c$, associated with each $\hat{y}$ appearing in the bin's interval. Three types of the bin may now appear,

- **Empty bins:** have no $\hat{y}_p$ appearing in their interval.
- **Pure bins:** have $\hat{y}_p$ appearing in their interval such that the majority of labels y_p are the same. A 'pure bin' assumes the label y_p that reflects the majority of the bin content and is declared when the following condition holds,

$$\frac{C_{bin} - y^*}{C_{bin}} < \beta \tag{1}$$

where C_{bin} is the count of the number of records appearing in the bin, $C(c)$ represents the number of records of each class in the bin, and $y^* = \max_c C(c)$.
- **Ambiguous bins:** imply that some $\hat{y}_p$ appear at an interval such that bin purity does not hold.

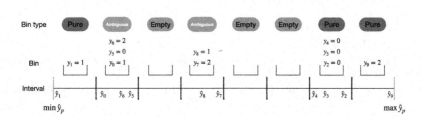

Fig. 1. Illustration for the relationship between program outputs ($\hat{y}_p$), intervals, bins, labels ($y_p = c$) and bin type.

If a pure bin is encountered for an individual during training (Step 12) that also has best fitness, then the next champion for including in the ensemble has been identified. Moreover, each time a new ensemble member is to be evolved, they have to improve on the fitness of the previous ensemble member (Step 13).

Algorithm 1. StackBoost($\langle X^t, Y^t \rangle$, New_Pop_size, Max_Boost_epoch, Max_GP_epoch). PP is the parent pool

---

1: $Best\_Fit \leftarrow 0$
2: $Ensemble \leftarrow []$ ▷ Initialize ensemble stack to null
3: **for** $i \leftarrow 1$ to Max_Boost_epoch **do**

4:   $best \leftarrow$ **False**
5:   $Pop \leftarrow Initialize(New_Pop_size)$ ▷ Initialize a new pop

6:   **for** $(j \leftarrow 1$ to Max_GP_epoch$)$ & $(best =$ **False**$)$ **do**   ▷ Evolve population
7:    $Fitness \leftarrow [\,GiniIndexFitness(Tree, X^t, Y^t)$ for $Tree \in Pop]$
8:    $Ranked\_Pop \leftarrow [\arg \mathrm{sort}(Fitness, Pop)]$
9:    $PP \leftarrow \mathrm{top}(Ranked\_Pop, \%Gap)$ ▷ Return parent pool

10:    **while** $Tree \in PP$ **do**   ▷ Test for champion
11:     $Histogram, Interval \leftarrow fitHist(Tree, X^t, Y^t)$
12:     **if** $(\exists Interval \in Histogram =$ Pure$)$ **then**
13:      **if** $(Tree.Fitness > Best\_Fit)$ **then**
14:       $best \leftarrow$ **True**   ▷ Exit early if champion found
15:       $Best\_Fit \leftarrow Best\_Fit$
16:       $Champion\_Histogram \leftarrow Histogram$
17:       $Champion \leftarrow Tree$ ▷ Record champion
18:      **end if**
19:     **end if**
20:    **end while**
21:    $Offspring \leftarrow \mathrm{Variation}(PP, \%Gap)$
22:    $Trees \leftarrow PP \cup Offspring$
23:   **end for**

24:   $\langle X', Y' \rangle \leftarrow \emptyset$
25:   **for** $Interval \in Champion\_Histogram$ **do**   ▷ Identify Pure bin content
26:    **if** $Interval =$ $Pure$ **then**
27:     $\langle X', Y' \rangle \leftarrow \mathrm{copy}(\langle x_p, y_p \rangle \in Interval)$ ▷ Identify correctly labelled data
28:    **end if**
29:   **end for**

30:   $\langle X^t, Y^t \rangle \leftarrow \langle X^t, Y^t \rangle / \langle X', Y' \rangle$ ▷ Define residual dataset
31:   $Ensemble.push(Champion)$ ▷ Add $Champion$ to ensemble stack
32:   **if** $\langle X^t, Y^t \rangle = \emptyset$ **then**   ▷ Case of early stopping
33:    **return** $Ensemble$
34:   **end if**
35: **end for**
36: **return** $Ensemble$

Algorithm 2. Gini Index Fitness $(Tree, X, Y)$ returns gini index weighted by model complexity.

Require: $Total(i) \leftarrow \#$ records mapped to interval 'i'
Require: $Count(i, c) \leftarrow \#$ class 'c' mapped to interval 'i'
Require: $\#Inst(c) \leftarrow \#$ records of class 'c' in Y
Require: $\%Used\_bins$ is the % of bins with a non-zero count
1: **for** $i \leftarrow 1$ to NumBin **do**
2:     **for** $c \leftarrow 1$ to NumClass **do**
3:         **if** $Total(i) \neq 0$ **then**
4:             $hist(i, c) \leftarrow \frac{Count(i,c)}{Total(i) \times \#Inst(c)}$
5:         **else**
6:             $hist(i, c) \leftarrow 0$
7:         **end if**
8:         $GiniIndex \leftarrow GiniIndex + hist(i, c)^2 \times \#Inst(c)$
9:     **end for**
10: **end for**
11: **return** $GiniIndex + \alpha(\%Used\_bins)$

Note that multiple Pure bins can appear, where this does not place any constraint on the labels representing different Pure bins. All training records corresponding to the 'Pure bins' are then identified (Step 27) and removed to define a candidate residual dataset (Step 30).

Finally, the champion individual is then 'pushed' to the list of ensemble members (Step 31). Note that the order in which champions are pushed to the list is significant (Sect. 3.2). At the next iteration of the algorithm, an entirely new population of programs is evolved to specifically label what the members of the ensemble currently cannot label. Moreover, 'early stopping' might appear if the residual dataset is empty (Step 32).

3.2 Evaluating an Ensemble Stack Post Training

Post-training, we need to define how the ensemble operates collectively to produce a single label; whereas, during training, programs are constructed 'independently', without direct reference to other members of the ensemble. Algorithm 3 summarizes how an ensemble is deployed as a 'stack' and evaluated following the order in which each ensemble member was originally added to the ensemble, i.e. the stack chooses programs under a first-in, first-out order (Step 3). Program i may only suggest a label for bins that were originally identified as 'Pure bins'. If the program produces a $\hat{y}_p$ corresponding to any other type of bin (Empty or Ambiguous), then the next program, $i + 1$, is called on to suggest its mapping $\hat{y}_p$ for record p (Step 16). At any point where a program maps the input to a 'Pure bin', then the corresponding label for that bin can be looked up. If this matches the actual label, y_p, then the ensemble prediction is correct (Step 13). This also implies that the entire ensemble need not be evaluated in order to make a prediction.

The first-in, first-out deployment of programs mimics the behaviour of 'falling rule lists', a model of deployment significant to interpretable machine learning [22]. Thus, the most frequent queries are answered by the classifier deployed earliest. As the classes of queries become less frequent (more atypical) prediction is devolved to classifiers appearing deeper in the list (queue). The proposed evolutionary ensemble learner adds programs to the stack in precisely this order.

Algorithm 3. StackEvaluation(X, Y, `Ensemble`, `stack_depth`)

1: $\#correct \leftarrow \#error \leftarrow 0$	
2: **for** $\langle x_p, y_p \rangle \in X, Y$ **do**	▷ Loop over all data partition
3: $Initialize.Stack(\text{Ensemble})$	▷ Initialize stack, using original 'fifo' order
4: $n = \text{stack\_depth}$	▷ Initialize to max ensemble members
5: **repeat**	
6: $Tree \leftarrow pop(Stack)$	▷ Pop a tree, oldest first
7: $\langle bin_p, \hat{y}_p \rangle \leftarrow Tree(x_p)$	▷ Execute Tree on record x_p
8: **if** bin_p is Pure **then**	▷ Test for a 'Pure' bin type
9: $n \leftarrow 0$	▷ Pure bin, so use this Tree for prediction
10: **if** $\hat{y}_p \neq y_p$ **then**	▷ Case of prediction not matching class
11: $\#error + +$	▷ Update classification performance
12: **else**	
13: $\#correct + +$	
14: **end if**	
15: **else**	
16: $n - -$	▷ Not a pure bin, so next Tree
17: **end if**	
18: **until** n is 0	▷ No further trees in Ensemble
19: **end for**	
20: **return** $\langle \#error, \#correct \rangle$	

3.3 Using an Extremely Large Number of Bins

One approach to parameterizing BStacGP is to force the number of bins to be very low (2 or 3). The intuition of is that this rewards BStacGP discovering a mapping of records to bins that develops at least one bin to be pure. This approach will be adopted later for the 'small scale' benchmark. Under high cardinality datasets the opposite approach is assumed. The insight behind this is that there is sufficient data to support multiple pure bins such that we maximize the number of training records mapped to pure bins by the same program. This will also result in the fastest decrease in data cardinality during training.

Taking this latter approach to the limit, we assume the number of bins is set by the size of a floating point number or 2^{32}. During training, bins are again identified as pure, ambiguous or empty. However, given the resolution of the bins, most training data will likely be mapped to a 'pure' bin, reducing the number of boosting epochs necessary to build the ensemble. Under test conditions, given that the bins have a high resolution and the data has not previously been encountered, it is likely that some test data will be mapped to 'empty' bins. Hence, under test conditions, the test data is labelled by the bin that it is closest to. If the closest bin is not pure but ambiguous or empty, then the next tree in the stack is queried.

4   Experimental Methodology

A benchmarking comparison will be made against a set of five datasets from a recent previous study [27]. This enables us to establish to what degree the proposed approach is competitive with five state-of-the-art approaches for GP ensemble classification. The datasets appearing in this comparison are all widely used binary classification tasks from the UCI repository,[2] as summarized by Table 1. The training and test partitions are stratified to provide the same class distribution in each partition and the training/test split is always 70/30%.

The algorithms appearing in this comparison take the following form:

- **2SEGP:** A bagging approach to evolving GP ensembles in which the underlying design goal is to maintain uniform performance across the multiple bootstrap bags. Such an approach was demonstrated to be particularly competitive with other recent developments in evolutionary ensemble learning [27]. In addition, the availability of a code base enables us to make additional benchmarking comparisons under a high cardinality dataset.
- **eGPw:** Represents the best-performing configuration of the cooperative coevolutionary approach to ensemble learning from [21]. Specifically, benchmarking revealed the ability to discover simple solutions to binary classification problems while being competitive with Random Forests and XGBoost.
- **DNGP:** Represents an approach to GP ensembles in which diversity maintenance represents an underlying design goal [28]. Diversity maintenance represents a reoccurring theme in evolutionary ensemble learning, where the motivation is to reduce the correlation between ensemble members. The framework was reimplemented and benchmarked in the study of [27].
- **M3GP:** Is not explicitly an evolutionary ensemble learner but does evolve a set of programs to perform feature engineering [18]. The underlying objective is to discover a mapping to a new feature space that enables clustering to separate between classes. M3GP has been extensively benchmarked in multiple contexts and included in this study as an example of what GP classification can achieve without evolutionary ensemble learning being the design goal [4,18].

A second benchmarking study is then performed on a large cardinality dataset describing an intrusion detection task. This dataset is the union of the normal and botnet data from the CTU-13 dataset [10], resulting in hundreds of thousands of data records (Table 1). In this case, we compare the best two

---

[2] https://archive-beta.ics.uci.edu.

Table 1. Properties of the benchmarking datasets. Class distribution reflects the distribution of positive to negative class instances.

Dataset	# Features (d)	Cardinality ($|\mathcal{D}|$)	Class distribution (%P / %N)
Small benchmarking datasets			
BCW	11	683	35/65
HEART	13	270	45/55
IONO	33	351	65/35
PARKS	23	195	75/25
SONAR	61	208	46/54
Large benchmarking dataset			
CTU	8	801132	55/45

evolutionary ensembles from the first study with C4.5 [20] and XGBoost [5]. The latter represent very efficient non-evolutionary machine learning approaches to classification.

Table 2. BStacGP parameterization. *Gap* defines the size of the parent pool. β is defined by Eq. (1). α weights the fitness regularization term, Algorithm 2. `NumBin` appears in Algorithm 3 and the remaining parameters in Algorithm 1. Instruction set takes the form of the arithmetic operators $\langle +, -, \div, \times \rangle$

Benchmarking Study	Small Scale		Large Scale	
Parameter	fast	slow	fast	slow
`Max_Boost_epoch`	1000		10	
`Max_GP_epoch`	30		3	6
`New_Pop_Size`	30	1000	30	
Gap	10	300	10	
`NumBin`	2		2^{32}	
Bin Purity (β)	0.99		0.6	0.75
Regularization (α)	0.0		0.4	
Num. Trials	40			

Table 2 summarizes the parameters assumed for BStacGP. Two parameterizations are considered in each benchmarking study: 'slow and complex' versus 'fast and simple'. One insight is that higher data cardinality can imply a higher bin count. We therefore use the lowest bin count (2) and a high purity threshold (0.99) as a starting point under the Small Scale benchmarking study. Such a combination forces at least one of the 2 bins to satisfy the high bin purity threshold. We then differentiate between 'slow and complex' versus 'fast and simple' scenarios by increasing the population size (with the parent pool increasing proportionally). The value for `Max_Boost_epoch` is set intentionally high, where in practice such an ensemble size is not encountered due to early stopping being triggered (Algorithm 1, Step 32). Under the large scale benchmark, the largest bin count was assumed, where this does not imply that this number of bins needs to contain values, but it does mean that the distribution has the most resolution.

5 Results

Two benchmarking studies are performed.[3] The first assumes a suite of 'small scale' classification tasks (Sect. 5.1) that recently provided the basis for comparing several state-of-the-art GP evolutionary ensemble learners [27]. The second study reports results on a single large cardinality dataset using the best two evolutionary ensemble learners from the first study, and two non-evolutionary methods (Sect. 5.2). Hereafter, the proposed approach is referred to as BStacGP.

5.1 Small Scale Classification Tasks

Tables 3 and 4 report benchmarking for the five small datasets. In the previous benchmarking study, 2SEGP and M3GP were the best-performing algorithms on these datasets [27]. Introducing the proposed Boosted Stack ensemble (BStacGP) to the comparison changes the ranking somewhat. That said, all five GP formulations perform well on the BCW dataset (>95% under test), whereas the widest variance in classifier performance appears under HEART and SONAR. Applying the Friedman non-parametric test for multiple models to the ranking of test performance fails to reject the null hypothesis (all algorithms are ranked equally). Given that these are all strong classification algorithms this is not in itself unexpected.

Table 3. Classifier accuracy on the *training partition* for small binary datasets. Bold indicates the best-performing classifier on the dataset

Dataset	BStacGP		2SEGP	DNGP	eGPw	M3GP
	slow	fast	[27]	[27]	[27]	[27]
BCW	0.994	**0.995**	**0.995**	0.979	0.983	0.971
	± 0.006	± 0.005	± 0.005	± 0.010	± 0.008	± 0.002
HEART	**1.000**	0.985	0.944	0.915	0.907	0.970
	± 0.0	± 0.015	± 0.022	± 0.021	± 0.025	± 0.017
IONO	**0.993**	0.983	0.976	0.955	0.884	0.932
	± 0.008	± 0.017	± 0.017	± 0.015	± 0.032	± 0.042
PARKS	0.982	**0.996**	0.948	0.931	0.923	0.981
	± 0.018	± 0.004	± 0.011	± 0.057	± 0.042	± 0.024
SONAR	0.999	**1.000**	0.966	0.924	0.924	**1.000**
	± 0.001	± 0.0	± 0.034	± 0.043	± 0.034	± 0.012

---

[3] Laptop with Intel i7 10700k CPU, 4.3 GHz single core.

Table 4. Classifier accuracy on the *test partition* for small binary datasets. Bold indicates the best-performing result

Dataset	BStacGP		2SEGP	DNGP	eGPw	M3GP
	slow	fast	[27]	[27]	[27]	[27]
BCW	**0.96**	0.957	**0.965**	0.959	0.956	0.957
	± 0.017	± 0.022	± 0.018	± 0.019	± 0.018	± 0.014
HEART	0.803	0.796	**0.815**	**0.815**	0.790	0.778
	± 0.094	± 0.052	± 0.062	± 0.049	± 0.034	± 0.069
IONO	**0.924**	0.901	0.896	0.901	0.830	0.871
	± 0.027	± 0.047	± 0.047	± 0.026	± 0.057	± 0.057
PARKS	**0.951**	0.937	0.936	0.917	0.822	0.897
	± 0.017	± 0.013	± 0.012	± 0.055	± 0.064	± 0.051
SONAR	0.76	0.728	0.738	0.730	0.762	**0.810**
	± 0.083	± 0.131	± 0.067	± 0.063	± 0.060	± 0.071
Av.Rank	2.0	3.8	2.7	3.2	5.0	4.3

BStacGP and 2SEGP are the two highest ranked algorithms on training and test. With this in mind we consider the average solution complexity and time to evolve solutions. In the case of complexity, the average number of nodes in the Tree structured GP individuals comprising an ensemble is counted, Fig. 2. It is apparent that BStacGP is able to discover solutions that are typically an order of magnitude simpler. Figure 3 summarizes the wall clock time to conduct training. Both BStacGP and 2SEGP are implemented in Python. BStacGP typically completes training an order of magnitude earlier than 2SEGP. In summary, the process by which BstackGP incrementally only evolves classifiers against the 'residual' misclassified data results in a significant decrease in computational cost and model complexity.

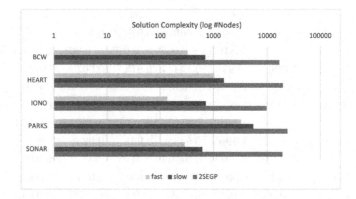

Fig. 2. Solution Complexity for BStacGP and 2SEGP on small scale classification tasks. 'fast' and 'slow' represent the two BStacGP parameterizations.

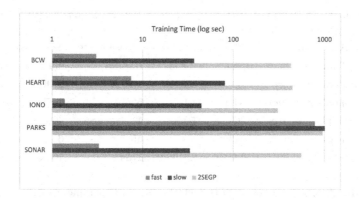

Fig. 3. Training time for BStacGP and 2SEGP on small scale classification tasks. 'fast' and 'slow' represent the two BStacGP parameterizations.

5.2 Large Scale Classification Task

BStacBP and 2SEGP accounted for 4 out of 5 of the best classifier performance under the test partition of the 'Small Scale' task (Table 4). With this in mind, we now compare the performance of BStacBP and 2SEGP under the high-cardinality CTU dataset with the non-evolutionary classifiers of C4.5 and XGBoost. Table 5 reports performance for the average (over 40 trials) of BStacBP, C4.5 and XGBoost and a single run of 2SEGP. Specifically, the high run time cost of 2SEGP precluded performing multiple trials on this benchmark.[4] The best-performing algorithm under test is XGBoost, however, BStacGP is within 1.3% of this result. Conversely, 2SEGP returns a 10% lower accuracy, a result in part due to 20 generations taking 4hrs to perform, so limiting the ability to optimize parameters.

In the case of solution complexity (as measured by node count), BStacGP returns solutions that are 2 to 3 orders of magnitude simpler than any comparator algorithm. From the perspective of the computational cost of performing training, C4.5 is the fastest. However, of the ensemble algorithms, BStacGP is twice as fast as XGBoost and 3 orders of magnitude faster than 2SEGP.

A second experiment is performed using the CTU dataset, the hypothesis, in this case, being that constraining C4.5 and XGBoost to the same complexity as BStacGP will have a significant impact on their classification performance. Put another way, the comparator algorithms will not be able to discover solutions with similar complexity to BStacGP without significantly compromising their classification accuracy.

Table 6 summarizes performance under the low complexity condition. It is now apparent that BStacGP is able to maintain solution accuracy on this task as well as further reduce the computation necessary to identify such a solution. This implies that BStacGP has the potential to scale to tasks that other formulations of evolutionary ensemble learning fail to scale to while maintaining

---

[4] 2SEGP parameterization: pop. size 500, ensemble size 50, max. tree size 500.

Table 5. CTU dataset with BStacGP using 'slow' parameters (Table 2)

Algorithm	BStacGP	2SEGP	Decision Tree	XGBoost
Train Accuracy	0.987	0.8437	**0.9986**	0.9716
Test Accuracy	0.953	0.8419	0.9625	**0.9681**
number of nodes	**157.9**	8454	52801	26792
number of trees	2.87	50	**1**	300
avg. tree depth	**6.03**	–	55	**6**
time (sec)	23.09	13160.92	**1.54**	46.92

Table 6. CTU dataset with BStacGP using 'fast' parameters (Table 2)

Algorithm	BStacGP	Decision Tree	XGBoost
Train Accuracy	**0.985**	0.914	0.8796
Test Accuracy	**0.952**	0.915	0.8795
number of nodes	**47.55**	59	75
number of trees	2.25	**1**	5
avg. tree depth	3.8	8	**3**
time (sec)	11.28	0.83	**0.41**

state-of-the-art classification performance. Thus, as dataset cardinality increases algorithm efficiency has an increasing impact, as it is simply not possible to tune parameters, an important practical property to have.

Post training, BStacGP solutions can be queried. For example, the 1st program from a typical BStack stack ensemble provided $\approx 57\%$ of the labels. The 2nd $\approx 24\%$, the 3rd $\approx 15\%$ and the 4th $\approx 4\%$. This illustrates the 'fall through' nature of BStacGP operation in which most of the data is labeled by a single GP program. The complexity of trees associated with each stack level are respectively 138, 250, 305 and 407 nodes, i.e. increases with position in the stack.

6 Conclusion

An approach to evolutionary ensemble learning is proposed that employs boosting to develop a stack of GP programs. Key components of the framework include: 1) interpreting the program output as a distribution; 2) quantizing the distribution into intervals and therefore 'binning' the number of records mapped to an interval; 3) making predictions on the basis of 'bin purity'; and 4) removing records from the training partition corresponding to correctly classified instances. The combination of these properties incrementally reduces the cardinality of the training partition as classifiers are added to the ensemble, and explicitly focuses the role of the next program on what previous programs could not classify. Moreover, the resulting ensemble is then deployed as a 'Stack'. This is important because it now means that only part of the ensemble is responsible for providing a label, thus improving the explainability of the resulting ensemble.

The accuracy of the BStacGP framework on small cardinality datasets previously employed for benchmarking is empirically shown to be comparable to state-of-the-art evolutionary ensemble learners. Moreover, training time and model simplicity is significantly improved. This property is shown to be key to scaling BStacGP much more efficiently to a large cardinality dataset containing hundreds of thousands of records. Indeed the results are competitive with non-evolutionary methods.

Future work will scale BStacGP to multi-class classification and continue to investigate scalability and solution transparency.

Acknowledgements. This research was enabled by the support of the Natural Science and Engineering Research Council (NSERC) of Canada Alliance Grant.

References

1. Agapitos, A., Loughran, R., Nicolau, M., Lucas, S.M., O'Neill, M., Brabazon, A.: A survey of statistical machine learning elements in genetic programming. IEEE Trans. Evol. Comput. **23**(6), 1029–1048 (2019)
2. Badran, K.M.S., Rockett, P.I.: Multi-class pattern classification using single, multidimensional feature-space feature extraction evolved by multi-objective genetic programming and its application to network intrusion detection. Genet. Program Evolvable Mach. **13**(1), 33–63 (2012)
3. Brameier, M., Banzhaf, W.: Evolving teams of predictors with linear genetic programming. Genet. Program Evolvable Mach. **2**(4), 381–407 (2001)
4. Cava, W.G.L., Silva, S., Danai, K., Spector, L., Vanneschi, L., Moore, J.H.: Multidimensional genetic programming for multiclass classification. Swarm Evol. Comput. **44**, 260–272 (2019)
5. Chen, T., Guestrin, C.: XGBoost: a scalable tree boosting system. In: Proceedings of the ACM SIGKDD International Conference on Knowledge Discovery and Data Mining, pp. 785–794. ACM (2016)
6. Curry, R., Lichodzijewski, P., Heywood, M.I.: Scaling genetic programming to large datasets using hierarchical dynamic subset selection. IEEE Trans. Syst. Man, Cybern. - Part B **37**(4), 1065–1073 (2007)
7. Dietterich, T.G.: An experimental comparison of three methods for constructing ensembles of decision trees: bagging, boosting, and randomization. Mach. Learn. **40**(2), 139–157 (2000)
8. Fahlman, S.E., Lebiere, C.: The cascade-correlation learning architecture. In: Advances in Neural Information Processing Systems, vol. 2, pp. 524–532. Morgan Kaufmann (1989)
9. Folino, G., Pizzuti, C., Spezzano, G.: Training distributed GP ensemble with a selective algorithm based on clustering and pruning for pattern classification. IEEE Trans. Evol. Comput. **12**(4), 458–468 (2008)
10. García, S., Grill, M., Stiborek, J., Zunino, A.: An empirical comparison of botnet detection methods. Comput. Secur. **45**, 100–123 (2014)
11. Gathercole, C., Ross, P.: Dynamic training subset selection for supervised learning in genetic programming. In: Davidor, Y., Schwefel, H.-P., Männer, R. (eds.) PPSN

1994. LNCS, vol. 866, pp. 312–321. Springer, Heidelberg (1994). https://doi.org/10.1007/3-540-58484-6_275

12. Iba, H.: Bagging, boosting, and bloating in genetic programming. In: Proceedings of the Genetic and Evolutionary Computation Conference, pp. 1053–1060. Morgan Kaufmann (1999)

13. Imamura, K., Soule, T., Heckendorn, R.B., Foster, J.A.: Behavioral diversity and a probabilistically optimal GP ensemble. Genet. Program Evolvable Mach. 4(3), 235–253 (2003)

14. Lichodzijewski, P., Heywood, M.I.: Managing team-based problem solving with symbiotic bid-based genetic programming. In: Proceedings of the Genetic and Evolutionary Computation Conference, pp. 363–370. ACM (2008)

15. Lichodzijewski, P., Heywood, M.I.: Symbiosis, complexification and simplicity under GP. In: Proceedings of the Genetic and Evolutionary Computation Conference, pp. 853–860. ACM (2010)

16. McIntyre, A.R., Heywood, M.I.: Classification as clustering: a pareto cooperative-competitive GP approach. Evol. Comput. 19(1), 137–166 (2011)

17. Muni, D.P., Pal, N.R., Das, J.: A novel approach to design classifiers using genetic programming. IEEE Trans. Evol. Comput. 8(2), 183–196 (2004)

18. Muñoz, L., Silva, S., Trujillo, L.: M3GP – multiclass classification with GP. In: Machado, P., et al. (eds.) EuroGP 2015. LNCS, vol. 9025, pp. 78–91. Springer, Cham (2015). https://doi.org/10.1007/978-3-319-16501-1_7

19. Potter, M.A., Jong, K.A.D.: Cooperative coevolution: an architecture for evolving coadapted subcomponents. Evol. Comput. 8(1), 1–29 (2000)

20. Quinlan, J.R.: C4.5: Programs for Machine Learning. Morgan Kaufmann, Burlington (1993)

21. Rodrigues, N.M., Batista, J.E., Silva, S.: Ensemble genetic programming. In: Hu, T., Lourenço, N., Medvet, E., Divina, F. (eds.) EuroGP 2020. LNCS, vol. 12101, pp. 151–166. Springer, Cham (2020). https://doi.org/10.1007/978-3-030-44094-7_10

22. Rudin, C.: Stop explaining black box machine learning models for high stakes decisions and use interpretable models instead. Nat. Mach. Intell. 1(5), 206–215 (2019)

23. Sipper, M., Moore, J.H.: Symbolic-regression boosting. CoRR abs/2206.12082 (2022)

24. Song, D., Heywood, M.I., Zincir-Heywood, A.N.: Training genetic programming on half a million patterns: an example from anomaly detection. IEEE Trans. Evol. Comput. 9(3), 225–239 (2005)

25. Soule, T.: Voting teams: a cooperative approach to non-typical problems using genetic programming. In: Proceedings of the Genetic and Evolutionary Computation Conference, pp. 916–922. Morgan Kaufmann (1999)

26. Thomason, R., Soule, T.: Novel ways of improving cooperation and performance in ensemble classifiers. In: Proceedings of the Genetic and Evolutionary Computation Conference, pp. 1708–1715. ACM (2007)

27. Virgolin, M.: Genetic programming is naturally suited to evolve bagging ensembles. In: Proceedings of the Genetic and Evolutionary Computation Conference, pp. 830–839. ACM (2021)

28. Wang, S., Mei, Y., Zhang, M.: Novel ensemble genetic programming hyper-heuristics for uncertain capacitated arc routing problem. In: Proceedings of the Genetic and Evolutionary Computation Conference, pp. 1093–1101. ACM (2019)

29. Wu, S.X., Banzhaf, W.: Rethinking multilevel selection in genetic programming. In: Proceedings of the Genetic and Evolutionary Computation Conference, pp. 1403–1410. ACM (2011)

Adaptive Batch Size CGP: Improving Accuracy and Runtime for CGP Logic Optimization Flow

Bryan Martins Lima[✉][ID], Naiara Sachetti[ID], Augusto Berndt[ID],
Cristina Meinhardt[ID], and Jonata Tyska Carvalho[ID]

Federal University of Santa Catarina - UFSC, Florianópolis, Brazil
{bryan.l,naiara.sachetti}@grad.ufsc.br, augusto.berndt@posgrad.ufsc.br,
{cristina.meinhardt,jonata.tyska}@ufsc.br

Abstract. With the recent advances in the Machine Learning field, alongside digital circuits becoming more complex each day, machine learning based methods are being used in error-tolerant applications to solve the challenges imposed by large integrated circuits, where the designer can obtain a better overall circuit while relaxing its accuracy requirement. One of these methods is the Cartesian Genetic Programming (CGP), a subclass of Evolutionary Algorithms that uses concepts from biological evolution applied in electronic design automation. CGP-based approaches show advantages in the logic learning and logic optimization processes. However, the main challenge of CGP-based flows is the extensive runtime compared to other logic synthesis strategies. We propose a new strategy to tackle this challenge, called Adaptive Batch Size (ABS) CGP, in which the CGP algorithm incrementally improves the fitness estimation of the candidate solutions by using more terms of the truth table for evaluating them along the evolutionary process. The proposed approach was evaluated in nine exemplars from the IWLS 2020 contest, in which 3 exemplars are from the arithmetic domain, and six are from image recognition domain, specifically three from the CIFAR-10 dataset and three from the MNIST dataset. The results show that ABS presented an accuracy increase of up to 8.19% and decreased the number of candidate solutions evaluations required by up to 84.56%, in which directly affects the runtime of the algorithm. Furthermore, for all circuits, no significant accuracy reduction was observed while a significant reduction in the number of evaluations was achieved.

Keywords: Logic synthesis · Cartesian Genetic Programming (CGP) · Evolutionary algorithms · Approximate Computing

1 Introduction

Logic optimization is an initial task when converting an abstract specification of a digital circuit in terms of logic gates. It focus on reducing the number

G. Pappa et al. (Eds.): EuroGP 2023, LNCS 13986, pp. 149–164, 2023.
https://doi.org/10.1007/978-3-031-29573-7_10

of logic elements, including nodes and logic depth, used for producing a specific input-output mapping (function). These optimizations will influence future steps in the design flow, improving circuit area, delay and energy consumption [23]. Traditional logic optimization methods simplify a Boolean function, exploring exact logic minimization techniques like the Algebraic method, the Karnaugh map technique [12], or the Quine-McCluskey method [22]. However, the main limitation of the traditional logic optimization methods is the number of inputs that they can deal with. For instance, the Quine-McCluskey method is limited to functions with up to 15 variables [6]. As for real-world applications, Espresso is used for the simplification of circuits with many inputs reaching faster results by exploring sub-optimal heuristic methods [5,25]. These fast simplification methods provide a trade-off for computing performance at the cost of output quality.

Some new logical optimization flows address fast logic optimization based on machine learning approaches like decision trees [1]. However, many of them fail to deal with scaling the logic function complexity. For example, decision tree solutions, in general, must expand all the input combinations, which becomes prohibitive for large inputs. To deal with the expensive rising of complexity on logic functions, neural networks [19] and evolutionary algorithms such as Cartesian Genetic Programming (CGP) [3,4] have been recently studied. An important drawback of these approaches is the large runtime compared to the traditional logic synthesis approaches.

Most of these new optimization flows are particularly interesting when applied to Approximate Computing. The traditional methods, such as Espresso, struggle when problems have an incomplete specification, as this method was designed to tackle exact minimization. In this scenario, the novel optimization flows are becoming an alternative when specifications for problems do not require an exact circuit. In this case, using CGP with Approximate Computing paradigm can be a good alternative to traditional methods [4].

This paradigm have been mainly used to design power-efficient solutions for error-tolerant applications [26,29]. Examples of error-tolerant applications include those using image and sound data, video processing, sensors for Internet of Things, and neural networks. As these applications have less strict accuracy requirements of the implemented functions, it is possible to focus on a smaller circuit, which can improve power and delay [2].

Furthermore, we can use Approximate Computing to generalize a circuit based on a few selected samples. This was one of the goals of the International Workshop on Logic and Synthesis (IWLS) Contest in 2020 [23], in which multiple teams of different countries competed to generalize logic functions. The teams used multiple strategies, including Espresso [25], multi-layer perceptrons (MLP), random forests, lookup-table (LUT) networks [19] as well as CGP [4]. Overall, the presented results confirmed that sacrificing some accuracy was possible to achieve a significantly smaller circuit, as well as many of them used a combination of these strategies; therefore, none of these strategies dominated the others.

One of the promising flows proposed in the IWLS was CGP [4], which presented reasonable accuracy performance but with a long runtime for synthesizing the circuits. This flow can be used to optimize already synthesized circuits as well as synthesize a new circuit. In this context, we present a new technique called Adaptive Batch Size (ABS) CGP which aims to reduce the number of evaluations required for synthesizing and optimizing circuits, while maintaining or improving the accuracy compared to the standard CGP. The proposed strategy dynamically increases the number of selected terms of the truth table used for evaluating the individuals during the evolutionary process. By evaluating the proposed strategy on a subset of the IWLS contest, the results indicate a reduction of up to 84.6% on the number of evaluations needed compared to the standard CGP flow. Furthermore, some circuits presented increased accuracy of up to 8.2%, while no synthesized circuit presented an accuracy decrease.

This paper is organized as follows. Section 2 presents the Cartesian Genetic Programming algorithm. In Sect. 3 is described our methodology detailing the strategy proposed. In Sect. 4 there is a discussion on the data collected. Finally, Sect. 5 concludes this work and summarizes our findings.

2   Cartesian Genetic Programming

The CGP is a form of genetic programming that uses a graph representation to encode computer programs. It is called 'Cartesian' because it represents a program using a two-dimensional grid of nodes [18]. It was created by Julian F. Miller in 1999 to encode digital circuits [18]. It is capable of representing math equations, circuits and computer programs as a directed acyclic graph. The CGP is bio-inspired by concepts from genetics, that are used for building the meta-heuristic responsible for the optimization process and program/solution synthesis. For this, genes are the integers that form genotypes of a node, thus, they represent the input, operation and output of the given node.

There are many applications where CGP is used. It has been used in image filters and image processing [10,11,27], it can be used to encode Artificial Neural Networks [13,14] and to optimize the learning of Convolutional Neural Networks [28], as well as in cryptography field [20,21]. For further discussion on its applications see the works of [16,18].

2.1   Representation

Figure 1 shows the representation of a CGP individual as a 1-line array, instead of a 2-dimensional grid. The 1-line array was selected as it has faster convergence in the evolutionary search as demonstrated in [17]. However, both representations are equivalent for digital circuits, as they can be represented either way.

The 1-line array representation is an array of 4-tuples composed of two pairs, one pair for each input of the logic gate. It is noteworthy that the inputs are indexed as well as the individual nodes. Hence, there are 6 indexes representing 3 primary inputs and 3 nodes. The first pair (a, b), represents the first input, in which a represents if the input is inverted or not, and b representing from which

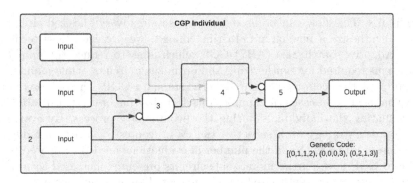

Fig. 1. CGP individual representation. Adapted from: [4]

index to get the input from; similarly, the second pair represents the second input.

Taking node 3 as an example, its first pair, the input is not inverted and it came from the primary input 1, which is indexed as 1; thus, the first pair is $(0, 1)$. For its second pair, the input is inverted and it came from the primary input 2, indexed as 2; therefore, the second pair is $(1, 2)$. Therefore, the complete genetic code for node 3 is $(0, 1, 1, 2)$. There are two main considerations for this representation. The first one is that the input of a given node can be a primary input, i.e. (0, 1, 2), or an output of another node, i.e. (3, 4, 5). It is important to notice that backward connections are not allowed so, for instance, node 4 can receive inputs from node 3, but not from node 5. The second one regards the meaning of the term functional node. Note that, in this simple example, node 4 is represented in gray color in Fig. 1. That is because it is not a functional node, i.e. it does not have a connection to the output of the CGP individual. The functional part of the CGP individual is called the *phenotype*, and the genetic code, is called *genotype*. Is noteworthy, that the number of primary inputs, primary outputs and the number of nodes is fixed through the whole evolutionary process.

It is important to note that we used AND and Inverter gates; thus, it represents a AND-Inverter graph or AIG. It is noteworthy that any logic function can be described with only AND and Inverter logic gates.

2.2 Evolutionary Process

Figure 2 presents a hypothetical CGP search that uses the evolutionary approach $(1 + 4)$, in which, at each generation, a parent circuit generates four offspring. These offspring circuits are a mutated copy of the parent's circuit, i.e., a copy that had some of its connections mutated by chance according to a given probability, the mutation rate. Direct connections are represented by a straight line and inverted connections are represented by dotted lines.

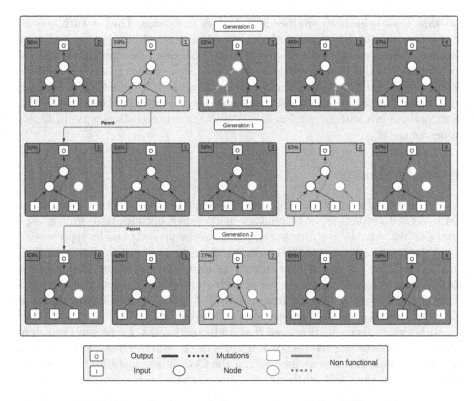

Fig. 2. Example of an evolutionary process of the CGP search.

As can be seen in Fig. 2, in generation 0, individuals are randomly generated, and each circuit has a different accuracy (fitness value). The parent circuit of the next generation is highlighted in green. The circuit with the highest accuracy is selected as the parent of the next generation. Therefore, regarding generation 0, the next parent is the circuit with 59% of accuracy. In generation 1, this parent generates one unmodified copy and four mutated copies, i.e., copies with a chance of having different connections. Mutations are highlighted in red. The same process repeats for generation 2 and so on, until reaching the maximum number of generations defined by the user.

In this example, for sake of clarity, we used a simpler representation of a CGP individual when compared to Fig. 1. We used this simpler representation to focus on the learning of the search rather then focus in the individuals. Moreover, it is important to note that each individual had 3 nodes; however, the number of nodes is generally hundreds or even thousands.

Furthermore, as demonstrated by [17], the CGP search tends more easily to escape from local-optima regions, so avoiding premature convergence, if phenotypically larger solutions are considered preferred candidates when analyzing the individual's fitness accuracy due to higher genetic variation. This happens since artificial evolution tends to select genotypes that are robust to mutations, i.e., that are less likely to produce maladaptive mutations. This, in turn, pushes in the direction of genotypes with tiny functional circuits, since the genotype becomes robust to any mutation that does not affect the functional part of the circuit, and this characteristic leads to premature convergence. The authors demonstrate that this limitation could be eliminated by selecting individuals with larger functional sizes when they have the same fitness. Therefore, in this work, individuals with larger functional sizes are selected when a tie in fitness values happens.

3   Methodology

In our implementation, we explore the utilization of AIGs, which are the state-of-the-art data structure for technology-independent optimizations during logic synthesis [24]. An AIG is a directed acyclic graph composed of nodes representing AND gates and edges representing inverted or directed connections. An AIG node is composed of exactly two inputs and an arbitrary number of outputs. An AIG may represent any logic function. In our CGP implementation, an AIG is represented as an individual from the CGP population, and CGP mutations concern modifying the AIG connections and inversions.

To better explain our proposed flow and experiments, we present four definitions before detailing two important parts of our work. Then, in Sect. 3.3 we present the hyperparameters used to run the CGP implementation and analyze their impact in the CGP synthesis.

3.1   Definitions

The following definitions describe fundamental concepts in our approach.

Definition 1 (Batch size): The actual number of terms used for the individuals' evaluation is called *batch size* (BS).

Definition 2 (Adaptive Interval (σ)): represents the interval, in number of generations, in which ABS CGP will use to monitor the evolution stagnation.

Definition 3 (Mutation): to explore the search space, the circuit needs to change. For this, to add variability to the system, mutations occur in the AIG nodes, as demonstrated by red drawings in Fig. 2. Mutations during generations happen according to the mutation rate value, which is given by the $1/5^{th}$ rule [8]. When a mutation happens, one of the values of the 4-tuple representing the inputs and input-inversions of a node is randomly changed. Notice that backward connections are not allowed, so for a given node i, the mutated value will be drawn from a uniform distribution ranging from 0 to $i - 1$.

Definition 4 (Fitness): in this work, fitness means the capability of a circuit to produce the correct output given a set of terms (accuracy). Meaning that Boolean signals are propagated and binary outputs achieved are checked if they match with the expected value.

3.2   Adaptive Batch Size CGP

Traditionally, CGP evaluates all the individuals every generation after they are created by the parent mutation. The evaluation process for each of them requires processing all the training batch inputs. This is done with a depth-first search along the AIG. In other words, the evaluation has a time complexity of $O(i*n*b)$, with i being the number of individuals in the CGP population, n the size of each individual, and b the batch size (BS) used for training.

Figure 3 presents a flowchart of a simplified version of the proposed mechanism for the CGP technique. First, ABS CGP starts with an initial batch size of β terms. The next step is choosing a criterion for defining and detecting evolution stagnation, and by doing so increasing the batch size when it happens. The initially chosen criterion was a Simple Moving Average (SMA) from the accuracy of the synthesized circuits considering a window of σ generations.

For the next step, there is a verification step that compares the current SMA to the previous SMA. By always saving the previous SMA calculated we can compare this value with the SMA of the current generation. The comparison of both values provides us with the information that CGP is actually learning the function at hand. Therefore, if these values are the same, there were no improvements in accuracy in the last $2*\sigma$ generations. This is the core of our strategy, if the CGP is not improving the circuit's accuracy, i.e. the estimation quality is stagnant, exposing the algorithm to more terms could help broaden its search. By using a BS smaller than the whole truth table, stochasticity in the evaluation process is inherently added. The smaller the BS value, the faster is the evaluation process and higher is the stochasticity. Therefore, there is a moment in which the individuals' fitness reached a point that, due to stochasticity, the algorithm cannot differ which are the best CGP individuals anymore. Our proposed strategy tries to solve this by increasing the number of terms when it detects that this situation occurs, reducing the stochasticity in the evaluation, and by doing so, improving the quality of the fitness estimation.

Before actually increasing the batch size, there is another verification checking if increasing the batch will not surpass the size of the whole data available. If this verification yields true, i.e. increasing the batch does not surpass the complete data, then the batch is increased by α terms and the SMA is calculated again after σ generations. If the previously mentioned verification yields false,

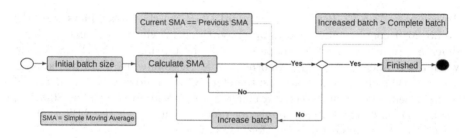

Fig. 3. Adaptive Batch Size CGP flowchart

Table 1. Exemplars details

Exemplar	Logic function	# of inputs
40, 41, 43	LSB of n-square root	16, 10, 18
80, 81, 84	MNIST	196
93, 98, 99	CIFAR-10	768

that means the algorithm is almost or already using the complete data available; thus, increasing the batch would surpass all data available. In this case, the whole data available is chosen as the batch size of the next generation, and there is no need for any more steps, as the CGP search is already using the maximum amount of data available. After this step, the CGP continues its search as the standard version.

It is important to note that the value of the BS directly impacts the runtime. As the search algorithm will utilize this value to measure the synthesized circuit's accuracy. Therefore, evaluating a circuit's accuracy with a BS of 128 is computationally cheaper than evaluating with a BS of 1024.

Furthermore, this work performed an improvement regarding the evaluation process using the CGP C++ implementation[1] used in [4]. The main difference in our version is when to evaluate the nodes accuracy, as this process is computationally costly. For this, we used the proposed strategy "Skip" in [9] which means skipping the evaluation of individuals in which the functional nodes of their parent were not mutated. This change was not only better for performance, but it was necessary to be possible to evenly compare the multiples hyperparameters used in CGP. The "Skip" strategy does not affect by any means the performance of the circuits synthesized by the algorithm, it only prevents unnecessarily evaluating nodes that certainly have the same accuracy as their parent, thus, it does not waste time evaluating an already known accuracy.

Hence, by avoiding unnecessary evaluations and dynamically increasing the number of terms for evaluating the candidate solutions during CGP search, we reduce the number of evaluations required for synthesizing circuits, and consequently reduce runtime.

3.3   Experimental Protocol

To compare ABS CGP with the standard CGP, we chose exemplars from the IWLS 2020 contest. Specifically, we chose those in which CGP had a bigger difference of accuracy when compared to the best teams in the IWLS contest when using different strategies. In other words, those in which CGP had a worse accuracy than the others strategies; therefore, CGP could still have a margin of improvement. Thus, the selected exemplars were: ex40, ex41, ex43, ex80, ex81, ex84, ex93, ex98, ex99. As ex41 was our proof of concept for the ABS strategy, we selected two similar exemplars for comparison.

---

[1] Source code available in: https://gitlab.com/gudeh/cgpv3

Table 2. Hyperparameters tested

Parameter	Value
Number of generations	50,000
Number of nodes	1,000
Number of seeds	10
Evolutionary Strategy	$(1 + 4)$
ABS CGP-specific	
Adaptive interval (σ)	100, 250, 500, 1000, 2500, 5000
Initial batch size (β)	64
Increase batch factor (α)	64
Standard CGP-specific	
Batch size	6400

Table 1 presents more details upon the exemplars chosen for validating the proposed algorithm. The nine exemplars are from arithmetic - ex40, ex41 and ex43 - and image recognition domains - MNIST [7] and CIFAR-10 [15]. It is noteworthy that the image recognition functions are the largest in the contest; furthermore, the contest only provided 6,400 inputs for all exemplars, which only represent a tiny portion of all possible combinations for MNIST and CIFAR, 2^{196} and 2^{768} combinations respectively.

Table 2 shows the hyperparameters used in our experiments. We used 50,000 generations to identify the initial impacts on the learning of the circuit. We used a number of nodes of 1,000, which means that the CGP has a maximum of 1,000 functional nodes. Moreover, we replicated each experiment ten times with different random seeds. As this algorithm is impacted by different sources of stochasticity, for instance, its initial randomly generated circuits, ten runs provided enough information for the statistical validation of the results. These variations in the accuracy caused by stochasticity will be noticeable on the box plots presented in Sect. 4.

For the ABS CGP-specific hyperparameters, we varied the AI parameter between 0.2% and 10% of the total evolutionary process of 50,000 generations to analyze its impact. Due to the high computational cost of optimizing all hyper-parameters, we decided to fix the values for the initial batch size and the increased batch factor, which were fixed to the value of 64. Investigating the impact of varying these values on the performance of the ABS CGP is relevant and planned as future work. To compare the proposed strategy to the standard CGP version the value of the BS was fixed with 6,400 terms, as it was the maximum value for number of terms for all exemplars of the IWLS contest. Along this, we used the same 50,000 generations with 1,000 nodes and ten seeds for the control case.

Finally, instead of using the algorithm runtime as one of the comparison metrics, we use the number of candidate solutions evaluations performed during the evolutionary process. This is a standard metric within the evolutionary computing community since it removes any hardware-related differences among studies. To measure the total number of evaluations we use the sum over the generations of the result of the multiplication between the BS and the offspring in each generation. Equation 1 presents the complete equation.

$$Evaluations = \sum_{i=1}^{i=N} BS_i * offspring \tag{1}$$

As we use Eq. 1 to calculate the number of evaluations, it is clear that the BS value is the one that impacts the number of evaluations the most. As the number of generations is fixed and the $offspring$ variable is independent of the ABS strategy, the only variable that is dependent on the proposed strategy is the BS value.

4   Results

Our results show that in most cases the ABS approach not only increased the accuracy but also led to a significant reduction in the number of evaluations performed during the evolutionary search. Figure 4 presents each circuit accuracy when using all available terms of the truth table versus our approach with ABS CGP. The results of ABS consider the best set of hyperparameters for each exemplar, considering the best AI values, presented in Table 3.

Table 3. Best AI values for each exemplar of Fig. 4 and Fig. 5

Exemplar	Adaptive interval
ex40	100
ex41	500
ex43	100
ex80	250
ex81	100
ex84	250
ex93	250
ex98	100
ex99	250

Figure 5 presents the number of evaluations to learn the circuits and Table 4 presents their average values from the ten different seeds used. Furthermore, Table 5 presents the gain/loss in accuracy and the number of evaluations for each exemplar between ABS CGP and the standard CGP using all data available. Exemplars from the arithmetic domain had the best-improved accuracy

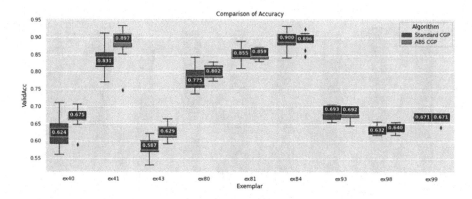

Fig. 4. Accuracy of Standard CGP and ABS CGP

Table 4. Average number of evaluations (in 10^7) of Fig. 5

Exemplar	Complete version	Adaptive version
ex40	123.00 ± 1.99	70.18 ± 10.2
ex41	126.17 ± 1.47	18.80 ± 2.15
ex43	122.10 ± 1.93	62.45 ± 10.4
ex80	121.67 ± 2.70	32.06 ± 3.02
ex81	123.11 ± 3.02	79.38 ± 5.64
ex84	122.67 ± 2.97	35.39 ± 5.95
ex93	119.18 ± 0.08	37.31 ± 3.85
ex98	119.12 ± 0.09	85.56 ± 3.85
ex99	119.15 ± 0.09	35.21 ± 3.12

of all nine exemplars tested, with a statistically significant accuracy increase of 7.74% on average (Mann-Whitney U Test, p-values of .02, .03 and .001 for exemplars ex40, ex41 and ex43, respectively), while decreasing 57.96% on average the number of evaluations performed (Mann-Whitney U Test, $p < .001$ for all exemplars). As for the exemplars from the image recognition domain, the accuracy improvement was 0.792% in average, which was not statistically significant (Mann-Whitney U Test, p-values of .16, .909, .71, .569, .87 and .939, for exemplars ex80, ex81, ex84, ex93, ex98 and ex99, respectively). Despite not achieving a significant improvement for the exemplars ex80 through ex99, the accuracy level obtained by the control case was maintained, while decreasing the evaluations performed by 58.015% on average for these exemplars (Mann-Whitney U Test, $p < .001$ for all exemplars). From these results, it seems that ABS CGP is capable of synthesizing circuits with better accuracy in the arithmetic domain. Although more exemplars of this type and further analysis are needed to better understand why this happens for this particular family of functions.

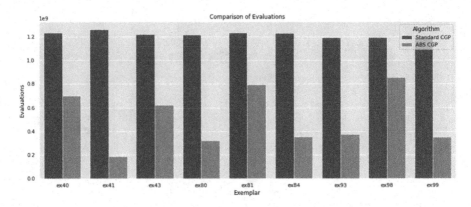

Fig. 5. Number of evaluations between ABS CGP and Standard CGP

Another noteworthy insight from these experiments are the values of AI. As we can see in Table 3, most of the AI values are in the lower range of the chosen AI parameters of Table 2. Only the AI values of 100 and 250 managed to reach a point in the evolutionary process in which all available lines were used for evaluating individuals in all the experiments. Furthermore, the results with AI above 500 had worse accuracy compared to the results presented herein. This indicates that more generations could be executed on the search since with these AI values the CGP search could not reach the complete batch size. Further investigation is required to analyze if by increasing the number of generations the evolutionary processes using these higher AI values present better results with respect to using lower AI values as they would have more time to reach the maximum size. Moreover, using a fixed size of α, which controls by how much the batch will be increased, may be too low. In our set of experiments we used a value of α representing only 1% of the truth table which could be optimal for some exemplars, but not others. Overall, this data indicates that there might be room for improvement if this hyperparameter would be optimized as well.

To further investigate the learning curve of the ABS CGP evolutionary search we checked how the fitness of the best individuals increased through generations. Besides collecting the partial accuracy of the circuit in each generation, we collected its validation accuracy as well, i.e. the accuracy using the validation set. Figure 6 presents this data for the initial evolutionary search for exemplar 41. The blue line represents the search using all terms available when using the standard CGP. The green line represents the search for the Adaptive version with AI set to 100. Hence, the first generations are using a smaller portion of the truth table, and the last ones are using all terms available. The orange line is representing the accuracy of the same Adaptive version of 100 AI while using all terms from the beginning. For this reason, the orange line and green line converge at the end.

It is noteworthy that our first assumption for the ABS version was that ABS CGP could improve faster at the beginning, even when using a rough fitness evaluation of the candidate solutions, i.e., using fewer lines for evaluating each

Table 5. Adaptive CGP gains over standard CGP

Exemplar	Accuracy gain/lost	Evaluation reduction
ex40	+8.186%[1]	−44.180%[1]
ex41	+7.890%[1]	−84.563%[1]
ex43	+7.143%[1]	−45.125%[1]
ex80	+3.443 %	−73.617%[1]
ex81	+0.550 %	−35.343%[1]
ex84	−0.399 %	−72.641%[1]
ex93	−0.144 %	−68.823%[1]
ex98	+1.217 %	−26.839%[1]
ex99	+0.029 %	−70.827%[1]
Average	3.102%± 3.344	−57.995% ± 19.110

[1] Statistically significant difference (Mann Whitney-U with $p < .05$).

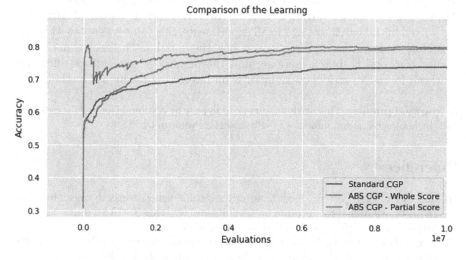

Fig. 6. Accuracy during learning for ex41

circuit, and performing the fine-tuning of circuits at the end, using the whole data available. Indeed, the learning of the ABS CGP (orange line) managed to achieve better accuracy in these initial evaluations compared to the complete version even with a worst fitness estimation than the version using all data available for evaluating individuals. Furthermore, ABS kept improving the circuits beyond the performance achieved by the standard version, providing a better final result.

5   Conclusion

Due to the increasing complexity of digital circuits novel techniques for logic synthesis are required. The combination of logic synthesis and machine learning has been gaining much attention in the literature due to the promising results regarding circuit area reduction and power consumption, mainly in approximate computing domains. Among these techniques, Cartesian Genetic Programming is capable of synthesizing better circuits than traditional techniques in certain domains, but at the cost of demanding a great computational power, which slows down the synthesis. This work proposes a novel technique called Adaptive Batch Size CGP seeking to alleviate the computational resources required by CGP, and by doing so, improve the runtime and accuracy of the synthesized solutions. Our results confirmed that by incrementally improving the fitness estimation along the evolutionary process, we can obtain significant improvements in the CGP convergence speed and in the quality of the synthesized circuits. For all experiments performed, the ABS CGP achieved better, or at least the same, accuracy as the base version of the algorithm while performing significantly fewer evaluations, and being significantly faster. In a number of cases of the presented results, the final BS did not reach a point in the evolutionary process in which all data available was used for evaluating the individuals. These results indicate that some particular combinations of data available, number of generations, and the AI were not optimal for certain cases. Moreover, the current version of ABS CGP has a fixed initial batch size and in all experiments, this value was set to 64 terms. Therefore, as future work, investigating how the initial batch size, the parameter α that controls the batch increase, and the number of generations affect the synthesis for each particular exemplar is an interesting research direction.

Acknowledgments. This work was financed in part by National Council for Scientific and Technological Development - CNPq and the Propesq/UFSC.

References

1. de Abreu, B.A., et al.: Fast logic optimization using decision trees. In: 2021 IEEE International Symposium on Circuits and Systems (ISCAS), pp. 1–5 (2021). https://doi.org/10.1109/ISCAS51556.2021.9401664
2. Barua, H.B., Mondal, K.C.: Approximate computing: a survey of recent trends— bringing greenness to computing and communication. J. Inst. Eng. (India): Series B **100**(6), 619–626 (2019). https://doi.org/10.1007/s40031-019-00418-8
3. Berndt, A., et al.: Accuracy and size trade-off of a cartesian genetic programming flow for logic optimization. In: Proceedings of the 34th Symposium on Integrated Circuits and Systems Design. SBCCI 2021 (2021). https://doi.org/10.1109/SBCCI53441.2021.9529968
4. Berndt, A.A.S., et al.: A CGP-based logic flow: optimizing accuracy and size of approximate circuits. J. Integr. Circ. Syst. **17**(1), 1–12 (2022). https://doi.org/10.29292/jics.v17i1.546
5. Brayton, R.K., Hachtel, G.D., Mcmullen, C.T., Sangiovanni-Vincentelli, A.L.: Logic Minimization Algorithms for VLSI Synthesis. Kluwer Int. Ser. Eng. Comput. Sci. **2**, 1–194 (1984). https://doi.org/10.1007/978-1-4613-2821-6
6. Coudert, O., Sasao, T.: Two-level logic minimization. In: Hassoun, S., Sasao, T. (eds.) Logic Synthesis and Verification. SECS, vol. 654, pp. 1–27. Springer, Boston (2002). https://doi.org/10.1007/978-1-4615-0817-5_1

7. Deng, L.: The MNIST database of handwritten digit images for machine learning research. IEEE Signal Process. Mag. **29**(6), 141–142 (2012). https://doi.org/10.1109/MSP.2012.2211477

8. Doerr, B., Doerr, C.: Optimal parameter choices through self-adjustment: applying the 1/5-th rule in discrete settings. In: Proceedings of the 2015 Annual Conference on Genetic and Evolutionary Computation, pp. 1335–1342 (2015). https://doi.org/10.1145/2739480.2754684

9. Goldman, B.W., Punch, W.F.: Analysis of cartesian genetic programming's evolutionary mechanisms. IEEE Trans. Evol. Comput. **19**(3), 359–373 (2014)

10. Harding, S.: Evolution of image filters on graphics processor units using cartesian genetic programming. In: 2008 IEEE Congress on Evolutionary Computation (IEEE World Congress on Computational Intelligence), pp. 1921–1928. IEEE (2008)

11. Harding, S., Leitner, J., Schmidhuber, J.: Cartesian genetic programming for image processing. In: Riolo, R., Vladislavleva, E., Ritchie, M., Moore, J. (eds.) Genetic programming theory and practice X. (GEVO), pp. 31–44. Springer, Cham (2013). https://doi.org/10.1007/978-1-4614-6846-2_3

12. Karnaugh, M.: The map method for synthesis of combinational logic circuits. Trans. Am. Inst. Electr. Eng. Part I: Commun. Electron. **72**(5), 593–599 (1953). https://doi.org/10.1109/TCE.1953.6371932

13. Khan, M.M., Ahmad, A.M., Khan, G.M., Miller, J.F.: Fast learning neural networks using cartesian genetic programming. Neurocomputing **121**, 274–289 (2013)

14. Khan, M.M., Khan, G.M., Miller, J.F.: Evolution of neural networks using cartesian genetic programming. In: IEEE Congress on Evolutionary Computation, pp. 1–8. IEEE (2010)

15. Krizhevsky, A., Hinton, G., et al.: Learning multiple layers of features from tiny images (2009)

16. Manazir, A., Raza, K.: Recent developments in cartesian genetic programming and its variants. ACM Comput. Surv. (CSUR) **51**(6), 1–29 (2019). https://doi.org/10.1145/3275518

17. Milano, N., Pagliuca, P., Nolfi, S.: Robustness, evolvability and phenotypic complexity: insights from evolving digital circuits. Evol. Intel. **12**(1), 83–95 (2019). https://doi.org/10.1007/s12065-018-00197-z

18. Miller, J.F.: Cartesian genetic programming: its status and future. Genet. Program Evolvable Mach. **21**(1), 129–168 (2019). https://doi.org/10.1007/s10710-019-09360-6

19. Miyasaka, Y., Zhang, X., Yu, M., Yi, Q., Fujita, M.: Logic synthesis for generalization and learning addition. In: 2021 Design, Automation Test in Europe Conference Exhibition (DATE), pp. 1032–1037 (2021). https://doi.org/10.23919/DATE51398.2021.9474169

20. Picek, S., Carlet, C., Guilley, S., Miller, J.F., Jakobovic, D.: Evolutionary algorithms for Boolean functions in diverse domains of cryptography. Evol. Comput. **24**(4), 667–694 (2016)

21. Picek, S., Jakobovic, D., Miller, J.F., Batina, L., Cupic, M.: Cryptographic Boolean functions: one output, many design criteria. Appl. Soft Comput. **40**, 635–653 (2016)

22. Quine, W.V.: A way to simplify truth functions. Am. Math. Mon. **62**(9), 627–631 (1955). https://doi.org/10.1080/00029890.1955.11988710

23. Rai, S., et al.: Logic synthesis meets machine learning: trading exactness for generalization. In: 2021 Design, Automation & Test in Europe Conference & Exhibition (DATE). IEEE (2021). https://doi.org/10.23919/DATE51398.2021.9473972

24. Riener, H., Haaswijk, W., Mishchenko, A., De Micheli, G., Soeken, M.: On-the-fly and DAG-aware: rewriting Boolean networks with exact synthesis. In: 2019 Design, Automation Test in Europe Conference Exhibition (DATE), pp. 1649–1654 (2019). https://doi.org/10.23919/DATE.2019.8715185
25. Rudell, R.L., Sangiovanni-Vincentelli, A.: Multiple-valued minimization for PLA optimization. IEEE Trans. Comput. Aided Des. Integr. Circ. Syst. **6**(5), 727–750 (1987). https://doi.org/10.1109/TCAD.1987.1270318
26. Scarabottolo, I., et al.: Approximate logic synthesis: a survey. In: Proceedings of the IEEE, pp. 1–19 (2020). https://doi.org/10.1109/JPROC.2020.3014430
27. Sekanina, L., Harding, S.L., Banzhaf, W., Kowaliw, T.: Image processing and CGP. In: Miller, J. (ed.) Cartesian genetic programming. NCS, pp. 181–215. Springer, Heidelberg (2011). https://doi.org/10.1007/978-3-642-17310-3_6
28. Suganuma, M., et al.: Evolution of deep convolutional neural networks using cartesian genetic programming. Evol. Comput. **28**(1), 141–163 (2020). https://doi.org/10.1162/evco_a_00253
29. Venkataramani, S., Kozhikkottu, V., Sabne, A., Roy, K., Raghunathan, A.: Logic synthesis of approximate circuits. IEEE Trans. Comput. Aided Des. Integr. Circ. Syst. **39**, 2503–2515 (2019)

Faster Convergence with Lexicase Selection in Tree-Based Automated Machine Learning

Nicholas Matsumoto[1], Anil Kumar Saini[1], Pedro Ribeiro[1], Hyunjun Choi[1], Alena Orlenko[1], Leo-Pekka Lyytikäinen[2], Jari O. Laurikka[3], Terho Lehtimäki[2], Sandra Batista[1(✉)], and Jason H. Moore[1(✉)]

[1] Cedars-Sinai Medical Center, Los Angeles, CA 90048, USA
{nicholas.matsumoto,anil.saini,pedro.ribeiro,hyunjun.choi,
alena.orlenko,sandra.batista}@cshs.org, jason.moore@csmc.edu
[2] Tampere University, Tampere, Finland
leo-pekka.lyytikainen@tuni.fi, terho.lehtimaki@uta.fi
[3] Sydänsairaala Hospital, Tampere, Finland
jari.laurikka@sydansairaala.fi

Abstract. In many evolutionary computation systems, parent selection methods can affect, among other things, convergence to a solution. In this paper, we present a study comparing the role of two commonly used parent selection methods in evolving machine learning pipelines in an automated machine learning system called Tree-based Pipeline Optimization Tool (TPOT). Specifically, we demonstrate, using experiments on multiple datasets, that lexicase selection leads to significantly faster convergence as compared to NSGA-II in TPOT. We also compare the exploration of parts of the search space by these selection methods using a trie data structure that contains information about the pipelines explored in a particular run.

Keywords: Parent Selection · NSGA-II · Lexicase · Convergence · Trie

1 Introduction

In evolutionary computation (EC) systems, just like many machine learning (ML) algorithms such as neural networks, the time and resources required to converge to an acceptable solution are important. For tasks such as classification and regression, faster convergence to an acceptable solution may form the basis of early stopping criteria in resource-constrained environments. However, even the notion of convergence for EC systems requires careful specification of what convergence means and how it will be measured. This also requires careful consideration of effects of specific aspects of the implementations of EC systems.

Parent selection methods are used in EC systems to select parents from the current generation that are used to produce the next generation. These methods differ in the way they use the fitness of the individuals on different objectives. Consequently, parent selection methods can affect various properties of

© The Author(s), under exclusive license to Springer Nature Switzerland AG 2023
G. Pappa et al. (Eds.): EuroGP 2023, LNCS 13986, pp. 165–181, 2023.
https://doi.org/10.1007/978-3-031-29573-7_11

the evolving population. In this paper, we conduct a case study on the effect of parent selection algorithms on convergence in Tree-based Pipeline Optimization Tool (TPOT) [16], an automated machine learning system that uses genetic programming (GP) to evolve machine learning pipelines for classification and regression tasks. Specifically, through experiments, we demonstrate that using lexicase selection, as compared to the default selection algorithm, Non-dominated Sorting Genetic Algorithm II (NSGA-II), leads to faster convergence without loss of accuracy on the holdout set. We use a variant of lexicase selection called automatic ϵ-lexicase selection that is used when the objectives are real-valued. We define convergence as the generation when the best model in the population (based on cross-validation accuracy on the training data) attains at least 99% of the cross-validation accuracy of the best model in the final generation of that GP run.

For our experiments we use synthetic and real clinical datasets. The DIverse and GENerative ML Benchmark (DIGEN) [18] provides 40 synthetic binary classification datasets that were designed to produce a diverse distribution of performance scores for different popular machine learning algorithms. The other one is the Angiography and Genes Study (ANGES) dataset that came from the study conducted at Tampere University Hospital, Finland. The study includes 925 patients and provides their clinical data, measured coronary arteries angiography, and metabolic profiling.

In our experiments with the DIGEN and ANGES datasets, we found that, on average, TPOT with lexicase selection converges multiple generations earlier compared to TPOT with NSGA-II, without sacrificing on the holdout set accuracy or changing the number of machine learning methods used in the best models of the final generations.

We also looked at the exploratory behavior of TPOT with the selection methods using an *exploration trie* that visualizes sequences of machine learning operators in pipelines TPOT explores. While the tries for both selection methods had similar depth, we found that, with NSGA-II, tries for TPOT are larger with more balanced branching. Lexicase selection, on the other hand, leads to smaller tries with more branching in certain directions. The selection methods, therefore, exhibit different behaviors when it comes to prefixes of sequences of machine learning operators used in pipelines. NSGA-II tends to explore pipelines with many different prefixes, but lexicase tends to focus on a few prefixes and explore around them.

Various sections in this paper are organized as follows. We start with related work in the GP literature in Sect. 2. Then after describing the TPOT system and the selection methods studied in this paper in Sect. 3, we describe the experimental design and the datasets used in Sect. 4. The results are presented in Sect. 5 and discussed in Sect. 6.

2   Related Work

Convergence. In computer science literature, the term convergence is defined in a variety of contexts. In neural networks, for example, parameters of the networks are said to be converging when the corresponding weights are not changing on account of gradient descent or other updates [15]. In genetic programming, convergence is sometimes defined in terms of diversity, i.e., if the proportion of unique individuals in the population is very low, the population is said to have converged [2,10]. The term 'uniqueness' may itself be defined in terms of the individuals which have the same outputs, same size, or any other characteristics. In this study, however, we define convergence in a somewhat related way as the generation after which there is no visible change in the performance of the best individual in the population.

Comparison Among Selection Methods. Multiple works analyze the effect of parent selection methods on certain properties of evolving populations such as their diversity (e.g., number of unique genotypes) and modularity. In Metevier et al. [13], the authors compared lexicase selection, tournament selection, and fitness proportionate selection on the success rate, the number of generations used to find a solution, and structural diversity. In Saini and Spector [19], the authors demonstrated that lexicase selection, compared to other selection methods, leads to a significantly greater number of individuals with looping instructions in the evolving population.

In multi-objective and many-objective optimization settings, lexicase selection and its variants have been compared to NSGA-II and other methods. In evolving gaits in quadrupedal animats, NSGA-II significantly outperforms lexicase selection when distance traveled, efficiency, and vertical torso movement were used as objectives during evolution [14]. When experimenting on many-objective optimization problems like DTLZ [4], lexicase selection outperforms NSGA-II, especially when the number of objectives is more than 5 [8].

Analyzing GP Runs. Various works [1,12] try to summarize information about GP runs in a graphical form. For example, McPhee et al. [12] presents a way to record information about genetic ancestry in a graph database, which basically means recording the individuals in a particular run which contributed to the material in the final solution. In this paper, we introduce a different method of recording information in exploration tries.

3   Methods

3.1   Review of TPOT

TPOT is a tree-based genetic programming system implemented in Python that searches for machine learning pipelines for classification and regression tasks. A given evolving individual is a machine learning pipeline including its methods and hyperparameters represented as a tree. Within pipelines, different machine learning methods or *operators* such as Random Forest, PCA, and RFE, may be applied in composition on the data. TPOT uses the Distributed Evolutionary Algorithm in Python (DEAP) [6] framework to evolve pipelines by using the variation operators and selection methods as implemented in DEAP. Using

DEAP's use of primitives and terminals, we describe primitives in TPOT as single machine learning modules or operators and the terminals as their respective hyperparameters. For classification tasks, TPOT currently supports 32 operators with corresponding hyperparameters.

3.2   Parent Selection Algorithms

The procedure to generate individuals for the next generation from the current evaluated individuals is given in Algorithm 1. From the current population, we generate the same number of individuals through crossover and mutation. For crossover, we search in the current population for the individuals which share at least one primitive using a function *ChooseEligibleIndividuals()*. Then after performing crossover, we choose the first child. If that child has already appeared as one of the offspring, we discard it, and instead apply mutation on a randomly chosen individual from the population. Note that *ChooseRandomly()* chooses the individuals uniformly at random. The mutation operation proceeds in the regular fashion.

After applying the variation operators, we input the parent and offspring individuals to one of the selection methods. The description of the different selection methods used in this study is given in the following sections.

Algorithm 1. Selection Procedure

1: **procedure** SELECT($curr\_pop, cross\_prob, mut\_prob, method, pop\_size$)
2:     $offspring=[]$
3:     **while** $size(offspring) \neq pop\_size$ **do**
4:         rand = rand()   ▷ Generate a random number from $[0,1)$
5:         **if** $rand < cross\_prob$ **then**
6:             $ind1, ind2$ = ChooseEligibleIndividuals(curr_pop)
7:             $ind3, ind4$ = Crossover(ind1, ind2)
8:             **if** $ind3$ is a duplicate from *offspring* **then**
9:                 $ind3$ = Mutation(ChooseRandomly(curr_pop))
10:             offspring.append($ind3$)
11:         **else**
12:             $ind1$ = ChooseRandomly(curr_pop)
13:             $ind2$ = Mutation(ind1)
14:             offspring.append($ind2$)
15:     **if** method=NSGA2 **then**
16:         parents = NSGA2(curr_pop + offspring, pop_size)
17:     **else**
18:         parents = ϵLex(curr_pop + offspring, pop_size)

NSGA-II: Non-dominated Sorting Genetic Algorithm II (NSGA-II) is a multi-objective evolutionary algorithm that has been widely used for optimization problems with multiple objectives [3].

For every selection event, as shown in Algorithm 2, we combine the parent and offspring populations and perform non-dominated sorting. The sorting will lead to the combined population getting organized into 'fronts', whereby individuals in front 1 dominate[1] individuals in front 2, and so on. Consequently, in Algorithm 2, *ParetoFrontsNonDominatedSort()* returns a list of lists of individuals. For all of the individuals in various fronts, we also calculate a metric called 'crowding distance' which is a measure of the density of individuals around a particular individual in the objective space (see [3] for more details).

Then, we start adding individuals from the fronts to the new population until its size reaches n individuals (with n being the size of population, *pop_size*). If the size of the first front is more than the population size, we sort the individuals by crowding distance and keep the first n individuals. Otherwise, we add the whole front to the population and look at the second front. If the total individuals in fronts 1 and 2 is more than n, we sort the second front by crowding distance and keep the first $(n - size(front1))$ individuals. Otherwise, we look at the subsequent fronts, and repeat the process as summarized in lines 4–8 in Algorithm 2.

Algorithm 2. NSGA-II

1: **procedure** NSGA2(pop, pop_size)
2:     $parents = [], i = 1$
3:     $fronts = $ ParetoFrontsNonDominatedSort(pop)   ▷ list of lists of individuals
4:     **while** $size(parents) + size(fronts[i]) \leq pop\_size$ **do**
5:         $parents = parents + fronts[i]$
6:         $i = i + 1$
7:     sort $fronts[i]$ using crowding distance
8:     $parents = parents + fronts[i][: pop\_size - size(parents)]$

Lexicase Selection. Lexicase selection [7,13] is a parent selection method used in genetic programming and other evolutionary computation techniques. Our implementation of lexicase selection is given in Algorithm 3. For every selection event, first, we randomly shuffle the list of objectives on which a given individual is evaluated. Then, the pool of candidates, which initially contains the whole population, is whittled down based on their performance on the objectives: the individuals that perform the best on the first objective are kept in the pool and others are removed. Then from this pool, the individuals that perform the best on the second objective are kept in the pool and others are removed, and so on. The process is repeated until we are left with only one individual in the pool, or, we are out of objectives. In the second case, we randomly choose one of the individuals as the parent.

ϵ-lexicase selection [9] and automatic ϵ-lexicase selection [9] are variants of lexicase selection that have been developed for the settings where individuals can

---

[1] Individual i_1 dominates i_2 if i_1 is better than or the same as i_2 on all objectives and strictly better than i_2 on at least one objective.

have real-valued errors or fitness values, as for example, in symbolic regression. In ϵ-lexicase selection, for a given objective, all the individuals within ϵ of the fitness of the best individual in the current pool are kept in the pool and the rest are removed. The value of ϵ is a parameter of the algorithm. It is specified by the user and is fixed during the whole process. Automatic ϵ-lexicase selection has the same process as ϵ-lexicase selection, but the value of ϵ is automatically determined by the algorithm: the median absolute deviation (MAD) of the fitness values of the individuals in the current pool on a selected objective. In other words, $\epsilon_t = median(|x_1 - median(x)|, |x_2 - median(x)|, ...)$, where $x_1, x_2, ...$ are the fitness values of the individuals in the *current pool* on objective t.

We ran experiments on both the regular lexicase and the automatic ϵ-lexicase selection methods. However, we include results for automatic ϵ-lexicase (simply called lexicase from here onwards) only since there was no substantial difference between the two in terms of convergence, accuracy on the holdout set, and the number of operators.

Algorithm 3. ϵ-Lexicase Selection

1: **procedure** ϵLEX(pop, pop_size)
2:      $parents = []$
3:      **while** $size(parents) \neq pop\_size$ **do**
4:          $curr\_pool = pop$
5:          $curr\_objectives =$ objectives sorted in a random order
6:          **for** obj in $curr\_objectives$ **do**
7:              $best\_val =$ best value on obj in $curr\_pool$
8:              $\epsilon =$ median absolute deviation of obj values for $curr\_pool$
9:              $curr\_pool =$ inds. from $curr\_pool$ with obj values within ϵ of $best\_val$
10:         **if** $size(curr\_pool \neq 1)$ **then**
11:             $parents = parents +$ one individual from $curr\_pool$ chosen randomly
12:         **else**
13:             $parents = parents + curr\_pool$

---

4   Experimental Set-Up

We begin by summarizing the datasets that we used for our experiments in the following subsection. In subsequent subsections, we give implementation details and parameters for our experiments, the metrics used to evaluate convergence, and finally, the construction and metrics of exploration tries used to examine the behavior of the selection algorithms.

4.1   Datasets

DIGEN: The DIverse and GENerative ML Benchmark (DIGEN) is a set of 40 synthetic, binary classification data sets [18]. Each dataset consists of an 800 sample training set and a 200 sample testing set. There are 10 features independently generated from a Gaussian distribution. The binary target is generated with a unique generative function. The set of 40 generative functions for the datasets were designed to produce a diverse distribution of performance scores and relative ranking for eight popular machine learning algorithms such as Decision Trees and Gradient Boosting.

After running TPOT with both selection methods on all 40 DIGEN sets, we noticed that within the initial population, TPOT achieved a high average balanced accuracy score for 27 DIGEN datasets on the training set, and there was little change in the later generations. For the present study, since we need a sufficient difference to determine the efficacy of each selection algorithm, we will focus on only 13 of the 40 DIGEN data sets which provided at least a 10% increase from the average balanced accuracy across all pipelines from the first generation to twentieth generation pipelines.

ANGES: The Angiography and Genes Study (ANGES) includes data on 925 Finnish subjects with coronary angiography, or specifically, the evaluation of the degree of coronary artery stenosis, and targeted metabolic profiling (for detailed study population description, see [17]). ANGES dataset contains 73 metabolic and clinical features, and one binary outcome for coronary artery disease (CAD) status, where patients were considered cases for the disease when any major coronary artery is detecting stenosis greater than 50% and controls otherwise. The dataset was split into training (75 percent) and testing (25 percent) sets prior to the analysis.

4.2   Implementation

As mentioned earlier, we conduct our experiments in an AutoML system called TPOT[2]. Except for the selection method, the implementation of TPOT was kept constant as we test the effects of NSGA-II and lexicase on the evolution of machine learning pipelines. For both methods, we use the following objectives:

1. Maximize the balanced accuracy calculated from the 10-fold cross-validation scores on the training data.
2. Minimize the number of machine learning operators used in a given pipeline.

During crossover on two individual pipelines, one of the shared primitives is chosen randomly and the subtrees at nodes corresponding to the chosen primitive in both individual trees are swapped.

For every mutation operation, one of the following mutation strategies are chosen with equal probability:

1. *mutInsert* - Insert a randomly chosen primitive into the tree.

---

[2] https://github.com/EpistasisLab/exploration-trie-tpot including code and supplementary material.

2. *mutShrink* - Remove an a primitive from the tree. If there is only one primitive in the pipeline, the removal will not take place; instead, the mutInsert or mutNodeReplacement technique would be applied with equal probability.
3. *mutNodeReplacement* - Replace a randomly selected node in the tree; if the node is a primitive one, it is replaced by a new randomly chosen primitive node, otherwise a terminal node is mutated by changing the values of hyperparameters in that node.

The rates of crossover and mutation are 0.1 and 0.9, respectively. This means effectively, 10% of offspring are produced by crossover and the rest using mutation operator (see lines 3–14 in Algorithm 1). Table 1 summarizes the parameters used for TPOT for each dataset for the experiments.

Table 1. Genetic Programming parameters.

Parameter	Values (ANGES data)	Values (DIGEN data)
Population size (initial gen.)	100	80
Population size (later gen.)	50	40
Number of generations	100	20
Number of runs per selection method	50	40
Mutation operator	mutInsert, mutShrink, mutNodeReplacement	mutInsert, mutShrink, mutNodeReplacement
Mutation rate	90%	90%
Crossover operator	one-point crossover	one-point crossover
Crossover rate	10%	10%

4.3   Evaluating Convergence

To evaluate convergence for a TPOT run on a single random seed, we first define our 'convergence point' to be the first generation at which any individual pipeline in the population reaches 99% of the best balanced accuracy found in any individual pipeline in the final generation of that GP run. Therefore, we will have a convergence point for each run launched for a particular data set and selection method combination. Note that we use only one of the objectives while defining convergence, since in most settings, accuracy on the training set is considered a primary objective with objectives such as size of the model considered secondary. Using only one objective while defining convergence does not change the basic behaviour of the algorithms studied here.

We record the balanced accuracy and the number of operators for each individual pipeline in each generation of a TPOT run. Then using the TPOT runs as our experimental samples, we test for any statistically significant difference in the convergence points, the balanced accuracy on the holdout set, and the number of operators in pipelines at the convergence points for the selection methods. The non-parametric Mann-Whitney-U test is used for these comparisons. Another way we compare the accuracy of the best performing classifiers from each selection method is by constructing binomial confidence intervals for the balanced accuracy of the classifier on the holdout set [5, 20].

4.4   Exploration of Pipelines

The convergence to a solution in genetic programming is often affected by the extent of the search space being explored. Therefore, we implement an *exploration trie* to investigate the explored search subspace in TPOT. The root node of the trie represents an empty sequence and all other nodes represent a machine learning operator. A path from the root to a node in the trie represents a sequence of machine learning operators explored during the run, and as with more general tries, any sequences that share the same prefix of operators will share the same path from the root in the trie. The trie represents the space of possible sequences of machine learning operators considered during the TPOT run and may not correspond directly to the evaluated pipelines with hyperparameters. Therefore, we are using trie graphs only to *compare* the exploration capabilities of the selection methods, instead of summarizing the pipelines explored by them.

Let us consider an example exploration trie. In Fig. 1, we trace the construction of the TPOT trie for a TPOT run that has explored the following set of sequences: { *LogisticRegression(X), LogisticRegression(PCA(X)), DecisionTreeClassifier(X), LogisticRegression(Normalizer(X))* }. First, starting from an empty sequence representing a root node, TPOT explores the sequence consisting of only logistic regression on the data, X. When the second sequence applies PCA to the data before logistic regression is explored, a node for PCA is added to the trie following logistic regression. When TPOT explores the decision tree classifier on the data, a new branch is added to the trie directly from the root. Finally, when TPOT explores a sequence that first normalizes the data and applies logistic regression, a node for a normalizer is added to the trie following logistic regression.

We use graph metrics on the trie such as nodal global efficiency, leaf-to-node ratio, and the number of unique nodes to better understand the evolutionary selective pressure that lexicase and NSGA-II provide.

1. **Nodal Global Efficiency:** To get a measure of the length of the unique sequences explored, we use the metric called the nodal global efficiency, g. It is the average inverse shortest distances between a given node i and all other nodes in the trie. If T is the set of nodes in trie excluding the node i, n is the number of nodes in the trie, and d_{ij} is the shortest distance between node i and j, the nodal global efficiency [11] for node i can be defined as: $g_i = \frac{1}{(n-1)} \sum_{j \in T} \frac{1}{d_{ij}}$. In this work, we use nodal global efficiency for the root node (simply called 'nodal global efficiency' in this paper). This metric has a value between 0 and 1, where 1 indicates the node is directly connected to all other nodes and a value closer to 0 indicates longer average shortest paths to other nodes (0 would mean the node is disconnected from all other nodes). We calculated the root nodal global efficiencies by constructing the

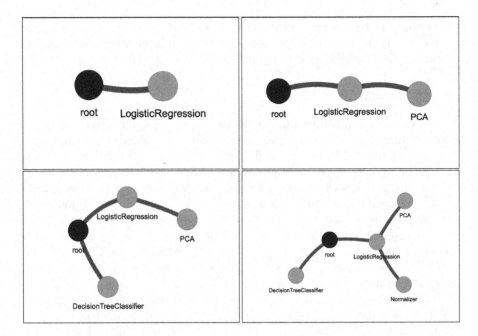

Fig. 1. Sequences inserted in order and the resulting trie graphs after each insertion: LogisticRegression (X), LogisticRegression (PCA (X)), DecisionTreeClassifier (X), LogisticRegression (Normalizer (X)). The root node is shown in red. (Color figure online)

adjacency matrices for tries assuming uniform weights on edges and using Dijkstra's algorithm [11].

2. **Number of trie nodes:** The number of total nodes excluding the root in the exploration trie is the total number of machine learning operators explored in sequences.

3. **Leaf-to-node ratio:** The leaf nodes represent the ends of branches in the trie or the endpoints of the sequences of operators. Accordingly, the leaf-to-node ratio measures the proportion of branches in the trie relative to the total number of nodes. The leaf-to-node ratio is between 0 and 1. The maximum value of leaf-to-node ratio means the root node is directly connected to all other nodes by single edges whereas the value is minimum if there is one linear sequence of all nodes. The leaf-to-node ratio captures the notion of branching in the trie.

5   Results

In Table 2, we show the mean convergence points for various selection methods averaged over 40 runs for each DIGEN dataset and over 50 runs for the ANGES dataset. The procedure to calculate the convergence points has been described in the previous section. To check whether the differences in the convergence points are statistically significant or not, we used a non-parametric test called Mann-Whitney-U test and reported the resulting p-values in the same table. We also report the accuracy of the models on the holdout set found at the convergence points. We show both the mean accuracy across the respective runs and the p-values for the Mann-Whitney-U test applied to the accuracy values on the holdout set. In Fig. 2, we report the balanced accuracy values of the best model (based on training accuracy) per generation across multiple runs for ANGES and select DIGEN datasets and for the models with the best accuracy, demonstrate comparable accuracy across methods on the holdout set with 95% confidence intervals.

Table 2. (a) Mean convergence points for various selection methods on different datasets. The third column shows the p-values obtained by applying the Mann-Whitney-U test on the distribution of values from both selection methods in the corresponding column. (b) Mean accuracy on the holdout set at convergence points for various selection methods on different datasets. The sixth column shows the p-values obtained by applying the Mann-Whitney-U test on the distribution of values from both selection methods in the corresponding columns.

Dataset	NSGA-II (a)	Lexicase (a)	p-value (a)	NSGA-II (b)	Lexicase (b)	p-value (b)
ANGES	16.44	10.10	8.85E−03	0.73	0.73	0.59
DIGEN-2	7.65	4.70	1.26E−18	0.95	0.92	0.06
DIGEN-4	8.97	5.97	3.36E−05	0.95	0.95	0.84
DIGEN-7	10.30	5.60	4.99E−07	0.97	0.96	0.90
DIGEN-14	10.10	5.60	1.86E−06	0.97	0.97	0.85
DIGEN-23	10.32	6.15	2.40E−05	0.96	0.94	0.28
DIGEN-24	4.25	2.47	1.38E−04	0.94	0.94	0.42
DIGEN-25	8.27	5.37	3.66E−05	0.95	0.96	0.31
DIGEN-27	5.42	2.82	4.08E−05	0.93	0.93	0.79
DIGEN-28	8.72	5.37	8.83E−06	0.93	0.94	0.46
DIGEN-30	10.70	8.95	4.79E−02	0.95	0.94	0.34
DIGEN-32	3.35	3.22	6.40E−01	0.93	0.94	0.98
DIGEN-35	5.92	3.45	1.41E−03	0.94	0.95	0.85
DIGEN-40	5.35	3.42	2.73E−03	0.94	0.91	0.31

5.1   DIGEN Datasets

For 12 of the 13 DIGEN datasets tested (all excluding DIGEN 32), the selection methods have statistically significant impact on the number of generations used to reach the best cross validation score. On average across all 13 datasets, the convergence point for lexicase is about 2.96 generations sooner. The largest absolute difference was in DIGEN 7 where the average convergence point for lexicase was 4.7 generations sooner than NSGA-II. The selection methods mostly did not have statistically significant differences in maximum balanced accuracy scores on the holdout set. The few datasets that were statistically significant still had differences of less than 0.01. There was also no statistically significant difference

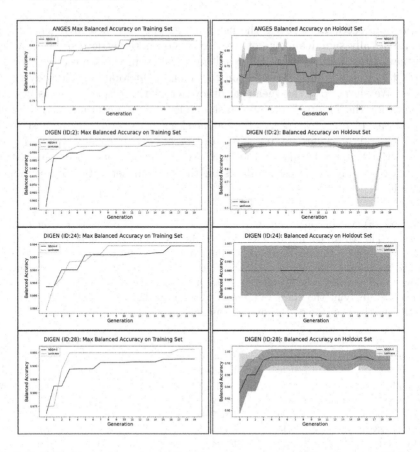

Fig. 2. Left: Best model based on training accuracy per generation across multiple runs of TPOT for ANGES (50 runs) and select DIGEN datasets (40 runs). Right: Holdout scores with 95% confidence intervals (±1.96 sd) for the best models based on training accuracy per generation for ANGES and DIGEN.

in the number of operators used for the best models at the convergence points (Table 1 in supplementary material). Both selection methods ranged between 2 and 4 operators depending on the datasets. As shown in Fig. 2, there does not appear to be any significant difference in the accuracy on the holdout set for lexicase and NSGA-II for DIGEN datasets.

For all experiments we calculated the median values of leaf-to-node ratio, nodal global efficiency, and the number of trie nodes for TPOT runs and these metrics are summarized in the supplementary material Table 2. As an example in Fig. 4, we plot the median values of leaf-to-node ratio, nodal global efficiency, and the number of trie nodes for TPOT runs on the DIGEN-24 and ANGES datasets. The plots for other DIGEN datasets look similar.

When we examine the TPOT tries for the selection variants, NSGA-II explored the largest set of sequences of machine learning operators. To illus-

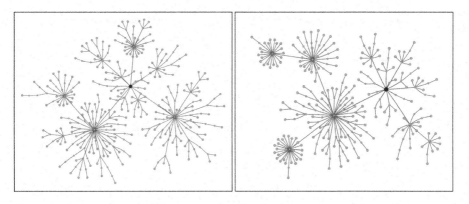

Fig. 3. Single Run of TPOT on DIGEN-24 dataset using NSGA-II (left), lexicase (right). Red node indicates the root node. (Color figure online)

trate this with an example in Fig. 3 and Fig. 4 for DIGEN-24, this can be viewed from the number of nodes in each respective graph. The length of sequences of operators observed was also greater for NSGA-II on average than for lexicase. Lexicase had slightly greater root nodal global efficiency compared to NSGA-II (as for example for the DIGEN-24 data set in Fig. 4) indicating shorter sequences of operators on average. However, the leaf-to-node ratio, the metric of relative branching, drastically differs in the methods. Lexicase maintains the leaf-to-node ratio at nearly 70% and NSGA-II permits the leaf-to-node ratio to drop to nearly 58% (as shown in Fig. 4).

5.2 ANGES Datasets

On the ANGES dataset the mean convergence point of lexicase of 10.10 generations was significantly shorter than the mean convergence point of NSGA-II at 16.44 generations while there was no statistically significant difference in mean balanced accuracy on the holdout set (as shown in Table 2) or the number of operators of the best models (Table 1 in supplemental material). When we consider the maximum balanced accuracy on the ANGES training set and the balanced accuracy of the best performing models on the ANGES holdout set in Fig. 2, both methods reach the same balanced accuracy on the training set and the same balanced accuracy on the holdout set with entirely overlapping confidence intervals at the final generations.

In terms of the exploration tries on the ANGES dataset in Fig. 4, the median number of nodes in the tries for NSGA-II were more than double the median number of nodes in the tries for lexicase whereas the leaf-to-node ratio for lexicase remained at nearly as 63% while for NSGA-II it fell to only 48%. The median root global efficiency of both methods declined to 33% for lexicase and 25% for NSGA-II indicating that both methods explores longer sequences of operators for the ANGES dataset. However, the greater leaf-to-node ratio of lexicase and fewer total nodes indicates that even on the ANGES dataset the exploration of

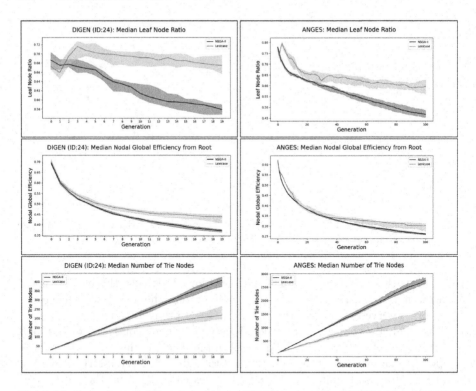

Fig. 4. Median values of different metrics on TPOT runs on the DIGEN-24 and ANGES datasets. From top to bottom: Leaf-to-node ratio, nodal global efficiency from the root node, total number of nodes in the trie.

the sequences of machine learning operators ends much sooner than in case of NSGA-II. In other words, NSGA-II has the tendency of lengthening sequences of ML operators of many different prefixes, but lexicase usually prefers fewer prefixes and lengthens them.

Interpretation of machine learning models is a crucial component of predictive analysis in biomedical studies. So in order to verify that TPOT with both lexicase and NSGA-II are using similar features in their best models, we perform the following analysis. We use permutation feature importance (PFI) analysis that generates the informative coefficients for ANGES models. We perform this analysis on the best models across all the runs based on the accuracy on the training dataset. While PFI can vary based on different initializations, PFI rank shows that most of the top 10 features in both models are the same. Both models selected features in their models that are known well-supported clinical risk factors or predictors for CAD. Detailed information about the top 10 features is given in supplementary material Sect. 3.

6 Discussion

When the selection methods are maximizing the balanced accuracy on cross-fold validation and minimizing the number of operators in pipelines, lexicase finds models of comparable balanced accuracy on the holdout set and the number of operators to those found by NSGA-II but in fewer generations.

To examine the behavior of the selection methods, we introduced the concept of an exploration trie that represents the explored search space of machine learning operator sequences for a TPOT run. Using graph metrics such as nodal global efficiency, leaf-to-node ratio, and total number of nodes in exploration tries, we can further understand the evolutionary selective pressure that lexicase and NSGA-II provide. Observing an overall lower leaf-to-node ratio and higher total number of nodes in exploration tries, NSGA-II explores a larger set of longer machine learning operator sequences compared to lexicase. Lexicase, which has an overall higher leaf-to-node ratio in exploration tries, creates more leaves around an optimal sequence of machine learning operators than is done in the case of NSGA-II. Comparatively, NSGA-II often keeps the best models for every possible number of operators. This means longer sequences of machine learning operators are considered for pipelines and consequentially longer pipelines are evaluated. Lexicase effectively chooses half its population to have minimal number of nodes while the other half is selected for the best accuracy. It does not automatically keep sequences with a greater number of machine learning operators.

In TPOT, evaluating machine learning pipelines is a resource demanding task. As the sequences of machine learning operators increase and corresponding pipelines include more operators, there are higher computational costs and longer evaluation times. In lexicase the selective pressures of fewer operators in pipelines and highly accurate solutions may lead to the evaluation of fewer pipelines of shorter length, which would in turn reduce the resource costs. NSGA-II, due to its diversity-preserving behavior, is more likely to evaluate a wider variety of pipelines of different lengths, which can often increase resource costs. With lexicase the reduction of average computational resources can permit more allocation of resources for increasing population size or increasing the number of generations if necessary.

Acknowledgements. This work is supported by National Institute of Health grants R01 LM010098 and R01 AG066833.

References

1. Burlacu, B., Affenzeller, M., Kommenda, M., Winkler, S., Kronberger, G.: Visualization of genetic lineages and inheritance information in genetic programming. In: Proceedings of the 15th Annual Conference Companion on Genetic and Evolutionary Computation, pp. 1351–1358 (2013)
2. Ciesielski, V., Mawhinney, D.: Prevention of early convergence in genetic programming by replacement of similar programs. In: Proceedings of the 2002 Congress

on Evolutionary Computation, CEC 2002 (Cat. No. 02TH8600), vol. 1, pp. 67–72. IEEE (2002)

3. Deb, K., Pratap, A., Agarwal, S., Meyarivan, T.: A fast and elitist multiobjective genetic algorithm: NSGA-II. IEEE Trans. Evol. Comput. **6**(2), 182–197 (2002)

4. Deb, K., Thiele, L., Laumanns, M., Zitzler, E.: Scalable test problems for evolutionary multiobjective optimization. In: Abraham, A., Jain, L., Goldberg, R. (eds.) Evolutionary Multiobjective Optimization. Advanced Information and Knowledge Processing, pp. 105–145. Springer, London (2005). https://doi.org/10.1007/1-84628-137-7_6

5. Dietterich, T.: Approximate statistical tests for comparing supervised classification learning algorithms. Neural Comput. **10**, 1895–1923 (1998)

6. Fortin, F.A., De Rainville, F.M., Gardner, M.A., Parizeau, M., Gagné, C.: DEAP: evolutionary algorithms made easy. J. Mach. Learn. Res. **13**, 2171–2175 (2012)

7. Helmuth, T., Spector, L., Matheson, J.: Solving uncompromising problems with lexicase selection. IEEE Trans. Evol. Comput. **19**(5), 630–643 (2014)

8. La Cava, W., Moore, J.H.: An analysis of ϵ-lexicase selection for large-scale many-objective optimization. In: Proceedings of the Genetic and Evolutionary Computation Conference Companion, pp. 185–186 (2018)

9. La Cava, W., Spector, L., Danai, K.: Epsilon-lexicase selection for regression. In: Proceedings of the Genetic and Evolutionary Computation Conference 2016, pp. 741–748 (2016)

10. Langdon, W.B.: Genetic programming convergence. Genet. Program Evolvable Mach. **23**(1), 71–104 (2022)

11. Latora, V., Marchiori, M.: Efficient behavior of small-world networks. Phys. Rev. Lett. **87**, 198701 (2001). https://doi.org/10.1103/PhysRevLett.87.198701, https://link.aps.org/doi/10.1103/PhysRevLett.87.198701

12. McPhee, N.F., Finzel, M.D., Casale, M.M., Helmuth, T., Spector, L.: A detailed analysis of a PushGP run. In: Riolo, R., Worzel, B., Goldman, B., Tozier, B. (eds.) Genetic Programming Theory and Practice XIV. GEC, pp. 65–83. Springer, Cham (2018). https://doi.org/10.1007/978-3-319-97088-2_5

13. Metevier, B., Saini, A.K., Spector, L.: Lexicase selection beyond genetic programming. In: Banzhaf, W., Spector, L., Sheneman, L. (eds.) Genetic Programming Theory and Practice XVI. GEC, pp. 123–136. Springer, Cham (2019). https://doi.org/10.1007/978-3-030-04735-1_7

14. Moore, J.M., McKinley, P.K.: A comparison of multiobjective algorithms in evolving quadrupedal gaits. In: Tuci, E., Giagkos, A., Wilson, M., Hallam, J. (eds.) SAB 2016. LNCS (LNAI), vol. 9825, pp. 157–169. Springer, Cham (2016). https://doi.org/10.1007/978-3-319-43488-9_15

15. Oh, H., et al.: Convergence-aware neural network training. In: 2020 57th ACM/IEEE Design Automation Conference (DAC), pp. 1–6. IEEE (2020)

16. Olson, R.S., Urbanowicz, R.J., Andrews, P.C., Lavender, N.A., Kidd, L.C., Moore, J.H.: Automating biomedical data science through tree-based pipeline optimization. In: Squillero, G., Burelli, P. (eds.) EvoApplications 2016. LNCS, vol. 9597, pp. 123–137. Springer, Cham (2016). https://doi.org/10.1007/978-3-319-31204-0_9

17. Orlenko, A., et al.: Model selection for metabolomics: predicting diagnosis of coronary artery disease using automated machine learning. Bioinformatics **36**(6), 1772–1778 (2020)

18. Orzechowski, P., Moore, J.H.: Generative and reproducible benchmarks for comprehensive evaluation of machine learning classifiers. Sci. Adv. **8**(47), eabl4747 (2022). https://doi.org/10.1126/sciadv.abl4747, https://www.science.org/doi/abs/10.1126/sciadv.abl4747

19. Saini, A.K., Spector, L.: Relationships between parent selection methods, looping constructs, and success rate in genetic programming. Genet. Program Evolvable Mach. **22**(4), 495–509 (2021)
20. Snedecor, G.W., Cochran, W.G.: Statistical Methods, 8th edn. Iowa State University Press (1989)

Using FPGA Devices to Accelerate Tree-Based Genetic Programming: A Preliminary Exploration with Recent Technologies

Christopher Crary$^{(\boxtimes)}$ ⓘ, Wesley Piard ⓘ, Greg Stitt ⓘ, Caleb Bean ⓘ, and Benjamin Hicks ⓘ

University of Florida, Gainesville, FL 32611, USA
ccrary@ufl.edu

Abstract. In this paper, we explore the prospect of accelerating tree-based genetic programming (TGP) by way of modern field-programmable gate array (FPGA) devices, which is motivated by the fact that FPGAs can sometimes leverage larger amounts of data/function parallelism, as well as better energy efficiency, when compared to general-purpose CPU/GPU systems. In our preliminary study, we introduce a fixed-depth, tree-based architecture capable of evaluating type-consistent primitives that can be fully unrolled and pipelined. The current primitive constraints preclude arbitrary control structures, but they allow for entire programs to be evaluated every clock cycle. Using a variety of floating-point primitives and random programs, we compare to the recent TensorGP tool executing on a modern 8 nm GPU, and we show that our accelerator implemented on a 14 nm FPGA achieves an average speedup of 43×. When compared to the popular baseline tool DEAP executing across all cores of a 2-socket, 28-core (56-thread), 14 nm CPU server, our accelerator achieves an average speedup of 4,902×. Finally, when compared to the recent state-of-the-art tool Operon executing on the same 2-processor CPU system, our accelerator executes about 2.4× slower on average. Despite not achieving an average speedup over every tool tested, our single-FPGA accelerator is the fastest in several instances, and we describe five future extensions that could allow for a 32–144× speedup over our current design as well as allow for larger program depths/sizes. Overall, we estimate that a future version of our accelerator will constitute a state-of-the-art GP system for many applications.

Keywords: Tree-based genetic programming · Field-programmable gate arrays · Hardware acceleration

1 Introduction

During any given time, the development of AI has been constrained and influenced by the computing technologies available [9,10,27]. Nevertheless,

This material is based upon work supported by the National Science Foundation under Grant Nos. CNS-1718033 and CCF-1909244.

G. Pappa et al. (Eds.): EuroGP 2023, LNCS 13986, pp. 182–197, 2023.
https://doi.org/10.1007/978-3-031-29573-7_12

novel applications of pre-existing technologies still happen, and they can drive research fields in whole new directions, occasionally prompting some form(s) of widespread adoption. For instance, the massive adoption of deep neural networks in the past decade was clearly enabled by developments regarding GPUs [9,10,27]. In this instance, computing advancements markedly extended the practical reach of neural networks, which then kickstarted a wave of popularity and research ventures. Importantly, in this domain and many others, the demands of ever-increasing performance and energy efficiency has now manifested into the broad development of *domain-specific hardware accelerators* [9]. However, in the wide-ranging domain of genetic programming (GP), there seems to exist only a few instances of specialized hardware accelerators [7,8,15,16,22], which is especially surprising given that GP is an "embarrassingly parallel" procedure.

Generally speaking, although general-purpose CPU/GPU systems can be made to effectively exploit some of the parallelism opportunities inherent to GP (e.g., by evaluating multiple data points, multiple operations, or multiple candidate solutions in parallel), the frequent, dynamic changes in control flow caused by GP (e.g., when evaluating different operations within a single program) generally limits how effective a general-purpose computing platform can perform [3,4,21]. Within this paper, we focus on the original tree-based GP (TGP) [12], and we explore how we may overcome the aforementioned limitations of CPU/GPU systems by way of an accelerator specialized to the evaluation phase of TGP, implemented with a modern field-programmable gate array (FPGA). In brief, FPGAs are programmable computing systems in which specialized digital circuitry can be synthesized from different levels of abstraction, without recourse to integrated circuit development.

Overall, as depicted in Fig. 1, our preliminary accelerator leverages a specialized, full tree of generic computing resources to compute any program relevant to a GP primitive set, as long as the depth of the program is not larger than the depth of the tree, the latter of which is defined by the user. By then pipelining the generic resources, the accelerator can generate an output for an entire program expression *every clock cycle* after some initial latency. To further increase throughput, the accelerator also dynamically compiles programs for the tree while evaluating, so that the tree may switch between programs *within a single clock cycle*. Importantly, such forms of parallelism have not been achieved via general-purpose CPU/GPU architectures.

We compare the performance of our architecture with the evaluation engines given by three actively maintained, open-source tree-based GP software tools: *DEAP* [6], *TensorGP* [1], and *Operon* [3].[1] From each tool, we use the evaluation engine—and no evolution engine—to execute a large set of randomly generated programs for various amounts of fitness cases (i.e., sample points), and we estimate evaluation performance in terms of *node evaluations per second (NEPS)*. For each software-based tool, we utilize a dual-socket server populated with two 2.6 GHz (3.7 GHz Turbo), 14-core (28 thread), 14 nm Intel Xeon Gold 6132 CPU packages, and we additionally use an 8 nm Nvidia RTX 3080 GPU (10 GB)

---

[1] Software: https://github.com/christophercrary/conference-eurogp-2023.

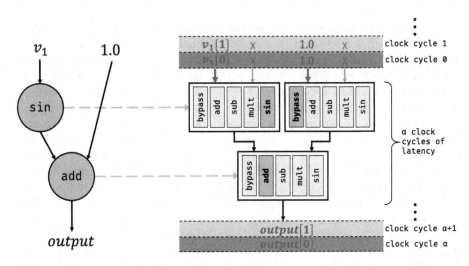

Fig. 1. A portrayal of how our GP accelerator can parallelize the evaluation of different data points and different solutions *every clock cycle* via a reconfigurable tree pipeline. Each node of the pipeline can perform any function within the GP primitive set, as well as a bypass, which allows for arbitrary program shapes.

for TensorGP. To implement our hardware accelerator, we utilize a 14 nm Intel Stratix 10 SX 1SX280HN2F43E2VG FPGA provided by an Intel Programmable Acceleration Card (PAC) through the Intel FPGA DevCloud service. We compile the accelerator by way of Quartus Pro 19.2.0, Build 57.

When compared to DEAP [6], a popular baseline for GP software tools, our accelerator achieves an average speedup of 4,902×. Compared to TensorGP [1], a recent general-purpose GP software tool targeting both CPU and GPU systems, our architecture achieves an average speedup of 61.5× in regard to CPU execution and 43× in regard to GPU execution. Finally, when compared to Operon [3], a recent state-of-the-art GP tool tailored to symbolic regression [13], our single-FPGA accelerator executes about 2.4× slower on average when compared to the same 2-processor CPU system, although there are several instances in which our accelerator performs the fastest. Despite not achieving an average speedup over every tool tested, we describe five future extensions that could allow for a 32–144× speedup over our current design. Separately, we note that it has been widely shown that FPGAs can often provide power and energy improvements when compared to CPU/GPU systems, sometimes by multiple orders of magnitude [18,20,23,24]. Although we do not provide power or energy estimates in this paper, if we can experience any of such improvements when compared to other GP tools, this should enable us to implement more energy-efficient (and, thus, potentially more cost-effective) GP systems than what has been presented in previous work [25]. Overall, we estimate that a future version of our accelerator will constitute a state-of-the-art GP system for many applications.

The remainder of the paper is organized as follows. Section 2 describes related work. Section 3 details our architecture. Section 4 describes our design of experiments. Section 5 presents results for our experiments. Section 6 discusses limitations of our current architecture and planned future extensions that should address the limitations and allow for state-of-the-art performance. Finally, Sect. 7 presents conclusions.

2   Related Work

In the context of CPU/GPU systems, there exist numerous works that discuss mechanisms for accelerating tree-based GP—see [5] for a recent review, as well as [1–4, 14, 21]—although there are comparatively few works that consider the use of FPGA devices [7, 22]. Compared to prior work [7, 22], our accelerator has several important contributions. Most significantly, our system dynamically compiles programs from a compressed prefix notation into configuration data for a reconfigurable pipeline, whereas previous work used a simpler, less flexible mechanism by which larger, fixed-size programs must be compiled. Ultimately, our compressed prefix notation allows for significantly reduced communication times as well as significantly reduced size requirements for on-chip RAM. Also, with the ability to dynamically compile arbitrary expressions directly on the target device, future extensions of our design can accelerate other GP stages without continued hardware/software communication. Besides dynamic compilation, we also explore the use of a higher-end FPGA device, multiple primitive sets, a range of fitness case amounts, different tree sizes, and 32-bit floating point, all while comparing to a range of modern GP tools.

Apart from tree-based GP, there exists some prior work on the FPGA acceleration of certain GP variants, e.g., Linear GP [15], Cartesian GP [16], and Geometric Semantic GP [8], although we note that the differences in evaluation schemes warrants a separate architecture dedicated to tree-based GP. Lastly, we note that the application area of evolvable hardware [28] has also leveraged FPGA devices, although this has been with the primary intention of evolving circuits, rather than accelerating the GP procedure via a single circuit.

3   Accelerator Architecture

In this section, we detail our accelerator architecture for the evaluation phase of tree-based GP. We focus on evaluation since it is the primary bottleneck for TGP. Eventually, we will investigate acceleration of the entire GP process.

The accelerator architecture currently consists of four major components, as shown in Fig. 2. The *program memory* (Sect. 3.1) stores candidate program solutions, where each candidate is encoded in a language defined by the specification of a particular primitive set. The *program compiler* (Sect. 3.2) reads program expressions from program memory and dynamically compiles them into configuration information for the *program evaluator*, which we implement as a reconfigurable *function tree pipeline* (Sect. 3.3). This function tree pipeline executes a compiled expression for all relevant fitness cases, resulting in a new output for the entire program *every clock cycle* after some initial latency. Finally, the *fitness evaluator* (Fig. 2d) compares the output of a current program to the relevant target data by way of some metric (*root-mean-square error* in this paper), which allows for other stages of GP to optimize the individual.

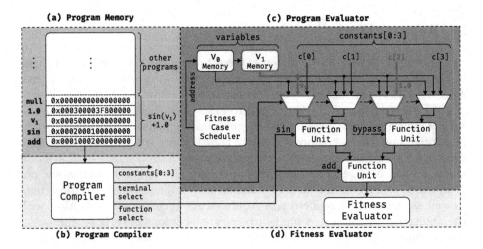

Fig. 2. High-level overview of the accelerator architecture. The accelerator stores programs (e.g., $\sin(v_1) + 1.0$) in (a) *program memory*, which are dynamically compiled by (b) *the program compiler* into configuration data for (c) *the program evaluator*. The program evaluator uses a reconfigurable function tree pipeline to execute a compiled expression for a set of fitness cases, resulting in a set of outputs to which (d) *the fitness evaluator* compares a set of desired outputs.

3.1   Program Memory

The architecture currently implements program memory with on-chip RAM resources and memory-mapped I/O. For a primitive set $P = F \cup V \cup C$, with function set F, variable terminal set V, and a set of 32-bit constant terminals C (e.g., all single-precision floating-point values), we define a 64-bit machine code for program nodes as follows:

1. The most-significant 16 bits of the machine code represent an opcode which specifies either the type of primitive or the *null word*, the latter of which is used to indicate the end of a program expression within memory. The null word is assigned opcode 0, each function is assigned an opcode in the range $[1, |F|]$, each constant is assigned opcode $|F|+1$, and each variable is assigned an opcode in the range $[|F| + 1 + 1, |F| + 1 + |V|]$.
2. The least-significant 32 bits of the machine code specify a constant value, which is only relevant if the opcode indicates that the node is a constant.

3. The remaining 16 bits specify the depth of a node within the context of a program, which is relevant to the program compiler (Sect. 3.2).

We encode program expressions via a prefix (i.e., Polish) notation. In essence, such a representation flattens tree-based programs into a linear structure [19]. For example, Fig. 2a shows how our architecture could support the program $\sin(v_1) + 1.0$ by way of a simple primitive set consisting of addition (+), sine (sin), two variable terminals (v_0 and v_1), and the set of all single-precision floats.

3.2 Program Compiler

The program compiler reads program expressions from program memory and dynamically compiles them into configuration information for the program evaluator. Currently, the program compiler is implemented as a finite-state machine (FSM) that continually writes configuration information into a configuration buffer, which is omitted from Fig. 2 due to space constraints. Such buffering enables the program compiler to generate configurations for a program in advance while the program evaluator is processing fitness cases for an existing program.

The program compiler's configuration data contains three major components (Fig. 2b, c): 1) *function select* values that configure individual function units within the function tree of the program evaluator (Sect. 3.3), *terminal select* values that dictate whether a variable or constant terminal is connected to the corresponding function tree input, and 3) *constant* values that specify the bits of any constant terminals feeding into the tree.

To compile a program, the program compiler conducts a pre-order traversal on a model of the relevant function tree, so that compilation can happen in parallel to program evaluation. We determine a model for the tree at compile time, based on the specified depth and branching factor of the tree, the latter of which is determined by the maximum function arity of the chosen primitive set.

Ultimately, depending on the shapes/sizes of programs being compiled and the number of fitness cases that are to be streamed into a function tree, the cost of compiling a program may be completely amortized such that there is no dead cycles in between evaluating consecutive programs. Fortunately, for *any* function tree structure, there will always be some threshold for the number of fitness cases such that, for any number of fitness cases above this threshold, compilation will be completely amortized. Separately, since the program compiler FSM needs relatively few resources (currently, less than 2% of all area for our target device), we can extend our architecture to support multiple compiler instantiations. With this ability, multiple programs could be compiled in parallel—perhaps to effectively support multiple function trees, or perhaps to ensure that the cost of compiling a single tree can be completely amortized. For the experiments in this paper, we support the compilation of one program at a time, and we incorporate a multiple-buffering approach, following the above.

3.3   Program Evaluator

The program evaluator (Fig. 2c) is a reconfigurable *function tree* that serves two purposes: 1) provide configurable resources that enable the program compiler to implement arbitrary expressions specified by program memory; and 2) provide a pipeline that enables streaming of fitness case data such that program outputs can be computed *every clock cycle*.

The motivation behind the function tree is that, with tree-based GP, every program expression can be represented as a tree. Therefore, if the accelerator provides a function tree containing pipelined generic resources capable of computing the functions of the relevant primitive set (i.e., a *function unit*), then the function tree can produce outputs for entire program expressions *every clock cycle* after some initial latency. In this paper, we consider a single tree structure, but we plan to support multiple tree structures in future work.

For the program evaluator, the user must specify the relevant primitive set and the depth of the underlying function tree, which define 1) the maximum function arity, 2) the operations supported by each function unit, and 3) the possible program shapes/sizes. A function tree with depth d can compute arbitrary programs that adhere to both 1) a maximum depth of $d + 1$—where the extra level accounts for terminal nodes—and 2) the syntax of the relevant primitive set. To be able to implement any program not represented by a full tree, a special bypass function is used to feed the leftmost input of a function unit directly to its output whenever that node within the tree is not to be used by a program. In regard to function primitives, we currently support any form of computation that can be unrolled and pipelined.

In addition to a function tree, the program evaluator also contains *variable memories*, which support variable terminals. The variable memories store fitness cases for every feature of the relevant training data. The particular data for each variable can be set at runtime, using memory-mapped I/O.

4   Design of Experiments

In this section, we detail our design of experiments, where the overall goal of these experiments was to compare our architecture (Sect. 3) with the core evaluation engines given by three tree-based GP software tools: *DEAP* [6], *TensorGP* [1], and *Operon* [3]. The computing technologies we used are listed in Sect. 1.

4.1   Comparison Metrics

We estimate and compare *median node evaluations per second (NEPS)*[2] values for each evaluation engine in the context of different combinations of program sizes, numbers of fitness cases, and primitive sets. For each software tool and for

---

[2] Frequently, the statistic of *GP operations per second (GPops)* is used when comparing the runtime performance of GP tools, but we use NEPS to emphasize that our runtimes do not include time taken for evolution.

each combination of parameters (detailed below), we conducted an experiment in which we evaluated 32 *program bins*, each consisting of 512 distinct random programs, and we estimated a median runtime for each program bin by measuring a certain number of evaluation runtimes and then taking the sample median. With a sample median runtime, we calculated an estimate for the true median NEPS value by dividing the total number of node evaluations for an experiment by the sample median runtime. Due to time constraints and significant performance differences between each of the GP tools, we used a different total number of executions for some tools when calculating sample median runtime. For Operon and TensorGP, the two fastest software tools, we ran the set of experiments 11 times. However, for DEAP, in which the set of experiments executed in about 44 hours (due to poor scaling at larger numbers of fitness cases), we ran each experiment just once. Running each DEAP experiment once seemed justified by the fact that any fluctuations in runtime due to other system processes were likely insignificant when compared to the processes used by the experiments, as indicated by the narrow $75^{th}/25^{th}$ percentile regions for the runtimes of TensorGP, given below. Lastly, we note that it was unnecessary for the accelerator experiments to be run more than once, since the circuitry created for the system had deterministic behavior.

4.2 Primitive Sets

Three distinct primitive sets were chosen. These primitive sets were inspired by recent work from Nicolau et al. [17], and, as such, were respectively named nicolau_a, nicolau_b, and nicolau_c. The first primitive set contained functions with the self-explanatory names add, sub, and mul, as well as a function by the name of aq, for "analytical quotient", defined by $aq(x_1, x_2) = x_1/\sqrt{1 + x_2^2}$, which is meant to behave similarly to divide, but without the asymptotic conditions at zero [17]. The second primitive set contained the same functions as the first, but also included sin and tanh. Lastly, the third primitive set contained the same functions as the second, but also included exp, log, and sqrt, where log and sqrt were "protected" in the typical GP sense [12,19]. We chose these specific primitive sets since they are relevant to symbolic regression [12,13,19], our primary target domain.

For a primitive set containing function set F, $|F| - 1$ terminal variables and one ephemeral random constant were employed so that the program generator (Sect. 4.3) would consistently construct programs in which the proportion of functions/terminals was approximately 0.5, so that the average runtime of a particular primitive set was not dictated by having more of one primitive type.

4.3 Program Generation

For each primitive set, a set of 32 program bins was constructed, each containing 512 random programs with sizes in some fixed range, where the particular range was dependent on the bin and primitive set, as described further below. The maximum possible program depth/size was chosen to be the largest that the target FPGA could support while also supporting up to 100,000 fitness cases for each of the relevant variable terminal memories (Sect. 3.3). To determine these values, the maximum possible function tree depth for each primitive set was manually determined through multiple hardware compilations—ultimately, depth values 8, 6, and 6 were respectively chosen for nicolau_a, nicolau_b, and nicolau_c. For a maximum possible function tree depth d, it was possible to support a program depth of up to $d + 1$ (Sect. 3.3), which corresponded to a maximum possible program size of $2^{d+2} - 1$, since every primitive set contained functions with arity of at most two. For a maximum size s, the range of program sizes $[1, s]$ was subdivided into 32 bins.

To randomly generate program expressions for each set of bins—which were kept the same for each GP tool—we utilized DEAP [6]. We chose DEAP for this task because it was simple to extend. DEAP offered, by default, several classic GP program initialization algorithms: *full*, *grow*, and *ramped half-and-half* [12,19]. Unfortunately, via the original version of these algorithms, the size of a generated program was completely random beyond a specified depth constraint, which made it too cumbersome to generate 512 distinct random programs for the bin structures established above. To circumvent this issue, we created a modified version of the *grow* method that allowed for the specification of a minimum/maximum program size, from which a random value was chosen in a uniform manner. Overall, choosing 512 distinct random programs for each bin structure meant that 16,384 programs were used to evaluate each of the three primitive sets, which corresponded to a total of 49,152 random programs.

4.4 Fitness Cases

For each primitive set, we used five amounts of fitness cases: 10, 100, 1,000, 10,000, and 100,000. For each number of fitness cases, we randomly generated input/target data in the range $[0, 1)$, and we used the same data for each of the evaluation engines. We note that using random data should elucidate the fact that our performance results are relevant to *any* GP application that can utilize the 1) chosen primitive sets, 2) maximum number of variables, and 3) maximum number of fitness cases, which, as shown in [13], allows for many.

5 Results

Figure 3 compares the performance of each evaluation engine in terms of sample median NEPS values, for six of the fifteen combinations of primitive set and number of fitness cases. For each combination, we plot results for five GP tool setups: 1) DEAP, 2) TensorGP with CPU, 3) TensorGP with GPU, 4) Operon, and 5) our FPGA-based hardware accelerator. More specifically, for each plot representing a GP tool, a sample median NEPS value is marked for each program bin containing 512 programs, with the particular number of fitness cases used for each program changing between sub-figures. In addition, for the tools in which experiments were run more than once (i.e., TensorGP and Operon), the 75th and

25th percentiles for runtime are plotted above/below each sample point; only a few of such percentile regions are noticeable, meaning that most runtimes vary little between multiple runs. Overall, due to space constraints, we include plots for just six experiments, but we chose these six in particular since they best conveyed the most important general trends for our accelerator, detailed below.

Overall, our accelerator mostly performed second-best behind Operon, but in several instances our accelerator obtained the highest performance, e.g., for the larger programs and larger number of fitness cases with the nicolau_a primitive set (Fig. 3a, b), and for the smaller programs and smaller numbers of fitness cases across all primitive sets (e.g., Fig. 3c, e). In some other instances, our accelerator performed very similarly to Operon, e.g., for the medium-sized programs and medium-sized numbers of fitness cases with nicolau_b (Fig. 3d). In general, the speedups we achieved stemmed from the fact that our accelerator had *constant throughput* once programs were compiled for the program tree. For larger numbers of fitness cases (e.g., 10K and 100K), compilation was completely amortized after the first program (Sect. 3.2), which allowed for maximal throughput. Interestingly, although the program tree structures for primitive sets nicolau_b and nicolau_c utilized the same depths/sizes—which should potentially allow for identical runtime—the hardware synthesis tool had to utilize a lower clock frequency for nicolau_c in order to support more complex primitives, which allowed for nicolau_b to have better performance. A similar discrepancy in clock frequency also explains why Fig. 3a lists better performance than Fig. 3b.

Table 1. Average NEPS speedups for various fitness case thresholds. For a given threshold value, the average is calculated from all results regarding thresholds less than or equal to this value. The last row represents an overall average.

	Average FPGA Speedup			
Fitness Case Threshold ($\leq$)		**Tool**		
	DEAP	**TensorGP (CPU)**	**TensorGP (GPU)**	**Operon**
10	569×	1210×	1408×	0.375×
100	741×	1197×	1384×	0.357×
1,000	2372×	1039×	1123×	0.290×
10,000	4611×	312.0×	318.5×	0.432×
100,000	4902×	61.5×	43.0×	0.423×

Beyond Operon, our accelerator was able to consistently outperform a modern GPU system running TensorGP, where our results for TensorGP align with results previously listed [1]. Interestingly, DEAP sometimes performed better than TensorGP for the smallest number of fitness cases (e.g., Fig. 3e), although TensorGP scaled much better with larger numbers of fitness cases. All in all, we

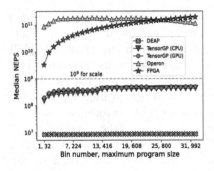

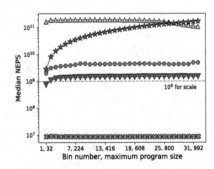

(a) Primitive set nicolau_a, 10K fitness cases. **Max value: 199 billion NEPS**.

(b) Primitive set nicolau_a, 100K fitness cases. Max value: 197 billion NEPS.

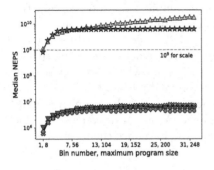

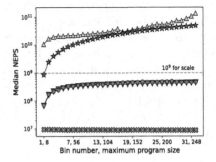

(c) Primitive set nicolau_b, 100 fitness cases. Max value: 18.3 billion NEPS.

(d) Primitive set nicolau_b, 10K fitness cases. Max value: 136 billion NEPS.

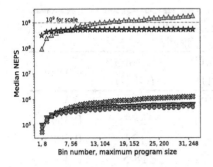

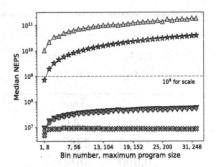

(e) Primitive set nicolau_c, 10 fitness cases. Max value: 1.86 billion NEPS.

(f) Primitive set nicolau_c, 1K fitness cases. Max value: 188 billion NEPS.

Fig. 3. Sample median node evaluations per second (NEPS) vs. program bin number and maximum program size, for six different combinations of primitive set and number of fitness cases. For Operon and TensorGP plots, the $75^{th}/25^{th}$ percentiles for runtime are plotted above/below each sample point, which are noticeable in only a few instances; for example, see the first bin of (a). Note that the legend from (a) applies to all subfigures. Also, note the use of a log scale.

note from the given plots that a single, pipelined program tree by way of our accelerator could keep up with and sometimes outpace a two-socket, 28-core, 56-thread CPU system running the state-of-the-art Operon tool, and we consistently outperformed a modern GPU system running the recent, high-performance TensorGP tool. On average, across all fifteen experiments and all random programs, the FPGA was $4,902\times$ faster than DEAP, $61.5\times$ faster than TensorGP executing with the CPU, $43\times$ faster than TensorGP executing with the GPU, and $2.4\times$ slower than Operon. For some more specific trends in regard to number of fitness cases, Table 1 presents an average NEPS speedup for the FPGA in the context of all experiments with 1) 10 fitness cases, 2) 100 fitness cases or less, 3) 1,000 fitness cases or less, 4) 10,000 fitness cases or less, and 5) 100,000 fitness cases or less, where the values provided by 5) are used to represent an overall average for the conducted experiments, already listed above. To calculate one of these averages for a particular tool and fitness case threshold, we first divided a sum of node evaluations by a sum of median runtimes, with the values in each sum stemming from all experiments regarding the particular tool and fitness case threshold. Then, to compute the relevant speedup, we divided a corresponding average for the FPGA by the average computed for the relevant tool. Note that using the median runtime from each bin in these calculations was appropriate, given that we wanted to establish a typical runtime value for each bin of each experiment. For more details, please refer to our code.

6 Current Limitations and Potential Optimizations

Below, we list three limitations of our initial accelerator architecture, and then we present five potential optimization strategies that could alleviate the three limitations and allow for an updated accelerator to achieve a speedup over our current design by 32–$144\times$ as well as support for larger program depths/sizes.

6.1 Current Limitations

Comprehensive Support. For a function tree (Sect. 3.3) to be able to support arbitrary programs, *every* function unit must support *all* function primitives defined by the primitive set. Therefore, depending on the number of function primitives and the types of low-level device resources utilized for these primitives, the maximum depth/size of function trees—and, thus, programs—can be restricted. For our experiments that utilized primitives relying on floating-point operations, we were ultimately constrained by the number of floating-point DSP and embedded memory resources available within the target FPGA device.

Exponential Growth. For an m-ary function tree (where m is the maximum function arity of the primitive set) with $m > 1$, the amount of area needed to implement the tree grows exponentially with increasing tree depth. Namely, for $m > 1$ and a function tree depth of d, $\frac{1-m^{d+1}}{1-m}$ generic function units are needed for the tree, which can prevent up to $\frac{100}{m}\%$ of some device resource(s) from being used when maximizing d. (For $m = 1$, only $d + 1$ function units are needed.)

Low Resource Utilization. If each function unit in the tree is capable of computing *every* function primitive, then for $|F|$ function primitives, the utilization of each function unit in terms of these high-level primitives is $\frac{1}{|F|}$. The utilization of low-level device primitives (e.g., floating-point DSPs) can be significantly lower, depending on the function primitive implementations.

6.2   Potential Optimizations

We note that the following five optimizations are independent from one another[3], and, thus, if all could be achieved simultaneously, a speedup between $2\cdot2\cdot2\cdot2\cdot2 = 32\times$ and $2\cdot6\cdot3\cdot2\cdot2 = 144\times$ could be achieved over our current accelerator.

Use Compacted Trees. To be able to more effectively leverage device resources as well as support larger program depths/sizes, we plan to explore various "compacted tree" architectures. Ideally, such an architecture would allow for the use of all resources that are currently unused due to exponential growth in area—either through the use of a single, more efficient compute engine or through multiple compute engines—and such an architecture would also offer native support for larger depths/sizes. One option may be to construct a unified parallel/sequential tree structure, similar to what has been developed for tree-based accumulators [26]. Another possibility may be to design a linear architecture that natively handles flattened tree (e.g., prefix/postfix) representations. If either option could result in a fully-pipelined architecture, the latter may be able to more effectively map flattened programs onto function unit resources, but such an architecture would seemingly require state memory (e.g., a temporary stack) to be included in the pipeline in order to maximize throughput, which would likely infer significantly more memory resources than a spatially parallel tree representation.

For the current study, if we could leverage all resources not currently used, we could improve upon our performance results by upwards of $2\times$ (Sect. 6.1).

Multiplex Function Unit Resources. Function unit primitives experience poor utilization due to the fact that they are implemented with independent IP blocks. This issue could be improved upon by implementing a function unit via a single IP block that multiplexes a minimal amount of some devices resource(s), e.g., floating-point DSPs. Such an "overlay" could free up a significant amount of resources, allowing for further parallelization of program evaluation. For example, in the context of the most complex primitive set used for this paper, nicolau_c, the most expensive primitive was tanh, which utilized $\frac{13}{42} \approx 31\%$ of all DSPs allocated for each function unit. Thus, with an appropriate overlay, $\frac{29}{42} \approx 69\%$ of all DSPs for function units could be recovered. Carrying out a similar process for all primitive sets used in this paper, about 50% of all DSPs

---

[3] A possible exception could occur when dealing with timing optimization, since the resulting clock frequency may unexpectedly get better or worse with design changes.

utilized for function units could be recovered on average, which could translate into an average speedup by up to 2×. However, in general, for a primitive set containing $|F|$ functions, an optimal overlay could allow for up to an $|F| \times$ speedup if all functions were to utilize the same amount of low-level resources. Therefore, in our case, where we currently utilize an average of approximately six functions, we estimate that we could achieve a speedup of 2–6× by using optimized (or alternate) functions.

Design for Higher Clock Frequencies. For our accelerator, throughput (i.e., performance) is directly proportional to clock frequency. With modern FPGAs, it is not uncommon for designs to achieve clock frequencies in the range 400–850 MHz after optimizing for timing [18,23,24]. Our current accelerator has not been fully optimized, and, as such, we achieved an average clock frequency of 178 MHz across the fifteen hardware compilations performed for this paper. We estimate that we can achieve up to a 2–3× higher average clock frequency once we further optimize for timing (and potentially move to a newer device), which would allow for an average speedup over our current design by up to 2–3×.

Use a Higher-End FPGA Device. With a more modern, higher-end FPGA implemented on newer process-node technology (e.g., [11]), we should be able to support at least 2× more floating-point DSP resources and 1.5× more embedded memory resources, in addition to higher clock frequencies. With 2× more DSP resources, we expect that we can further parallelize our current floating-point computations by up to 2×, which should allow for up to a 2× speedup.

Double-Buffer GP Runs. When our accelerator enters the context of a full GP system, including evolution, we expect that we can execute two GP runs simultaneously, by evolving one population whilst evaluating another. Such an optimization would generally not make sense for a typical GP system (with exception to possibly a combined CPU/GPU system), since any additional compute cores would likely be used to further parallelize program evaluation. If the total time taken for evolution and device communication can be less than the total time taken for evaluation, then this optimization should allow our accelerator to achieve an additional speedup by up to 2×.

7   Conclusion

In this paper, we leveraged a modern FPGA device to implement a hardware accelerator that more closely aligns with the computing model of tree-based GP when compared to CPU/GPU solutions. Specifically, the presented architecture dynamically compiles program trees onto a reconfigurable function tree pipeline that can generate outputs for entire program expressions every clock cycle and transition between separate programs within a single cycle.

We showed that our accelerator on a 14 nm FPGA achieves an average speedup of 43× when compared to a recent open-source GPU solution implemented on 8 nm process-node technology, and an average speedup of 4,902×

when compared to a popular baseline GP software tool running parallelized across all cores of a 2-socket, 28-core (56-thread), 14 nm CPU server. Despite our single-FPGA accelerator being 2.4× slower on average when compared to a recent state-of-the-art GP software tool executing on the same 2-processor CPU system, we described future extensions that could provide a 32–144× speedup over our current design.

References

1. Baeta, F., Correia, J., Martins, T., Machado, P.: Exploring genetic programming in TensorFlow with TensorGP. SN Comput. Sci. **3**(2), 1–16 (2022). https://doi.org/10.1007/s42979-021-01006-8
2. Banzhaf, W., Harding, S., Langdon, W.B., Wilson, G.: Accelerating genetic programming through graphics processing units. In: Worzel, B., Soule, T., Riolo, R. (eds.) Genetic Programming Theory and Practice VI. Genetic and Evolutionary Computation, pp. 1–19. Springer, Boston (2009). https://doi.org/10.1007/978-0-387-87623-8_15
3. Burlacu, B., Kronberger, G., Kommenda, M.: Operon C++: an efficient genetic programming framework for symbolic regression. In: Proceedings of the 2020 Genetic and Evolutionary Computation Conference Companion, GECCO 2020, pp. 1562–1570. Association for Computing Machinery, New York (2020). https://doi.org/10.1145/3377929.3398099
4. Chitty, D.M.: Fast parallel genetic programming: multi-core CPU versus many-core GPU. Soft. Comput. **16**(10), 1795–1814 (2012). https://doi.org/10.1007/s00500-012-0862-0
5. Chitty, D.M.: Faster GPU-based genetic programming using a two-dimensional stack. Soft. Comput. **21**(14), 3859–3878 (2016). https://doi.org/10.1007/s00500-016-2034-0
6. Fortin, F.A., De Rainville, F.M., Gardner, M.A.G., Parizeau, M., Gagné, C.: DEAP: evolutionary algorithms made easy. J. Mach. Learn. Res. **13**(1), 2171–2175 (2012)
7. Funie, A.-I., Grigoras, P., Burovskiy, P., Luk, W., Salmon, M.: Run-time reconfigurable acceleration for genetic programming fitness evaluation in trading strategies. J. Signal Process. Syst. **90**(1), 39–52 (2017). https://doi.org/10.1007/s11265-017-1244-8
8. Goribar-Jimenez, C., Maldonado, Y., Trujillo, L., Castelli, M., Gonçalves, I., Vanneschi, L.: Towards the development of a complete GP system on an FPGA using geometric semantic operators. In: 2017 IEEE Congress on Evolutionary Computation (CEC), pp. 1932–1939 (2017). https://doi.org/10.1109/CEC.2017.7969537
9. Hennessy, J.L., Patterson, D.A.: Computer Architecture: A Quantitative Approach, 6th edn. Morgan Kaufmann Publishers Inc., San Francisco (2017)
10. Hooker, S.: The hardware lottery. Commun. ACM **64**(12), 58–65 (2021). https://doi.org/10.1145/3467017
11. Intel: Intel Agilex™ M-Series FPGA and SoC FPGA Product Table (2015). https://cdrdv2.intel.com/v1/dl/getContent/721636
12. Koza, J.R.: Genetic Programming: On the Programming of Computers by Means of Natural Selection. MIT Press, Cambridge (1992)
13. La Cava, W., et al.: Contemporary symbolic regression methods and their relative performance. In: Vanschoren, J., Yeung, S. (eds.) Proceedings of the Neural Information Processing Systems Track on Datasets and Benchmarks, vol. 1 (2021)

14. Langdon, W.B., Banzhaf, W.: A SIMD interpreter for genetic programming on GPU graphics cards. In: O'Neill, M., et al. (eds.) EuroGP 2008. LNCS, vol. 4971, pp. 73–85. Springer, Heidelberg (2008). https://doi.org/10.1007/978-3-540-78671-9_7

15. Martin, P.: A hardware implementation of a genetic programming system using FPGAs and Handel-C. Genet. Program Evolvable Mach. 2(4), 317–343 (2001). https://doi.org/10.1023/A:1012942304464

16. Miller, J.F.: Cartesian genetic programming: its status and future. Genetic Programm. Evolvable Mach. 21(1), 129–168 (2020). https://doi.org/10.1007/s10710-019-09360-6

17. Nicolau, M., Agapitos, A.: Choosing function sets with better generalisation performance for symbolic regression models. Genet. Program Evolvable Mach. 22(1), 73–100 (2020). https://doi.org/10.1007/s10710-020-09391-4

18. Nurvitadhi, E., et al.: Can FPGAs beat GPUs in accelerating next-generation deep neural networks? In: Proceedings of the 2017 ACM/SIGDA International Symposium on Field-Programmable Gate Arrays, FPGA 2017, pp. 5–14. Association for Computing Machinery, New York (2017). https://doi.org/10.1145/3020078.3021740

19. Poli, R., Langdon, W.B., McPhee, N.F.: A Field Guide to Genetic Programming. Lulu Enterprises Ltd., UK (2008)

20. Putnam, A., et al.: A reconfigurable fabric for accelerating large-scale datacenter services. IEEE Micro 35(3), 10–22 (2015). https://doi.org/10.1109/MM.2015.42

21. Robilliard, D., Marion-Poty, V., Fonlupt, C.: Genetic programming on graphics processing units. Genet. Program Evolvable Mach. 10(4), 447–471 (2009). https://doi.org/10.1007/s10710-009-9092-3

22. Sidhu, R.P.S., Mei, A., Prasanna, V.K.: Genetic programming using self-reconfigurable FPGAs. In: Lysaght, P., Irvine, J., Hartenstein, R. (eds.) FPL 1999. LNCS, vol. 1673, pp. 301–312. Springer, Heidelberg (1999). https://doi.org/10.1007/978-3-540-48302-1_31

23. Stitt, G., Gupta, A., Emas, M.N., Wilson, D., Baylis, A.: Scalable window generation for the Intel Broadwell+Arria 10 and high-bandwidth FPGA systems. In: Proceedings of the 2018 ACM/SIGDA International Symposium on Field-Programmable Gate Arrays, FPGA 2018, pp. 173–182. Association for Computing Machinery (2018). https://doi.org/10.1145/3174243.3174262

24. Tan, T., Nurvitadhi, E., Shih, D., Chiou, D.: Evaluating the highly-pipelined Intel Stratix 10 FPGA architecture using open-source benchmarks. In: 2018 International Conference on Field-Programmable Technology (FPT), pp. 206–213 (2018). https://doi.org/10.1109/FPT.2018.00038

25. Veeramachaneni, K., Arnaldo, I., Derby, O., O'Reilly, U.-M.: FlexGP. J. Grid Comput. 13(3), 391–407 (2014). https://doi.org/10.1007/s10723-014-9320-9

26. Wilson, D., Stitt, G.: The unified accumulator architecture: a configurable, portable, and extensible floating-point accumulator. ACM Trans. Reconfigurable Technol. Syst. 9(3) (2016). https://doi.org/10.1145/2809432

27. Wright, L.G., et al.: Deep physical neural networks trained with backpropagation. Nature 601(7894), 549–555 (2022). https://doi.org/10.1038/s41586-021-04223-6

28. Yao, X.: Following the path of evolvable hardware. Commun. ACM 42(4), 46–49 (1999). https://doi.org/10.1145/299157.299169

Memetic Semantic Genetic Programming for Symbolic Regression

Alessandro Leite[(✉)][iD] and Marc Schoenauer[iD]

TAU, Inria Saclay, LISN, Univ. Paris-Saclay, Gif-sur-Yvette, France
{alessandro.leite,marc.schoenauer}@inria.fr

Abstract. This paper describes a new memetic semantic algorithm for symbolic regression (SR). While memetic computation offers a way to encode domain knowledge into a population-based process, semantic-based algorithms allow one to improve them locally to achieve a desired output. Hence, combining memetic and semantic enables us to (a) enhance the exploration and exploitation features of genetic programming (GP) and (b) discover short symbolic expressions that are easy to understand and interpret without losing the expressivity characteristics of symbolic regression. Experimental results show that our proposed memetic semantic algorithm can outperform traditional evolutionary and non-evolutionary methods on several real-world symbolic regression problems, paving a new direction to handle both the bloating and generalization endeavors of genetic programming.

Keywords: Genetic Programming · Memetic Semantic · Symbolic Regression

1 Introduction

For a given dataset (X, y), symbolic regression (SR) aims to find a function $f(X) : \mathbb{R}^n \mapsto \mathbb{R}$ that represents the underlying relationship between the input features (X) and an output (y). Over the last few years, genetic programming (GP) [14] has gained the attention of the machine learning (ML) community due to its capacity to learn both the model structure and its parameters without making assumptions about the data [26,34]. Moreover, the symbolic aspects of its solutions and their flexible representation enable it to learn complex data relationships. These properties have made it a candidate solution to replace neural networks, which are usually considered black-box and, consequently, hard to understand and explain. Symbolic regression is usually implemented through genetic programming (GP).

Traditional GP-based methods rely on the outcome of a program to decide how well it solves the task, ignoring intermediate results such as the semantics of its subtrees [22,27]. However, one can consider them to guide the search during its exploration process and, thus, to generalize on unseen data, and to favor short expressions that are usually easier to understand and analyze by the users. Furthermore, semantics can contribute to improving subtrees' reuse based not only on the performance of the whole tree but also on their effectiveness in approximating a desired output. Semantic backpropagation (SB) algorithm [9,27] has shown to be an effective strategy for dealing with such endeavors. Semantic backpropagation tries to find a set of subtrees that better approximate the desired outputs for a given tree's node in a supervised setting. In other words, it computes the desired outputs for each node on the path from the root regarding the target semantics and the semantics of the other subtrees in the tree.

At the same time, several mixtures of evolutionary and non-evolutionary methods have been proposed over the last few years. One example is memetics algorithms [26] that provide an effective way to compensate for the capability of global exploration of general evolutionary methods with the increased exploitation that can be obtained through local search. In this context, *this paper proposes an evolutionary multi-objective algorithm that combines both memetics and semantic backpropagation algorithms for symbolic regression problems.*

Different from traditional semantic backpropagation operators (e.g., random desired operator (RDO) [27]), our *memetic semantic GP for symbolic regression (MSGP)* approach (Sect. 4) only tunes the real-valued constants after a suitable tree has been found for the problem. Likewise, it computes them through linear scaling (LS) (i.e., regression) [11] and not randomly, and at each iteration, as implemented by the RDO operator [27]. Linear scaling aims to minimize the mean squared error (MSE) of a tree by performing a linear transformation on its outputs [11,12]. Consequently, it frees GP from this time-consuming task, allowing it to focus exclusively on the shape of the tree that fits the structure of the data rather than on trying to find a scale that approximates the target output. Last but not least, linear scaling helps in dealing with GP bloat problem [11,12].

Additionally, instead of trying to build a library with all possible precomputing subtrees up to a maximum height or a dynamic one, which increases the computing cost and interpretability due to the bloat problem, MSGP relies on a fixed library with a randomly generated population of subtrees up to a given height.

Experimental results (Sect. 6) on various real-world benchmark datasets show that MSGP either outperforms or is equal to traditional machine learning methods (e.g., decision tree (interpretable) and random forest (black-box method) [2]), and established GP-based methods (e.g., *gplearn*). Likewise, our approach leads to short expressions which improve the interpretability of the model without including any new parameter to be specified by the users. MSGP's code is available at gitlab.inria.fr/trust-ai/memetics/msgp.

2   Semantic GP

In GP, semantics describes the behavior of a program on a specific dataset. In other words, it is the outputs' vector for the fitness cases of a problem [23].

More formally, in a supervised setting, assume the data is a set $\mathcal{D}$ made of N fitness cases: $\mathcal{D} = \{(X_1, y_1), (X_2, y_2), \ldots, (X_N, y_N)\}$, where $X_i \in \mathbb{R}^n$ and $y_i \in \mathbb{R}^1$ are the inputs, and the corresponding desired outputs. The semantics (s) of a program p is the vector of outputs values computed by p from the set of all fitness cases $\mathcal{D}$, defined as [27]:

$$s(p) = [p(X_1), p(X_2), \ldots, p(X_N)] \tag{1}$$

Similar operations can be performed for every node of a given tree: the semantics can be computed from the tree's terminals up to the tree's root sequentially, defining the semantics for every node (i.e., subtree) of a GP tree.

Semantic backpropagation algorithms [9,27] try to find the subtrees whose semantics better approximate the desired outputs (d_i^N) of a node $\mathcal{N} \in p$. A prerequisite is that one can compute the desired outputs for every node in p, conditional on the target output and the semantics of the other nodes in the program. This operation can be done downward from the root node (where the desired outputs are the target values o_i of the problem definition given in the initial dataset). For all the other nodes, this is done by performing the inverse operation of the function implemented in the node, assuming that the semantics of all other nodes are fixed: from the target values are the root, semantic backpropagation recursively computes the desired output for a node $\mathcal{N}$ at depth D as [15,27]:

$$d_i^N = F_{A_{D-1}}^{-1}(d^{A_{D-1}}, S_d) \tag{2}$$

where, A represents the ancestor of $\mathcal{N}$ at depth D_i, S the siblings of A, and F^{-1} comprehends the inverse of the function implemented by node A.

It is fundamental to highlight the difference between the semantics of subtrees and the semantics of contexts. On the one hand, in the **semantics of subtrees**, the semantics of a node $\mathcal{N}$ only depends on its output for each fitness case, which means that if nodes $\mathcal{N}_1$ and $\mathcal{N}_2$ have the same semantics and a program p contains the former, replacing it with the latter will not change the semantics of p. On the other hand, in the **semantics of contexts**, given a node $\mathcal{N} \notin p$, it is usually hard to know how it will impact the semantics of the entire program (i.e., tree) since such information is conditioned to the semantics of the node that will be replaced, as well as the semantics of its ancestors and siblings [22]. In some contexts, it can remain the same (i.e., a fixed context independent of the replaced node) or change (i.e., variable context). Consequently, the semantics of a node is uniquely defined by the function it implements and the value of its arguments, and they are independent of the position in the tree. In contrast, context semantics depend on the function implemented by the immediate parent, the parent semantics, and the semantics of the siblings [22]. As a result, local improvements may degrade the global performance.

---

[1] We are focusing in this work on the specific case of SR, but X_i and y_i could belong to some other spaces, for instance, discrete spaces in the case of classification or boolean functions.

2.1 Library Building and Searching

For a given desired output at a given node, we want to search for a tree that better approximates these outputs than the current subtree. One can be achieved by building a library in a static or dynamic setting. In the static setting, the semantics of all possible subtrees with a maximum height are pre-computed, and redundant semantics are pruned to keep only one tree for each unique semantics. In the dynamic setting, also known as population-based, new trees are added based on the observed subtrees of every generation [5,27]. This strategy keeps only the subtrees with the smallest number of nodes if different ones exist in the population with the same output. Moreover, both strategies ignore the subtrees with constant outputs.

Once a library has been built, the search process looks for the individuals whose outputs o are the closest to the desired semantics d_i^j based on some distance metric (e.g., Euclidean or Minkowski distance). It means, finding a minimal distance d_i^j that minimizes $|d_i^j - o_i|^k, \forall k \in \{1, 2, \ldots, N\}$ [27]. Additionally, if the distance value remains the same, whatever the subtree, its value is defined as zero (i.e., $| * -o_i|^k = 0$). Finally, as subtrees with constant outputs are ignored, the search process checks if a constant semantics could reduce the distance between the tree outputs and the desired ones.

3 Memetic Algorithms

Memetic algorithms (MA) combine population-based search strategies with local search heuristics inspired by the concept in genetics [7]. They have been used across different domains due to their capacity to establish a good balance between exploration and exploitation when finding a solution for a complex optimization problem [6,24]. A meme, in this case, represents transferable knowledge built through local refinement procedures, which can be seen as a form of domain-specific expert knowledge on how a solution can be better improved [24]. For an optimization viewpoint, prototypical memetic algorithms comprise three main phases named creation, local improvement, and evolution (Algorithm 1)[2]. A population of randomly created individuals is set up in the creation phase. Then, each individual is locally improved up to a predefined level in the improvement phase. Finally, the evolution phase is the usual phase of evolutionary algorithms that starts by selecting individuals based on their fitness and combining/mutating them through variation operators (e.g., crossover and mutation), enabling them to share information in a cooperative manner. The last two phases repeat until they meet a stopping criterion [24].

---

[2] Though other types of hybridization between evolutionary computation (EC) and local search have been proposed, like using the local search as pre- or post-processor, as a mutation operator, among others that are beyond the focus of this work.

Algorithm 1. Memetic algorithm
___
create a population of individuals
repeat
    improve some or all individuals with some local search algorithm
    select, then combine and/or mutate the individuals
until stopping criteria

---

4   Memetic Semantic for Symbolic Regression

This section describes how semantic backpropagation and memetic algorithms are combined to evolve GP models for symbolic regression problems that are interpretable and have a lower learning error.

Although semantic backpropagations and memetic algorithms have been separately used on SR problems, combining them can improve the interpretability and the generalization efficiency of GP-based model. While SB helps one in finding programs (i.e., trees) with the approximated desired output, MA contribute to improving them by considering the semantics of their parts (i.e., subtrees).

In a standard SB-based approach, once a subtree is selected to replace a node in a tree, it adds some constants to enable the tree to output the desired output. Consequently, as evolution proceeds, the trees often undergo excessive growth, known as bloat, which penalizes the search process, drastically increases the evaluation cost, and hinders the generalization of the trees (i.e., the accuracy on unseen data). We handle these issues by using LS [11] to search for constants that correct the residual errors of the tree. As a result, at each iteration, the SB and memetic algorithmss can concentrate on the structure of the tree, leaving the scaling of the coefficients to linear scaling.

Given a dataset composed of N independent samples (X_i) with m independent input variables $(X_i = [x_{i,1}, x_{i,2}, \ldots, x_{i,m}])$ and a corresponding target output (y_i), the task of symbolic regression comprises in finding a tree $(\mathcal{T}(.))$ that minimizes the distance to an output (y) [13,31]. Such tree $\mathcal{T}(.)$ usually includes a set of predefined functions and terminals (a.k.a constants and input variables).

Hence, using the mean squared error (MSE) as the distance metric (a.k.a fitness function) for $\mathcal{T}(.)$, and denoting $\hat{y}$ the outputs of tree $\mathcal{T}$, the task of symbolic regression is to find a tree $\mathcal{T}(.)$ that minimizes $MSE(\mathcal{T})$ defined as:

$$MSE(\mathcal{T}) \equiv MSE(y, \hat{y}) = \frac{1}{N} \sum_{i=1}^{N} (y_i - \hat{y}_i)^2 \tag{3}$$

4.1   Algorithm

Given a tree $\mathcal{T}$ with a set of subtrees $\mathcal{S} = \{s_1, s_2, \ldots, s_n\}$ and a library $\mathcal{L}$ composed of l individuals (i.e., small subtrees), the goal of the proposed memetic semantic GP for symbolic regression (MSGP) algorithm (Algorithm 2) is to interactively improve $\mathcal{T}$ by checking for each subtree $s_i \in \mathcal{S}$ if there exists a subtree $s^* \in \mathcal{L}$ whose semantics are closest to the desired ones for s_i.

The starting tree $\mathcal{T}$ is usually created randomly. However, it can also be, for instance, the output of another GP-based SR approach to make it simpler and consequently easier to understand by the users. In other words, MSGP does not make any assumption about the size or nature of the initial tree when it uses the library $(\mathcal{L})$ to search the nodes that can better replace the one of a tree. Consequently, the size and heterogeneity of the library $\mathcal{L}$ can play an important role.

Further, the memetic part of MSGP, linear scalings computes a scaled version of the MSE [11] with a computing cost that is linear with the dataset size N, (i.e., $\mathcal{O}(n)$):

$$\text{MSE}^{a,b}(y, \hat{y}) = \frac{1}{N} \sum_{i=1}^{n} (y_i - (a + b\hat{y}_i))^2 \tag{4}$$

With a and b defined as:

$$a = \bar{y} - b\bar{\hat{y}} \tag{5}$$

$$b = \sum_{i=1}^{N} \frac{(y_i - \bar{y}_i)(\hat{y}_i - \bar{\hat{y}})}{(\hat{y}_i - \bar{\hat{y}})^2} \tag{6}$$

These coefficients a and b are then added to the final tree. Moreover, the algorithm ignores trees with constant outputs.

Finally, to avoid consuming computing resources to an already optimal tree, an early stopping strategy can be adopted in practice.

Algorithm 2. Memetic semantic for symbolic regression

Require: Initial tree $(\mathcal{T})$, library $(\mathcal{L})$
Require: # *epochs*, and fitness cases $(\mathbf{X}, y)$

1: $\mathcal{T}' \leftarrow clone(\mathcal{T})$
2: Evaluate($\mathcal{T}'$)
3: **while** $e \leq epochs$ **do**
4: $\quad \mathcal{T}^* \leftarrow lti(\mathcal{T}', \mathcal{L}, \mathbf{X}, y)$ $\qquad\qquad\qquad$ ▷ local tree improvement (Algorithm 3)
5: $\quad \mathcal{T}^* \leftarrow$LS($\mathcal{T}^*, y$) ▷ linear scaling [11] computes the coefficients of $\mathcal{T}^*$ (Eqs. (5) and (6))
6: $\quad \mathcal{T}' \leftarrow$ best($\mathcal{T}^*, \mathcal{T}'$)
7: **return** $\mathcal{T}'$

4.2 Local Tree Improvement

Local tree improvement (LTI) algorithm (Algorithm 3) identifies the subtrees with equal or better semantics than a randomly selected subtree of a given tree. It performs an exhaustive search in a library $\mathcal{L}$ with a set of pre-computed semantics using the ancestor's semantics of a subtree as the target process. Hence, given a library $\mathcal{L}$ and a tree $\mathcal{T}$, LTI finds a subtree $s^\star$ in $\mathcal{L}$ that minimizes

$$\arg\min_{s^\star \in \mathcal{L}} \min_{t \in \{t_i, \ldots t_n\}} d(t, s(s^\star)) \tag{7}$$

where, t is the desired semantics and s represents the semantics of a subtree $s^\star \in \mathcal{L}$. During the search process, the algorithm keeps track of the semantics distance between the already analyzed subtrees to avoid replacing them several times. The error function computes the distance between the semantics of the subtrees, which in this case, comprises the semantics of the candidate subtree and the ancestor's semantics of the selected subtree. If no local improvement was identified, the algorithm randomly replaces a subtree in $\mathcal{T}$ by the one also randomly selected from the library, which in this case, can be seen as a muta-tion operation. One can observe that local enhancement may degrade global criteria (e.g., accuracy, height, and generalization). Thus, it is up to the supe-rior level to keep or ignore the new proposed tree. A further investigation may evolve each proposed tree during a predefined number of generations and then crossover them using the LTI algorithm.

We consider a static library composed of trees up to a certain height and with heterogeneous semantics. Only the smallest tree is included in the library when two candidate ones have the same semantics. Moreover, we also individually include the features of the problems, as well as the operators (i.e., functions), into the library.

5   Experimental Setup

We evaluated the proposed algorithm on different real-world regression dataset benchmarks. They have a heterogeneous number of features and sample sizes, as depicted in Table 1. Moreover, they are commonly used in the GP litera-ture [19,36] as overfitting the training set occurs either when complex models are learned or when models are built using discontinuous functions. Furthermore, Dow Chemical and Tower datasets are recommended as benchmarks [37]. They come from the UCI machine learning repository (archive.ics.uci.edu) and from the repository (shortest.link/8n9V) provided by Martins et al. [21].

Table 2 includes the parameters settings to define the library, the initial tree, and the one used in standard GP experiments. We use analytical quotient (AQ) instead of protected division to avoid discontinuous behaviors [25], but keeping the same general properties of division. Likewise, the literature has shown that using it helps generalize at prediction time [4,25,36]. It is defined as:

$$AQ(x_1, x_2) = \frac{x_1}{\sqrt{1 + x_2^2}} \tag{8}$$

Algorithm 3. Local tree improvement

Require: Tree $(\mathcal{T})$, library $(\mathcal{L})$
1: COMPUTE-SEMANTICS$(\mathcal{T})$ $\triangleright \forall$ node $N \in \mathcal{T}$
2: $S \leftarrow$ SUBTREES$(\mathcal{T})$
3: $\tau \leftarrow$ SORT(S) $\triangleright$ by error ascending and height descending
4: $s \leftarrow$ RANK-SELECT(τ)
5: $best[s] \leftarrow \emptyset$
6: **for all** $s^\star \in \mathcal{L}$ **do**
7:     **if** $(s, s^\star) \in k$ **then**
8:         continue
9:     $e \leftarrow$ ERROR$(s, s^\star)$
10:    **if** $e < min\_error$ **then**
11:        $best[s] \leftarrow (s^\star, 0)$
12:        $min\_error \leftarrow e$
13:    **else if** $e == min\_error$ **then**
14:        $best[s]$ $best[s] \leftarrow \cup \{(s^\star, e)\}$
15:    **if** $|best[s]| > 0$ **then**
16:        $k \leftarrow k \cup \{(s, s^\star)\}$
17:        $\mathcal{T}^\star \leftarrow$ CROSSOVER(ancestor$(s), s^\star, \mathcal{T})$
18:        SEMANTICBACKPROGRATION(ancestor$(s), s^\star, \mathcal{T}^\star)$
19:        **return** $\mathcal{T}^\star$
20: $s \leftarrow$ RANDOM(τ)
21: $s^\star \leftarrow$ RANDOM$(\mathcal{L})$
22: $\mathcal{T}^\star \leftarrow$ CROSSOVER(ancestor$(s), s^\star, \mathcal{T})$
23: SEMANTICBACKPROGRATION(ancestor$(s), s^\star, \mathcal{T}^\star)$
24: $k \leftarrow k \cup \{(s, s^\star)\}$
25: **return** $\mathcal{T}^\star$

As baseline, we considered evolutionary (i.e., GP) and non-evolutionary (i.e., decision tree (DT) and random forest (RF) [2]) approaches. We relied on *gplearn* [33] as the GP-based model, and on the *scikit-learn* [28] implementation of decision tree and random forests as they are commonly used in the GP and machine learning literature [8,30,32]. While decision trees are normally considered interpretable models, *random forests* are defined as black-box. Nevertheless, the latter often outperforms the former. Consequently, they are usually employed by practitioners across different domains. For these models, we used the default parameter values defined by *scikit-learn*. Each experiment comprised 30 independent runs, and the median of the results is reported. MSGP was implemented in `python` using the `DEAP` library [10]. Finally, we run the experiments on a MacBook Pro with one Apple M1 processor (8 cores) and 16 GB of RAM memory.

Table 1. Regression datasets benchmarks considered by this work

Name	Acronym	# Features	# Samples
Airfoil	AF	5	1503
Boston housing	BH	13	506
Concrete compressing strength	CCS	8	1030
Dow chemical	DC	57	1066
Energy cooling	EC	8	768
Energy heating	EH	8	768
Tower	TW	25	4999
Wine red	WR	11	1599
Wine white	WW	11	4898
Yacht hydrodynamics	YH	6	308

Table 2. Parameter settings

Parameter	Value
Function set	$\{+, -, \times, \div (AQ)\}$
Terminal set	Features
Initial tree height	2
# epochs	$1e4$
Max time	$500\,s$
# Trials	30
Loss function	MSE
Library size	200
Initialization	Ramped H&H [1–2]
Train-validation-test-split	50%-25%-25%
Data normalization	$L2$

6   Results

Table 3 shows the error on the testing set, and the symbolic expressions out-putted by MSGP. The experiments reveal that MSGP outperformed the baseline methods except for the Boston housing (BH) dataset. Additionally, the expressions are short, showing that our proposed method could handle GP bloating on the selected problems. Moreover, it required less than 2000 epochs to find the best trees on most datasets, as depicted in Fig. 1. An early stopping strategy or other automatic options can be added to avoid running without further improvement. However, one can keep to users the decision of which tree to pick up for a given problem as an option to help them explore alternative solutions, as they all have distinct semantics.

Table 3. Performance on the test set for each benchmark dataset

DS	DT	RF	GP	MSGP	Expression
AR	82.88	75.89	30587	**24.50**	$-90.09 * (X_1 + X_3 - X_4 + X_5) + 129.9$
BH	**10.96**	75.89	36.9	28.55	$1598 * X_6 * (-X_{13} + X_6) + 19.47$
CCS	583.32	349.41	441.13	**145.85**	$2052.89 * (X1 + X8) * (X5 + X6) + 15.52$
DC	0.59	0.49	0.1	**0.07**	$-166.12 * (X_{17} - X_{49} - \dfrac{X_{49}}{X_4}) - 7.16$
EC	101.27	115.46	179.51	**14.68**	$558.67 * ((X_1 * X_7) + X_2 + X_5) - 56.23$
EH	143.78	94.88	162.33	**13.74**	$576.3 * (X_2 + X_3 * X_7 + X_5) - 60.97$
TW	10586.48	10214.52	33253	**3023.69**	$-27120.74 * (X_1 + X_{16} - X_{23} - X_6) + 346.49$
WR	2.48	1.53	0.7	**0.5**	$43.26 * X_{11} - 21.63 * X_2 - 21.63 * X_8 + 5.65$
WW	1.6	1.01	0.71	**0.63**	$127.95 - (125.5 * X_2)^2$
YH	791.61	699.25	100.77	**32.47**	$-28008.75 + 28014.61 * \dfrac{X_6}{X_2}$

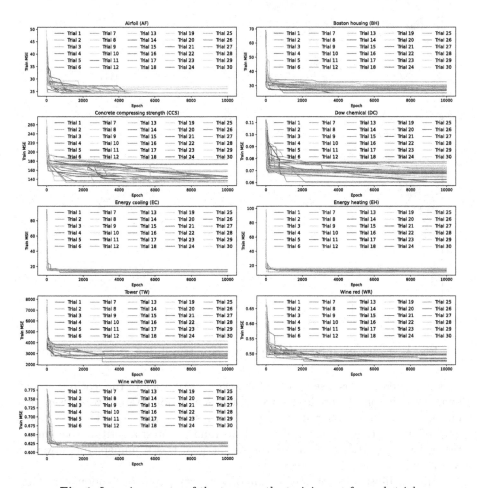

Fig. 1. Learning curves of the trees on the training set for each trial

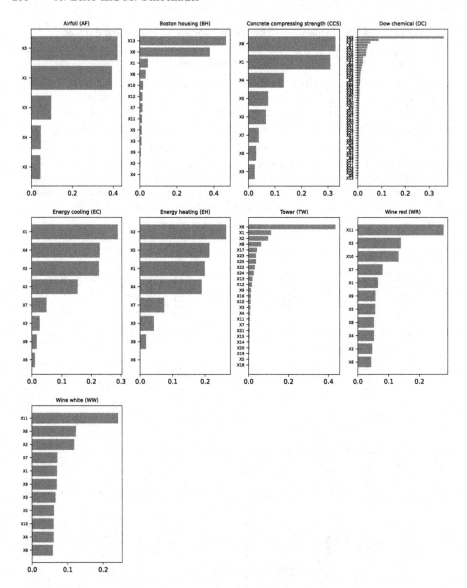

Fig. 2. Feature importance for each dataset obtained through random forest

Consequently, it is essential to check the contribution of the features chosen by the model from the prediction viewpoint. A common strategy comprises computing a score measure for all the input features. Examples include the Gini and mutual information measures. We used the random forest algorithm to compute the feature importance of the datasets. The goal was to understand if MSGP was relying on the relevant ones.

Figure 2 illustrates the feature importance computed by random forest for each dataset. We can observe that based on this metric, MSGP symbolic expressions (Table 3) include the important ones identified by random forest without needing external support. Indeed, we observed with some experiments that building a library with only the most important features degrades the model's performance and increases the output expressions' size. This behavior suggests that the proposed method requires some freedom to explore the search space. However, further analyses are still necessary to understand the reasons. Furthermore, complementary studies can be done to assess the performance of MSGP as a feature selection approach, which plays a fundamental role in post-hoc explanation methods, including SHapley Additive exPlanations (SHAP) [20] and Local Interpretable Model-Agnostic Explanations (LIME) [29].

In addition, the results showed that for some problems (e.g., CCS and DC), the algorithm stays stuck in a local minimum. Such behavior suggests the need to be able to identify and handle such behavior by, for instance, introducing some transformations to enhance exploration. Finally, another improvement comprises quantifying the impact of replacing a subtree with another one by considering the semantics of the tree. It means identifying which operations are necessary to change it to the desired one in an optimization setting.

7 Related Work

Several works have tried to handle the bloat problem of genetic programming. For example, Bleuler et al. [1] used SPEA2 to identify candidate solutions based on fitness and size. Experimental results showed that the proposed strategy could reduce GP bloat and speed up convergence. In [18], the authors proposed a pseudo-hill-climbing strategy to control trees' size during the crossover operation. In this case, the proposed approach discards an offspring if it either degrades the fitness or increases a tree's size. Although this approach can slow down GP bloating, it penalizes the running time of GP algorithms. In [3], the authors integrated a local search strategy into geometric semantic genetic programming (GSGP) to speed up convergence and limit overfitting. Experimental results showed that although the proposed strategy only required a few generations to find suitable candidate solutions, they were lost over the next generations leading to performance dropping. Other works have proposed semantic-based operators to handle both bloating and generalization issues. Some examples include [16,23,27,35] among others. Uy et al. [35] proposed to semantic-based crossover operators, named semantic aware crossover (SAC) and semantic similarity-based crossover (SSC). Their main difference is in the definition of the semantic distances, which, when exchanging two subtrees, must be different but not widely different. Similarly, Moraglio et al. [23] suggested a semantic crossover operator that creates offsprings with a weighted average of their parents' semantics. Notwithstanding, it cannot handle GP bloating properly without further simplification procedures. Geometric semantic crossover (AGX) [17] tries to handle these endeavors by replacing parents' subtrees with other ones semantically close to their parents' midpoint semantic. Random desired operator (RDO) [27] introduces back-propagation. In this case, after a crossover operation, the semantics of the new subtree is back-propagated as described in Sect. 2. Different studies have shown that RDO outperforms the other operators on both regression [27] and boolean [9] problems.

This work is closest to the RDO operator [27]. Nonetheless, we use a similar idea to improve a candidate solution's fitness locally instead of relying on crossover operation. Likewise, we use linear scaling [17] to compute the coefficients scale once a candidate solution was found. Finally, a memetic algorithm selects the solutions based on their fitness on a validation set and size. To the best of our knowledge, this is the first work to propose a memetic semantic algorithm for symbolic regression problems.

8   Conclusion

Symbolic regression (SR) searches for a set of mathematical expressions that better approximate a target variable of a given dataset. It is commonly implemented through genetic programming due to its characteristics in exploring the search space free of constraints' assumptions about the underlying data distribution. Nevertheless, GP-based approaches still face the challenge of overfitting the data expressed through complicated symbolic expressions (a.k.abloat). Semantic-based strategies [17,27] have been seen as a way to handle this issue. In this context, this paper proposed and evaluated a memetic semantic algorithm for symbolic regression (MSGP). The proposed approach combines a population-based search strategy with semantics-guided ones to output short symbolic expressions without penalizing the accuracy.

Experimental results demonstrated that in addition to favoring short and interpretable expressions, the proposed algorithm could outperform traditional machine learning models (i.e., decision tree (DT) and random forest (RF) [2]) and evolutionary one on different real-world datasets. Additionally, they demonstrated that the proposed algorithm only required a few iterations to identify the most predictive features. Further works include employing it to search for counterfactual outputs, as the counterfactual response can be framed as desired semantics. Another one comprises investigating new strategies to guide the construction of the semantics library to enhance the exploration and exploitation features of memetic semantic-based algorithms.

Acknowledgements. This research was partially funded by the European Commission within the HORIZON program (TRUST-AI Project, Contract No. 952060).

References

1. Bleuler, S., Brack, M., Thiele, L., Zitzler, E.: Multiobjective genetic programming: reducing bloat using SPEA2. In: Congress on Evolutionary Computation, vol. 1, pp. 536–543 (2001)
2. Breiman, L.: Random forests. Mach. Learn. **45**(1), 5–32 (2001)

3. Castelli, M., Trujillo, L., Vanneschi, L., Silva, S., Z-Flores, E., Legrand, P.: Geometric semantic genetic programming with local search. In: Annual Conference on Genetic and Evolutionary Computation, pp. 999–1006 (2015)
4. Chen, Q., Xue, B., Niu, B., Zhang, M.: Improving generalisation of genetic programming for high-dimensional symbolic regression with feature selection. In: IEEE Congress on Evolutionary Computation, pp. 3793–3800 (2016)
5. Chen, Q., Zhang, M., Xue, B.: Geometric semantic genetic programming with perpendicular crossover and random segment mutation for symbolic regression. In: Shi, Y., et al. (eds.) SEAL 2017. LNCS, vol. 10593, pp. 422–434. Springer, Cham (2017). https://doi.org/10.1007/978-3-319-68759-9_35
6. Chen, X., Ong, Y.S., Lim, M.H., Tan, K.C.: A multi-facet survey on memetic computation. IEEE Trans. Evol. Comput. **15**(5), 591–607 (2011)
7. Dawkins, R.: The Selfish Gene. Oxford University Press, Oxford (1976)
8. Ferreira, J., Pedemonte, M., Torres, A.I.: A genetic programming approach for construction of surrogate models. In: Computer Aided Chemical Engineering, vol. 47, pp. 451–456. Elsevier (2019)
9. Ffrancon, R., Schoenauer, M.: Memetic semantic genetic programming. In: Annual Conference on Genetic and Evolutionary Computation, pp. 1023–1030 (2015)
10. Fortin, F.A., De Rainville, F.M., Gardner, M.A.G., Parizeau, M., Gagné, C.: DEAP: evolutionary algorithms made easy. J. Mach. Learn. Res. **13**(1), 2171–2175 (2012)
11. Keijzer, M.: Improving symbolic regression with interval arithmetic and linear scaling. In: Ryan, C., Soule, T., Keijzer, M., Tsang, E., Poli, R., Costa, E. (eds.) EuroGP 2003. LNCS, vol. 2610, pp. 70–82. Springer, Heidelberg (2003). https://doi.org/10.1007/3-540-36599-0_7
12. Keijzer, M.: Scaled symbolic regression. Genet. Program Evolvable Mach. **5**(3), 259–269 (2004)
13. Korns, M.F.: A baseline symbolic regression algorithm. In: Riolo, R., Vladislavleva, E., Ritchie, M., Moore, J. (eds.) Genetic Programming Theory and Practice X. Genetic and Evolutionary Computation, pp. 117–137. Springer, New York (2013). https://doi.org/10.1007/978-1-4614-6846-2_9
14. Koza, J.R.: Genetic Programming: On the Programming of Computers by means of Natural Evolution. MIT Press, Massachusetts (1992)
15. Krawiec, K.: Semantic genetic programming. In: Krawiec, K. (ed.) Behavioral Program Synthesis with Genetic Programming. SCI, vol. 618, pp. 55–66. Springer, Cham (2016). https://doi.org/10.1007/978-3-319-27565-9_5
16. Krawiec, K., Lichocki, P.: Approximating geometric crossover in semantic space. In: 11th Annual conference on Genetic and Evolutionary Computation, pp. 987–994 (2009)
17. Krawiec, K., Pawlak, T.: Approximating geometric crossover by semantic backpropagation. In: 15th Annual Conference on Genetic and Evolutionary Computation, pp. 941–948 (2013)
18. Langdon, W.B., Poli, R.: Genetic programming bloat with dynamic fitness. In: Banzhaf, W., Poli, R., Schoenauer, M., Fogarty, T.C. (eds.) EuroGP 1998. LNCS, vol. 1391, pp. 97–112. Springer, Heidelberg (1998). https://doi.org/10.1007/BFb0055931
19. Liu, D., Virgolin, M., Alderliesten, T., Bosman, P.A.N.: Evolvability degeneration in multi-objective genetic programming for symbolic regression. In: Genetic and Evolutionary Computation Conference, pp. 973–981 (2022)
20. Lundberg, S.M., Lee, S.I.: A unified approach to interpreting model predictions. In: NeurIPS, pp. 4768–4777 (2017)

21. Martins, J.F.B., Oliveira, L.O.V., Miranda, L.F., Casadei, F., Pappa, G.L.: Solving the exponential growth of symbolic regression trees in geometric semantic genetic programming. In: Genetic and Evolutionary Computation Conference, pp. 1151–1158 (2018)

22. McPhee, N.F., Ohs, B., Hutchison, T.: Semantic building blocks in genetic programming. In: O'Neill, M., et al. (eds.) EuroGP 2008. LNCS, vol. 4971, pp. 134–145. Springer, Heidelberg (2008). https://doi.org/10.1007/978-3-540-78671-9_12

23. Moraglio, A., Krawiec, K., Johnson, C.G.: Geometric semantic genetic programming. In: Coello, C.A.C., Cutello, V., Deb, K., Forrest, S., Nicosia, G., Pavone, M. (eds.) PPSN 2012. LNCS, vol. 7491, pp. 21–31. Springer, Heidelberg (2012). https://doi.org/10.1007/978-3-642-32937-1_3

24. Moscato, P.: On evolution, search, optimization, genetic algorithms and martial arts: towards memetic algorithms. Technical report 826, Caltech Concurrent Computation Program, California Institute of Technology (1989)

25. Ni, J., Drieberg, R.H., Rockett, P.I.: The use of an analytic quotient operator in genetic programming. IEEE Trans. Evol. Comput. 17(1), 146–152 (2013)

26. Ong, Y.S., Lim, M.H., Neri, F., Ishibuchi, H.: Special issue on emerging trends in soft computing: memetic algorithms. Soft. Comput. 13(8), 739–740 (2009)

27. Pawlak, T.P., Wieloch, B., Krawiec, K.: Semantic backpropagation for designing search operators in genetic programming. IEEE Trans. Evol. Comput. 19(3), 326–340 (2014)

28. Pedregosa, F., et al.: Scikit-learn: machine learning in Python. J. Mach. Learn. Res. 12, 2825–2830 (2011)

29. Ribeiro, M.T., Singh, S., Guestrin, C.: Why should I trust you? Explaining the predictions of any classifier. In: SIGKDD, pp. 1135–1144 (2016)

30. Sathia, V., Ganesh, V., Nanditale, S.R.T.: Accelerating genetic programming using GPUs (2021)

31. Schmidt, M., Lipson, H.: Distilling free-form natural laws from experimental data. Science 324(5923), 81–85 (2009)

32. Sipper, M., Moore, J.H.: Symbolic-regression boosting. Genet. Program Evolvable Mach. 22(3), 357–381 (2021). https://doi.org/10.1007/s10710-021-09400-0

33. Stephens, T.: Genetic programming in python with a scikit-learn inspired API: gplearn (2016). github.com/trevorstephens/gplearn

34. Udrescu, S.M., Tegmark, M.: AI Feynman: a physics-inspired method for symbolic regression. Sci. Adv. 6(16), eaay2631 (2020)

35. Uy, N.Q., Hoai, N.X., O'Neill, M., McKay, R.I., Galván-López, E.: Semantically-based crossover in genetic programming: application to real-valued symbolic regression. Genet. Program Evolvable Mach. 12(2), 91–119 (2011)

36. Virgolin, M., Alderliesten, T., Witteveen, C., Bosman, P.A.: Improving model-based genetic programming for symbolic regression of small expressions. Evol. Comput. 29(2), 211–237 (2021)

37. White, D.R., et al.: Better GP benchmarks: community survey results and proposals. Genet. Program Evolvable Mach. 14(1), 3–29 (2013)

Grammatical Evolution with Code2vec

Michał Kowalczykiewicz$^{(\boxtimes)}$ ⓘ and Piotr Lipiński ⓘ

Computational Intelligence Research Group, Institute of Computer Science,
University of Wroclaw, Wrocław, Poland
{michal.kowalczykiewicz,piotr.lipinski}@cs.uni.wroc.pl

Abstract. This paper concerns using the code2vec vector embeddings of the source code to improve automatic source code generation in Grammatical Evolution. Focusing on a particular programming language, such as Java in the research presented, and being able to represent each Java function in the form of a continuous vector in a linear space by the code2vec model, GE gains some additional knowledge on similarities between constructed functions in the linear space instead of semantic similarities, which are harder to process. We propose a few improvements to the regular GE algorithm, including a code2vec-based initialization of the evolutionary algorithm and a code2vec-based crossover operator. Computational experiments confirm the efficiency of the approach proposed on a few typical benchmarks.

Keywords: Grammatical Evolution · Source Code Vector Embeddings · Code2vec

1 Introduction

Recent research in machine learning and software engineering focuses, among other things, on learning source code embeddings, i.e. some vector representations of the source code that move similarities of programs (e.g. some semantic properties) from the source code into a linear space. One of the most popular approaches is code2vec [3], based on a deep neural network and extending the word2vec approach [10] for vector representations of natural language words, that try to find a meaningful vector representation of snippets of source code in the form of continuous distributed vectors. It decomposes the source code into paths in abstract syntax trees, determines a vector representation of each path, and combines representations of paths into the vector representation of the entire snippet.

Learning source code embeddings becomes a common technique in many intelligent software engineering tools for facilitating the life of developers, including automatic detecting of source code duplicates or similarities, automatic suggesting of source code reusing or rewriting or intelligent support for source code improvements.

This paper studies the possibilities of using the vector embeddings of the source code to improve automatic source code generation in evolutionary

G. Pappa et al. (Eds.): EuroGP 2023, LNCS 13986, pp. 213–224, 2023.
https://doi.org/10.1007/978-3-031-29573-7_14

approaches, such as Grammatical Evolution (GE) [11]. GE is one of the popular evolutionary approaches to constructing a code of a function performing given operations, usually defined by an objective function that evaluates the constructed function. Focusing on a particular programming language, such as Java in our research, and being able to represent each Java function in the form of a continuous vector in a linear space, GE gains some additional knowledge on similarities between constructed functions in the linear space instead of semantic similarities, which are harder to process.

The paper proceeds as follows: Sect. 2 explains Grammatical Evolution and code2vec. Section 3 outlines new proposed methods for initialisation and selection in GE. Section 4 outlines the approach used to examine the proposed methods with experiments. Section 5 discuss the possible further work in this area.

2   Background

2.1   Grammatical Evolution

Grammatical Evolution (GE) was first introduced by M. O'Neill, JJ Collins and C. Ryan in [14]. It is an evolutionary algorithm, inspired by genetic processes from nature applied to evolution of programs. This method is able to satisfy the goal of finding the executable program in arbitrary language which achieves good fitness value for the specified objective function.

GE usually use the Backus-Naur Form (BNF) notation to describe the grammar for all possible programs in the search space. Programs (phenotypes) are mapped from binary strings (genotypes). The method has two parts: initialization and the creation of iterative offspring. During the initialization phase, chromosomes are generated to form the initial population. Then, until the maximum number of iterations is reached, offspring are created from the current population to form the next one. The creation of new chromosomes is done using genetic operators such as mutation, crossover, and selection. These operators guide the algorithm towards the final solution - the program that evaluates to the highest value for the given objective fitness function. An overview of a GE is presented in Algorithm 1. GE has been applied to develop programs addressing various problems, ranging from robot control [4] and aircraft modeling [5] to the automatic design of video games [15].

The initialization phase is an open research topic and literature discusses multiple methods for generating the initial population. From simple random generation of chromosomes to the algorithms based on Derivation Trees, like Ramped Half-Half [7,13] and Position Independent Grow [7] methods. The selection operator is also a widely studied research topic. Besides the commonly used Tournament selection, alternative methods such as gender specific selection [1], correlative tournament selection [9] and Semantic-Clustering Selection [8] have been proposed in the literature.

Algorithm 1. GrammaticalEvolution(bnf, fitness, popSize, maxGeneration)

INPUT: *bnf* - Backus-Naur Form grammar, *fitness* - objective function, *popSize* - amount of chromosomes in population, *maxGeneration* - maximal number of generations

OUTPUT: chromosome with best *fitness*

    Population ← generate *popSize* chromosomes for initial population
    $N \leftarrow 1$
    while $N \leq maxGeneration$ **do**
        Phenotypes ← map *chromosomes* from *Population* into programs on *bnf* grammar
        Fitnesses ← calculate the *fitness* function for all *Phenotypes*
        NextPopulation ← with *Fitnesses* and genetic operators generate offspring of *Population*
        $N \leftarrow N + 1$
        Population ← *NextPopulation*
    end while
    return chromosome from *Population* with best *fitness*

2.2   Code2vec

Code2vec was introduced in [3]. It is a neural model inspired by the word2vec algorithm [10] applied to code. This approach transforms program to collection of paths on its abstract syntax tree (AST) and based on this representation model learns code embeddings (fixed-size vectors) in a multidimensional space. Created vectors allow to model the correspondence between code - similar snippets are represented by similar vectors. Model was trained on programs from open GitHub repositories. Implementation details may by find in [3].

Code2vec can be used for many software engineering tasks like semantic labeling of code snippets (see example in Listing 1.1 and predicted labels in Table 1), code retrieval, classification and clone detection. This method continues to be an active area of research, with ongoing efforts to enhance the model and broaden its capabilities (e.g. [2,6]). In the next section we describe the first application of code2vec in GE.

Listing 1.1. Java snippet with unknown function name

```java
private static int f(char[] chars) {
    int result = chars[0];
    for (int index = 0; index < chars.length; index++) {
        if (chars[index] > result) {
            result = chars[index];
        }
    }
    return result;
}
```

Table 1. Predicted label for code snippet 1.1

Predicted method	Probability
max	65.5%
getResult	7.44%
occurrences	4.57%
getNext	4.17%
min	3.26%

3   Methods

In this paper, we propose an improvement of GE with using some latent data structure discovered by the code vectorization technique, namely code2vec, that introduces and benefits from some hidden relations between source codes reflected in the appropriate vector representation. In order to use such an additional knowledge in GE, we propose a set of hybrid algorithms, changing the initial population creation (*ClusterBooster*) and the next generation selection (*ClusterSelection*). In this section, we explain in details both methods.

Given code2vec's capability of mapping code snippets to vector embeddings, we can consider the population in GE as a collection of vectors within a multidimensional space and perform a variety of operations on these vectors, such as findings close vectors based on some distance metric and group similar vectors via clustering. For the further analysis, we first define abstractions:

- *phenotype$_v$*: denotes the vector $v_1, v_2, ..., v_n$ being the embedding created by code2vec
- *fitness$_{cluster}$*: denotes the best fitness calculated among all *phenotype$_v$* vectors belonging to *cluster*

3.1   ClusterBooster

In this section, we propose the new approach to create the initial population for GE. This method extends the classical random initialization using the code2vec algorithm which provides more diverse set of chromosomes.

The goal of *ClusterBooster* is to generate N chromosomes to form the starting population of GE. Initially, $N \cdot K$ chromosomes are generated randomly from which N chromosomes are selected. Subsequently, *phenotype$_v$* vectors for all chromosomes are calculated, generating the vector space that is partitioned into N clusters. By definition, clusters split the whole space, means each *phenotype$_v$* will be assigned to one and only one cluster. The vectors and the associated chromosomes within each cluster are similar amongst themselves and diverse from chromosomes from different clusters. In a looping process, a single chromosome is randomly picked from each cluster to construct the initial population. Algorithm 2 presents the method. In Sect. 4, we test different values for *boost* parameter K to analyse performance of this solution.

Algorithm 2. ClusterBooster(N, K)

INPUT: N - number of initial chromosomes, K - boost parameter
OUTPUT: *Result* initial chromosomes for GE

$Chromosomes \leftarrow$ generate list of random $N * K$ chromosomes
$vPhenotypes \leftarrow$ map all $Chromosomes$ into $phenotype_v$ representation
$Clusters \leftarrow$ generate N clusters on all $vPhenotypes$
$Result \leftarrow []$
while $cluster \in Clusters$ **do**
    $vPhenotype \leftarrow$ random $phenotype_v$ from $cluster$
    $chromosome \leftarrow$ get chromosome from $Chromosomes$ with $vPhenotype$ representation
    $Result \leftarrow Result + chromosome$
end while
return $Result$

3.2 ClusterSelection

In this section, we introduce the extension to GE which use knowledge of chromosome's vector representation also for creating subsequent generations.

In GE, the initial population is generated, and then successive generations are created using genetic operators, as detailed in [11]. One common selection operator is the Tournament method, which randomly selects a group of chromosomes using a uniform distribution, and then chooses the best one from that group. *ClusterSelection* follows the general structure of Tournament method and adjusts chromosome selection probability by taking into account the historical fitness values calculated in clusters.

Clusters created in *ClusterBooster* algorithm are reused and extended with new vectors. After the initialization phase, all clusters have just one $phenotype_v$ and sequentially with each next generated population all $phenotype_v$ vectors are added to the corresponding cluster. During the run of GE, method learns which clusters are the most promising ones to contain the chromosome with global fitness maximum, by calculating $fitness_{cluster}$ on all already added chromosomes into clusters. With this knowledge in each generation we iteratively adjust the probability of chromosome selection, by choosing the chromosomes with roulette like probability, *ClusterRoulette*:

$$ClusterRoulette(chromosome) = \frac{fitness_{cluster_i}}{\sum_{j=1}^{n} fitness_{cluster_j}},$$

for each *chromosome* in the i-th cluster.

The goal of *ClusterSelection* is to choose the best n chromosomes for next generation creation. Selection is done by n tournaments and each tournament consists of randomly picked *tournamentSize* chromosomes with uniform distribution, or *ClusterRoulette* distribution. Switching between these variants is toggled by hyperparameter *matchPercent*. Algorithm 3 presents the method. In Sect. 4, we test different values for *matchPercent* to analyse performance of this solution.

Algorithm 3. ClusterSelection(Chromosomes, matchPercent, tournamentSize)

INPUT: *Chromosomes* - List of chromosomes from current population, *matchPercent* - percent of selection with historic fintess, *tournamentSize* - size of tournament

OUTPUT: *Result* - selected chromosomes

    $n \leftarrow |Chromosomes|$
    $position \leftarrow 1$
    $Result \leftarrow []$
    while $position \leq n$ **do**
        $choice \leftarrow 1$
        $Tournament \leftarrow []$
        while $choice \leq tournamentSize$ **do**
            if $position/n \leq matchPercent$ **then**
                $chromosome \leftarrow$ random chromosome from *ClusterRoulette* distribution
            else
                $chromosome \leftarrow$ get random chromosome from *Chromosomes*
            end if
            $choice \leftarrow choice + 1$
            $Tournament \leftarrow Tournament + chromosome$
        end while
        $chromosome \leftarrow$ choose best chromosome in $Tournament$
        $Result \leftarrow Result + chromosome$
    end while
    return $Result$

---

4   Experiments

In this section, we present our computational experiments and discuss their results, comparing the regular GE to the GE augmented with *ClusterBooster* and *ClusterSelection*.

4.1   Benchmarks

We created 4 typical GE regression benchmark problems, shown in Listing 1.2, 1.3, 1.4 and 1.5. The objective in all benchmarks is to predict and generate a program that evaluates to the global minimum, which is 0. The search space of all possible programs is defined by grammar expressed in BNF, shown in Listing 1.6.

Listing 1.2. Benchmark I

```
total = 0.0
for i in range(100):
    value = float(i) / float(100)
    total += abs(<expr> - pow(value, 3))
fitness = total
```

Listing 1.3. Benchmark II

```
total = 0.0
for i in range(10):
    value = float(i) / float(10)
    total += abs(<expr> - sin(value) + pow(value, 4))
fitness = total
```

Listing 1.4. Benchmark III

```
total = 0.0
for i in range(10):
    value = float(i) / float(10)
    total += abs(<expr> - sin(value) + log(value + 5))
fitness = total
```

Listing 1.5. Benchmark IV

```
total = 0.0
for i in range(1, 100):
    value = float(i) / float(100)
    total += abs(<expr> - log(value) + pow(value, 2))
fitness = total
```

Listing 1.6. Benchmarks Grammar

```
<expr>        ::= <expr> <biop> <expr> | <uop> <expr>
               | <real> | log(abs(<expr>)) | <pow>
               | sin(<expr>)| value | (<expr>)
<biop>        ::= + | - | * | /
<uop>         ::= + | -
<pow>         ::= pow(<expr>, <real>)
<plus>        ::= +
<minus>       ::= -
<real>        ::= <int-const>.<int-const>
<int-const>   ::= <int-const> | 1 | 2 | 3 | 4 | 5 | 6 |
                  7 | 8 | 9 | 0
```

4.2 Experimental Setup

All methods are implemented in Python. For GE we used PyNeurGen[1] library and scikit-learn [12] for clustering calculation. In both methods, the clustering described in Sect. 3.2 was performed by the Lloyd algorithm for k-means++ clustering. Also, small library which maps Python snippets, constrained to BNF 1.6, into Java code was created.

---

[1] https://github.com/jacksonpradolima/PyNeurGen.

4.2.1   Code2vec

In all our experiments we used code2vec model[2], created by the authors of [6], for code vectorization. Model was trained on dataset introduced in [2] and consist of: "9500 top-starred Java projects from GitHub that were created since January 2007. It contains 9000 projects for training, 200 for validation and 300 for testing. Overall, it contains about 16M examples"[3].

4.2.2   GE

Every population consists of 50 chromosomes and maximal number of generations is set to 100. All methods use GE's operators - mutation, crossover and Tournament for selection. We do not remove possibly duplicated chromosomes in population. These common parameters are displayed in Table 2.

Table 2. GE experimental setup

Params	Value
Population	50
Generations	100
Number of runs	100
Mutation Rate	0.025
Crossover Rate	0.2
Duplicates	Possible
Tournament size	2
Elitism	0.05

4.2.3   ClusterBooster and ClusterSelection

Methods share the same common parameters as outlined in Table 2. Both methods were run with different variants of hyperparameters, tested *boost* and *matchPercent* params are displayed in Table 3. For all *ClusterSelection* runs, the initial population was build by *ClusterBooster* with $K = 100$.

Table 3. ClusterBooster and ClusterSelection experimental setup

Method	Param	Value
ClusterBooster	boost(K)	10, 50, 100
ClusterSelection	matchPercent(M)	10%, 20%, 50%

---

[2] https://zenodo.org/record/3577367/files/m1-standard.tar.gz.
[3] https://github.com/tech-srl/code2vec.

4.3   Results

Results for above experiments are presented in this section. Each benchmark problem was evaluated on all methods in 100 runs. For Benchmark I, we analyze the generated initial population, followed by a comparison of performance and a discussion on the size of the generated code. For benchmarks II, III and IV we conduct the performance results analysis.

Table 4 presents the statistics of chromosomes selected during the initialization phase. In all runs, the best fitness chromosome from each initial population was chosen. The comparison reveals that *ClusterBooster* produced better initial chromosomes than GE with classical random initialization.

Table 4. Benchmark I: Initial population results

Method	Avg	Median	Std
Random initialization	10.0921	9.3730	6.6099
ClusterBooster(K=10)	7.7260	7.2981	5.0069
ClusterBooster(K=50)	7.9871	8.1243	4.3051
ClusterBooster(K=100)	8.1036	7.1645	4.7294

Table 5 presents the performance results from all runs. Both *ClusterBooster* and *ClusterSelection* generated statistically improved solutions compared to GE, with a smaller variation in results. Figure 1 shows a visualization of the averaged run for GE, *ClusterBooster*($K = 10$) and *ClusterSelection*($M = 10$).

Table 5. Benchmark I: Performance results

Method	Avg	Median	Std
GE	0.1928	0.0	1.1488
ClusterBooster(K=10)	0.0701	0.0	0.1996
ClusterBooster(K=50)	0.0736	0.0	0.2005
ClusterBooster(K=100)	0.0873	0.0	0.2395
ClusterSelection(M=0.1)	0.0314	0.0	0.1289
ClusterSelection(M=0.2)	0.0392	0.0	0.1837
ClusterSelection(M=0.5)	0.0374	0.0	0.2477

Table 6 shows amount of runs which converged in 20, 50, 80 and 100 iterations. The vast majority of runs for all methods resulted in convergence at the global minimum. However, *ClusterBooster* and *ClusterSelection* for all hyperparameters, converged at a faster rate, with over half of the runs ending in 20 iterations.

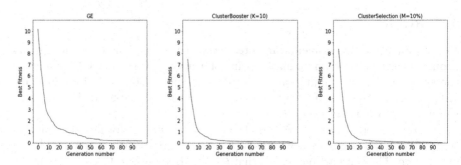

Fig. 1. Benchmark I: Average performance results

Table 6. Benchmark I: Successful runs with global minimum

Method	20	50	80	100
GE	50	70	75	76
ClusterBooster(K=10)	61	78	83	84
ClusterBooster(K=50)	57	76	82	82
ClusterBooster(K=100)	51	70	77	80
ClusterSelection(M=0.1)	59	77	80	84
ClusterSelection(M=0.2)	62	84	87	89
ClusterSelection(M=0.5)	67	88	92	92

Table 7. Benchmark I: Generated code-length comparison

Method	Avg	Median	Std
GE	213.2	212.0	4.5650
ClusterBooster(K=10)	211.44	210.0	2.3971
ClusterBooster(K=50)	212.01	210.0	3.4278
ClusterBooster(K=100)	212.14	210.0	4.1012
ClusterSelection(M=0.1)	212.9	212.0	4.2414
ClusterSelection(M=0.2)	212.49	210.0	8.1896
ClusterSelection(M=0.5)	212.09	210.0	3.8680

Table 7 shows the average length, calculated as the number of characters, of all generated programs. The results show a lack of significant difference among all methods, programs generated by *ClusterBooster* and *ClusterSelection* are similar to the ones created by GE.

Finally, Table 8 aggregates the performance results on all benchmarks for the hyperparameters with the best results from Table 5. Benchmarks II, III, and IV are more challenging than Benchmark I and none of the methods converged to the global minimum, resulting in a significant standard deviation in all cases. Despite this, methods *ClusterBooster* and *ClusterSelection* had better results than GE.

Table 8. Performance results summary

Method	Avg	Median	Std
Benchmark I			
GE	0.1928	0.0	1.1488
ClusterBooster(K=10)	0.0701	0.0	0.1996
ClusterSelection(M=0.1)	0.0314	0.0	0.1289
Benchmark II			
GE	0.9165	1.0969	0.3845
ClusterBooster(K=10)	0.8859	1.0392	0.3747
ClusterSelection(M=0.1)	0.7345	0.7341	0.3912
Benchmark III			
GE	1.6991	1.7462	0.5871
ClusterBooster(K=10)	1.6865	1.7462	0.5619
ClusterSelection(M=0.1)	1.5441	1.7462	0.6775
Benchmark IV			
GE	20.4731	22.6934	10.4753
ClusterBooster(K=10)	17.4413	18.7887	9.6822
ClusterSelection(M=0.1)	12.7584	12.2525	7.1833

5 Conclusion

In this paper, we proposed a novel idea to combine Grammatical Evolution with code2vec. Two extensions to GE were proposed, the conducted experiments proved that our methods have better average performance and the generated programs have similar code length as classical method. This study has indicated that usage of code embeddings represents a promising approach in GE. The next avenue of future work is to explore vectorization techniques in different problems in Genetic Programming.

Acknowledgment. This work was supported by the Polish National Science Centre (NCN) under grant OPUS-18 no. 2019/35/B/ST6/04379.

References

1. Affenzeller, M., Wagner, S., Winkler, S., Beham, A.: Genetic Algorithms and Genetic Programming: Modern Concepts and Practical Applications. Chapman and Hall/CRC (2009)
2. Alon, U., Brody, S., Levy, O., Yahav, E.: Code2seq: generating sequences from structured representations of code. arXiv preprint arXiv:1808.01400 (2018)
3. Alon, U., Zilberstein, M., Levy, O., Yahav, E.: Code2vec: learning distributed representations of code. Proc. ACM Program. Lang. **3**(POPL) (2019)

4. Burbidge, R., Walker, J.H., Wilson, M.S.: Grammatical evolution of a robot controller. In: 2009 IEEE/RSJ International Conference on Intelligent Robots and Systems, pp. 357–362. IEEE (2009)
5. Byrne, J., Cardiff, P., Brabazon, A., et al.: Evolving parametric aircraft models for design exploration and optimisation. Neurocomputing **142**, 39–47 (2014)
6. Compton, R., Frank, E., Patros, P., Koay, A.: Embedding Java classes with code2vec: improvements from variable obfuscation. In: Proceedings of the 17th International Conference on Mining Software Repositories, pp. 243–253 (2020)
7. Fagan, D., Fenton, M., O'Neill, M.: Exploring position independent initialisation in grammatical evolution. In: 2016 IEEE Congress on Evolutionary Computation (CEC), pp. 5060–5067. IEEE (2016)
8. Forstenlechner, S., Nicolau, M., Fagan, D., O'Neill, M.: Introducing semantic-clustering selection in grammatical evolution. In: Proceedings of the Companion Publication of the 2015 Annual Conference on Genetic and Evolutionary Computation, pp. 1277–1284 (2015)
9. Matsui, K.: New selection method to improve the population diversity in genetic algorithms. In: IEEE SMC 1999 Conference Proceedings. 1999 IEEE International Conference on Systems, Man, and Cybernetics (Cat. No. 99CH37028), vol. 1, pp. 625–630. IEEE (1999)
10. Mikolov, T., Sutskever, I., Chen, K., Corrado, G.S., Dean, J.: Distributed representations of words and phrases and their compositionality. In: Advances in Neural Information Processing Systems, vol. 26 (2013)
11. O'Neill, M., Ryan, C.: Grammatical evolution. IEEE Trans. Evol. Comput. **5**(4), 349–358 (2001)
12. Pedregosa, F., et al.: Scikit-learn: machine learning in Python. J. Mach. Learn. Res. **12**, 2825–2830 (2011)
13. Ryan, C., Azad, R.M.A.: Sensible initialisation in grammatical evolution. In: GECCO, pp. 142–145. AAAI (2003)
14. Ryan, C., Collins, J.J., Neill, M.O.: Grammatical evolution: evolving programs for an arbitrary language. In: Banzhaf, W., Poli, R., Schoenauer, M., Fogarty, T.C. (eds.) EuroGP 1998. LNCS, vol. 1391, pp. 83–96. Springer, Heidelberg (1998). https://doi.org/10.1007/BFb0055930
15. Shaker, N., Nicolau, M., Yannakakis, G.N., Togelius, J., O'neill, M.: Evolving levels for Super Mario Bros using grammatical evolution. In: 2012 IEEE Conference on Computational Intelligence and Games (CIG), pp. 304–311. IEEE (2012)

Short Presentations

Domain-Aware Feature Learning with Grammar-Guided Genetic Programming

Leon Ingelse[ID] and Alcides Fonseca[✉][ID]

LASIGE, Faculdade de Ciências da Universidade de Lisboa, Lisbon, Portugal
lingelse@lasige.di.fc.ul.pt, amfonseca@ciencias.ulisboa.pt

Abstract. Feature Learning (FL) is key to well-performing machine learning models. However, the most popular FL methods lack interpretability, which is becoming a critical requirement of Machine Learning. We propose to incorporate information from the problem domain in the structure of programs on top of the existing M3GP approach. This technique, named Domain-Knowledge M3GP, works by defining the possible feature transformations using a grammar through Grammar-Guided Genetic Programming. While requiring the user to specify the domain knowledge, this approach has the advantage of limiting the search space, excluding programs that make no sense to humans. We extend this approach with the possibility of introducing complex, aggregating queries over historic data. This extension allows to expand the search space to include relevant programs that were not possible before. We evaluate our methods on performance and interpretability in 6 use cases, showing promising results in both areas. We conclude that performance and interpretability of FL methods can benefit from domain-knowledge incorporation and aggregation, and give guidelines on when to use them.

Keywords: Interpretability · Domain-aware feature learning · Historical-data aggregation · Grammar-guided genetic programming

1 Introduction

One of the challenges of Machine Learning (ML) is to identify features that are relevant for predictors to consider. This step is considered critical to the application of ML in the real world.

Unfortunately, raw features obtained from sensors and datasets are not necessarily optimal. As such, these are generally processed through Feature Engineering (FE) methods, which can include removing irrelevant features and constructing new features by combining other features [15]. FE is resource intensive, as it requires "domain knowledge, intuition, and [...] a lengthy process of trial and error" [15].

FE can be done manually by the data scientist, using domain knowledge to identify relevant feature combinations. An example in the medical domain is

© The Author(s), under exclusive license to Springer Nature Switzerland AG 2023
G. Pappa et al. (Eds.): EuroGP 2023, LNCS 13986, pp. 227–243, 2023.
https://doi.org/10.1007/978-3-031-29573-7_15

the Body Mass Index (BMI), which combines sex, weight, and height. The BMI is more compact than the separate features, and contains the necessary health information, allowing for a quick consultation.

In some cases, the problem and data are so complex, that FE must be outsourced to learning algorithms, giving rise to the area of automated FE, here named FL. For example, in the areas of image recognition and natural language processing, deep-learning-based approaches have surpassed manually engineered features in performance. However, these features are typically less domain-specific, and, regarding many-layered models, are not considered interpretable [13]. Even simpler algorithmic methods like Support Vector Machines (SVM) and the widely used feature-dimensionality-reduction algorithm Principal Component Analysis [41] (PCA) are also not considered interpretable [13].

In domains where accountability is required, such as policy, medicine, defense and critical infrastructures, it is necessary to explain how decisions are made [1]. Interpretable features aid in understanding how decisions are made by ML algorithms. As such, feature interpretability falls within the Explainable AI (XAI) umbrella. Due to recent scandals in AI, such as racist and sexist models [42], there is a push for legislation to regulate the use of non-interpretable models [33].

We propose an ML solution that is aware of domain-specific knowledge and can create interpretable models. Our solution relies on Genetic Programming (GP) to build a classifier or regressor model for a given labeled dataset. We use the model produced by GP as a mapping of the feature set, similar to M3GP [28]. This mapping can be incorporated in ML *pipelines* as an FL method. GP finds interpretable models [7], and has a history of application for FL [4,5,8,13,35]. We introduce two FL methods that extend M3GP. The novelty of the first method is the encoding of domain knowledge in the FL process, and we name it Domain Knowledge M3GP (DK-M3GP). The second method uses the encoding of domain knowledge to learn features by dynamically aggregating over historical data, and we call it Domain Knowledge and Aggregation M3GP (DKA-M3GP). We use grammars to encode domain knowledge and aggregation into the solution space, through Grammar-Guided GP (GGGP) [38].

In this work, we research the effect of our methods on the search process for six use cases, when compared to normal M3GP, and other FL methods. We show that DK-M3GP improves the fitness in three of the six use cases, and worsens it in one use case. We argue that DK-M3GP improves the interpretability of learned features in three ways; (1) categorical features represented by numbers are converted back to their categorical values; (2) it blocks learning *nonsensical* features; and (3) learned features can be represented with branching conditions. Furthermore, DK-M3GP allows the introduction of aggregation in DKA-M3GP. Even though DKA-M3GP is computationally expensive, it improves the fitness of all use cases when compared to M3GP, and for five of the six use case when compared to DK-M3GP.

The rest of this paper is structured as follows. Section 2, we discuss other GP-based FL methods, especially those that touch upon domain knowledge and aggregation incorporation. In Sect. 3, we present the datasets, the grammars used for our methods, and Genetic Engine [17], the GGGP framework used throughout this work. Section 4, introduces the evaluation parameters, and Sect. 5, presents the results. Finally, in Sect. 6, we draw conclusions.

2   Related Work

There are plenty of FL methods, such as Principle Components Analysis [41] and Self-Organising Maps [22], and FL frameworks, such as FeatureTools [21] and AutoFeat [18]. In this work, we focus on GP-based FL methods (Sect. 2.1), and FL methods that incorporate domain knowledge and aggregation (Sect. 2.2).

2.1   Genetic-Programming-Based Feature Learning

Traditional GP (or standard GP) is the most basic GP-based FL method. Traditional GP evolves a mapping from the original feature set to a one-dimensional feature set. Normally, the function set consists of the following functions: addition, subtraction, multiplication, and *safe* division (division is made safe by mapping division by zero to a constant). The terminal set consists of the original features of the dataset, sometimes enriched with constant values. The terminal and function sets are used by the GP algorithm to create *building blocks*, referring to the parts that make up learned features. The mutation of an individual is done by changing a randomly selected tree node. The crossover of two individuals is done by swapping a randomly selected tree node of each individual.

A widely used GP-based FL method is Multidimensional Multiclass GP with multidimensional populations (M3GP) [28], a GP-based classification method that can also be used for FL [8] with excellent performance [2,8]. To classify a given dataset D, an M3GP individual maps the feature space of D to a new feature space.

In M3GP, an individual is a set of learned features, where each feature is assembled using a combination of the original features and the arithmetic operators $+$, $-$, $*$, and *safe* / (see Listing 1.1 for the grammar to learn a single feature). M3GP has two types of evolutionary operators; either the operation works on the sets of features; or on a single feature of the set. The mutation operator either adds or removes a single feature of the set, or it mutates a single feature in a traditional GP way. The crossover operator either swaps random subsets of the two feature sets, or it performs the crossover of two individual features in a traditional GP way, where each set supplies one feature. Each generation, either a crossover or a mutation is applied, with equal probability.

Originally, M3GP used the Mahalanobis measure for its fitness function, but a faster fitness function using a Decision Tree (DT) does not deteriorate performance [9]. More recently, autoencoders have replaced M3GP individuals to increase performance [36].

Later, M4GP [12] was presented, improving on M3GP. It introduces a stack-based program representation, multi-objective parent-selection techniques, and an archiving strategy. The stack-based program representation contributes to simpler models, making M4GP more efficient. The different multi-objective parent-selection techniques have different purposes. Where *lexicase* selection [34]

```
1 <Number>     ::= <Plus>
2              | <Multiply>
3              | <Variable>
4              | <Literal>
5 <Plus>       ::= <Number> + <Number>
6 <Minus>      ::= <Number> - <Number>
7 <Multiply>   ::= <Number> * <Number>
8 <Division>   ::= <Number> / <Number>
9 <Variable>   ::= x1 | x2 | x3 # Features of the dataset
```

Listing 1.1. Grammar for learning a single feature in M3GP.

rewards solutions that perform well on hard cases, *age-fitness Pareto survival* enhances the diversity of populations.

Feature Engineering Automation Tool (FEAT) [23] is a multi-objective, GP-based FL method that learns features for a linear regression model. It combines six different perturbation methods, four mutation variations, and two crossover variations. FEAT incorporates objective functions that aim to optimize three solution aspects; performance; the complexity of the solution; and the disentanglement of the solution. Performance is optimized using the mean square error. Complexity is optimized based on a function that sums the complexity weights of operators within a subtree. Here, the authors assign the complexity weights. Disentanglement is measured through two functions; one measures bivariate correlations and the other captures higher-order dependencies.

Like FEAT, there are other GP-based symbolic regression tools, such as Python's GP-GOMEA [37] and C++'s OPERON [11]. These tools perform very well on symbolic regression but do not directly support FL.

2.2 Domain-Aware Feature Learning and Aggregation Incorporation

Incorporating domain knowledge in GP-based FL can enhance performance and interpretability. For example, domain-knowledge-based operator definition has improved performance [24]. By integrating domain knowledge into a grammar, GGGP was used to improve performance and interpretability for FL [13]. Unfortunately, the above methods are not general, but domain-specific.

When dealing with time series or panel datasets, there is information contained in the data the model has already trained on that might benefit future data points. For example, look at the problems of daily bike occurrence prediction and a person's credit risk evaluation. The bike occurrence of the following day might correlate strongly with the bike occurrence of previous days. Similarly, the credit risk of a person might correlate with the risk of persons with similar characteristics, previously seen by the algorithm. An FL method that can generate features that involve historic data, can thus extract this information. In this paper, we refer to this concept as *aggregation of historic data* or just *aggregation*.

One attempt to incorporate aggregation into the FL process was done by Vectorial-GP (VE-GP) [6]. VE-GP includes vectors of (historic) data as features. These vectors can then be passed to a variety of vectorial functions, such as sum, mean, minimum, or maximum. However, VE-GP requires the relevant vectorial data to be available as a vector for each data point. Therefore, extra memory and manual curation of the relevant vectorial data are required.

Another attempt at automatically aggregating entities was done in AutoFE [32]. First, new features are exhaustively generated by combining terminals through a set of operators. AutoFE generates new features by (1) discretizing them, (2) combining features through standard arithmetic operators, and (3) feature aggregation after grouping on another discrete feature. The generated features then undergo an evolutionary-based FS process. This process indirectly optimizes a feature set by first expanding the feature set and then selecting relevant features. Furthermore, AutoFE does not allow for the combination of aggregated features with other features. Lastly, like with VE-GP, what is considered relevant historical data is statical.

3   Method

To improve the interpretability and performance of FL, we propose two methods. In Sect. 3.1, we introduce DK-M3GP, where the user defines the domain-specific operations allowed via a grammar. In Sect. 3.2, we introduce DKA-M3GP, which extends the first with aggregations over historical data.

3.1   Domain Knowledge M3GP

For practical use, tabular datasets used in supervised learning often use categorical features represented by numbers (CFRNs). For example, PMLB [30], a benchmark that collects problems from various fields, uses CFRNs. An example of CFRNs is encoding seasons by 1 for Winter, 2 for Spring, etc.

This representation comes with downsides: CFRNs are used to create *nonsensical* features, like multiplying the season with the temperature, which has no meaning for a domain expert. Features generated using CFRNs are therefore less interpretable than categorical counterparts [43].

In DK-M3GP, we propose the use of grammars to encode categorical features and what operations can interact between them. Furthermore, we encode relational properties between features that have semantic meaning in the domain.

Grammars can encode the terminal and function sets of GP that define the search space. However, instead of relying on a simple grammar, like Standard-GP or M3GP, we rely on user-defined grammars that encode domain specificities, as done in GGGP. To apply our method, the user should follow these steps:

Step 1: Identify categorical features. While the dataset may use a numeric representation, the user needs to identify each feature that is non-numeric and assign it a non-terminal, or a type.

```
 1  <BuildingBlock> ::= <Number> | <If>
 2  <Number>         ::= <Plus>
 3                    | <Multiply>
 4                    | <Variable>
 5  <Plus>           ::= <Number> + <Number>
 6  <Multiply>       ::= <Number> * <Number>
 7  <Variable>       ::= x1 | x2 | x3 # Features of the dataset
 8  <If>             ::= if <cond> then:
 9                         <BuildingBlock>
10                       else:
11                         <BuildingBlock>
12  <cond>           ::= <Equal>
13  <Equal>          ::= season == <season>
14  <season>         ::= Winter | Spring | Summer | Fall
```

Listing 1.2. Simplified example of a categorical-feature extension.

Step 2: Define a grammar that uses categorical features. To obtain a numeric feature from a combination of numeric and categorial features, the user must define functions that operate on categorical features, returning numeric features.

As an example, let us consider Listing 1.2, which treats seasons as categorical features. Seasons can be used to create features via the <Equal> expression, where the season of the instance is compared with one of the four seasons. This comparison is used in the <If> expression, to generate features with branches. The generated program is inherently more interpretable as season == Spring is more readable than x[13] == 1 [43]. In our evaluation, we extended this grammar with relational operations like <InBetween>, for ordered features, and <NotEqual>.

Note that ours is a generalizable method. In other words, the steps above can be done for any dataset that has CFRNs, where the grammar only differs on the categories defining each feature. Furthermore, other expressions and relational operations to handle the features can be introduced by the user.

3.2   Domain Knowledge and Aggregation M3GP

DKA-M3GP extends the previous method to allow the algorithm to use historic data for FL using aggregation (see Sect. 2.2 for a definition and an example). To apply DKA-M3GP, the user needs to include aggregation operations in the grammar. Listing 1.3 shows a very simple example for the bike example. A new feature can be created by considering the average of all past instances that pass a learned filter. One filter example might be only averaging winter days, when predicting the number of bikes recorded in a winter day.

Other possible aggregation methods that have shown to be beneficial are maximum, minimum, last, first, and sum [6].

Note that, like in AutoFE, the aggregation method in DKA-M3GP allows for filtering over a categorical feature before aggregating. This is not possible in VE-GP. DKA-M3GP also introduces features learned through aggregation back into

```
1  <BuildingBlock>  ::= <Number> | <If> | <Average>
2  <Average>        ::= historical_data.filter(<cond>)
3                                      .average()
```

Listing 1.3. Simplified example of aggregation, extending listing 1.2.

the search process, implemented in VE-GP, but not in AutoFE. Finally, unlike both VE-GP and AutoFE, the historic data that is aggregated is dynamic. DKA-M3GP evolves the window size of what is considered the historic data for each operation individually. In other words, each operation evolves what previous data it considers for its building block.

4 Evaluation

In this section we present the used datasets (Sect. 4.1), the implementation details of our methods (Sect. 4.2) and finally the experiment details (Sect. 4.3)

4.1 Datasets

We evaluated our FL methods on six use cases: two time-series regression problems; and four classification problems on *panel* datasets, detailed in Table 1.

Table 1. The used datasets and their characteristics.

Dataset	Problem type	Instances	Features	(#CFRNs)
Boom Bikes [10]	Regression	730	13	(7)
Credit risks [16]	Classification	1000	20	(13)
Caesarian [3]	Classification	80	5	(2)
Daily visitors† [29]	Regression	2167	4	(3)
Cleveland [14]	Classification	301	14	(6)
Horse colic† [16]	Classification	303	21	(9)

Because the original data had several features with a high percentage of missing data, the CFRN encoding would introduce erroneous information into the dataset, related to the representation of missing data when compared with existing values. To avoid this problem, we excluded features and instances with missing data. We label datasets that underwent preprocessing with †.

4.2 Implementation Details

We implemented all methods on top of Genetic Engine [17], a Python framework for GGGP that defines grammars using a Python internal DSL. Genetic Engine uses Python types to represent non-terminals, unlike the textual representation of grammars, and to define the relational properties as Python methods in those types. Genetic Engine also allows different GGGP implementations, like Grammatical Evolution (GE) [31], Structured GE (SGE) [25], and Context-Free Grammars GP (CFG-GP) [38]. As CFG-GP performs better than GE [20,26,39], and SGE has not shown a particular advantage over CFG-GP [26], in this work we use CFG-GP. We did not use the Probabilist Grammars [27] support, although it can be useful to model the probabilistic nature of the domain.

4.3   Experiment Details

We used the same parameter values for each GP-based algorithm; a maximum feature depth of 10; a population size of 200; a number of generations of 200; and an elitism of 2. For selection, we used a tournament selection with size 5. Crossover and mutation were done as in M3GP (Sect. 2.1), either occurring with a probability of 0.5.

Our M3GP implementation used a DT classifier or regressor, depending on the problem. We fitted the DT with a maximum depth of 4. We chose this depth for the DT to be small enough to be interpretable, and big for it to be effective. For the regression problems, we used the mean-squared error (MSE) as fitness metric. For the classification problems, we used the F1-score as fitness metric.

Besides comparing with M3GP, we also incorporated our methods within ML pipelines and compared their performance against traditional FL methods: traditional GP, PCA, and no FL. The ML pipeline consists of the FL method and a DT with a maximum depth of 4. Our FL methods use the same model within the fitness function as the one used to evaluate (namely DT). This creates an evaluation bias towards our methods. However, as we could change our fitness function to create bias for the problem at hand, this is a beneficial feature of our FL methods, and therefore reasonable to leave intact.

We tested with 25 % of the data, and we used 75 % of the data for training, in a time-series-aware manner [19]. Notice that we also use this method for the panel datasets to ensure the data used within the aggregation methods are only considered once they have been seen by the algorithm. We ran each experiment on 30 different seeds, to ensure statistical significance. The source code can be found on GitHub (https://github.com/Leoningel/FeatureLearningComparison).

5   Results

We evaluated our methods in three ways; (1) by comparing the fitness throughout the evolutionary process with M3GP; (2) by comparing the FL mapping of the best individual with other FL methods; and (3) on the interpretability of the found mapping.

Comparison with M3GP. The fitness during the evolutionary process of M3GP, DK-M3GP, and DKA-M3GP for each dataset is compared in Fig. 1. Except for Horse colic, we see that DK-M3GP performs on par with, or better than M3GP. Looking closer at the data of Horse colic, it only shows a single unordered categorical feature. Similarly, Boom Bikes and Caesarian, which do not benefit from domain-knowledge incorporation, have one and no unordered CFRN, respectively. As numbers have an inherent order, replacing ordered features with CFRNs is more logical. However, replacing unordered features with CFRNs includes an order where there is none, which introduces information not present in the original data. Cleveland has 2 unordered categorical features and sees benefits from incorporating domain knowledge. Credit risks benefits significantly from incorporating domain knowledge and has the highest percentage of unordered categorical features. Daily visitors is the exception, as it has no unordered features, but sees a strong fitness-progression improvement in DK-M3GP. However, further inspection of the Daily visitors data shows that the day of the week strongly influences the number of website visits in an unordered manner, namely, the first and last days see fewer visits than the other days. Even though the feature itself is ordered, its correlation with the target is unordered. As such, we hypothesize that the more unordered categorical features a dataset has, the more the evolutionary process benefits from the incorporation of domain knowledge.

Not only can domain-knowledge incorporation improve performance by itself, it also facilitates the incorporation of aggregation, as proposed in our method DKA-M3GP. DKA-M3GP shows a lot of potential, with big performance gains for Boom Bikes, Caesarian and Daily visitors. Furthermore, there is a slight improvement over DK-M3GP for Horse colic, whereas its performance is on par for Cleveland. For Credit risks DKA-M3GP shows a decrease in performance. As DKA-M3GP extends the search space with aggregation, this is possible when aggregation is irrelevant to the problem, as it will only distract the search. We hypothesize that for Boom Bikes, Caesarian and Daily visitors, historical data represent new data points well, and, as such, performance increases significantly with the incorporation of aggregation. For example, for Boom Bikes, the value for the *weather situation* column strongly influences the target, and the values of that column could be a good indication of future data. Further analysis shows that the *weather situation* column is the column that is most averaged on in the best individuals for Boom Bikes.

Comparison with Other FL Methods. To evaluate our methods as a FL method, we included it within an ML pipeline, and compared it to the functioning of other FL methods. We evaluated the methods using previously unseen data. See the results in Fig. 2, with statistical significance scores of the comparison of DKA-M3GP to each other method using the Mann-Whitney-Wilcoxon test [40]. The scores are given using stars (**** if $p \leq 0.0001$, *** if $p \leq 0.001$, ** if $p \leq 0.01$, * if $p \leq 0.05$, and ns if $p > 0.05$).

On unseen data, our methods show no clear performance benefits for the classification problems Caesarian, Credit risks, Cleveland and Horse colic. However, for the regression problems Boom Bikes and Daily visitors traditional GP outperforms our methods, and all other methods. One possible reason for our method to underperform, compared to traditional GP despite the better performance on training data is that it is learning faster, thus overfitting. Different

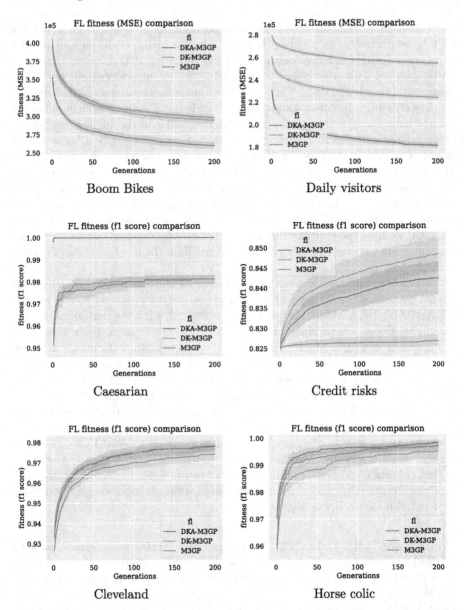

Fig. 1. Comparison of fitness throughout the evolutionary process for each dataset on the training data. The figures portray the average of 30 runs with different seeds

methods like lexicase or depth-penalization could be used to compensate for that overfitting and will be the focus of future work. Keep in mind that the higher performance of traditional GP comes at the cost of interpretability.

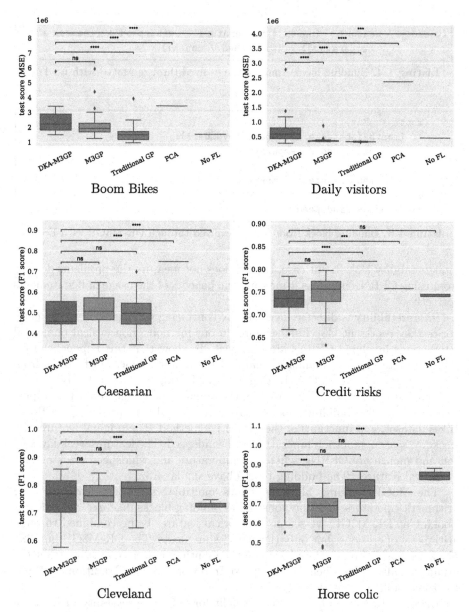

Fig. 2. Comparison of FL methods within ML pipeline on unseen data. The figures portray the average of 30 runs with different seeds.

Comparison on Interpretability. Apart from the performance benefits throughout the evolutionary process, our methods influence the interpretability of the found models. Measuring interpretability is not straight-forward and widely used metrics like the expression size (ES) do not cover all aspects of

```
1  (weekday / season) - (month * temperature) + windspeed + ((season *
         ↪ temperature) + (windspeed / month))
```

Listing 1.4. Building block from feature evolved through M3GP with ES 17.

```
1  if (month == May):
2      if (month inbetween (October, November)):
3              windspeed / temperature
4      else:
5              temperature * temperature
6  else:
7      windspeed + windspeed
```

Listing 1.5. Building block from feature evolved through DK-M3GP with ES 18.

interpretability [43]. In our case, the ES does not measure the problems derived from using CFRNs, nor does it measure the benefits of the usage of if statements correctly.

Interpretability is directly increased by converting CFRNs back to the categories they represent [43]. Furthermore, in our methods, nonsensical numerical interactions involving CFRNs are excluded. To show that this is beneficial for the interpretability, we manually analyzed the building blocks of features from both M3GP and DK-M3GP (Listing 1.4 and Listing 1.5, respectively), to try and understand the classifier, based on the domain knowledge about the respective problem. The building blocks consist of original features from the Boom Bikes dataset, and functions of the respective method. The original features are weekday, season, month, temperature and windspeed. The features weekday, season and month are CFRNs, whereas temperature and windspeed are standard numerical features. The building blocks have similar ES.

The M3GP building block combines multiple categorical features using arithmetic operators, creating nonsensical building blocks, like *weekday/season*. Looking at DK-M3GP, we see no categorical features being combined through arithmetic operations, but only through if statements. The DK-M3GP building block returns different values based on the month of a data point. Note that the arguments also holds when comparing to Traditional GP, as Traditional GP has the same building blocks as M3GP.

Using if statements has another benefit for the interpretability of features. Using if statements, each building block can be displayed in a tree-like format by branching conditions (Fig. 3).

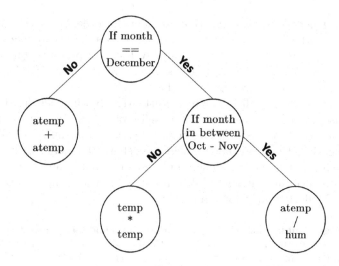

Fig. 3. The feature from listing 1.5 in a tree-like format.

6 Conclusion

In this paper, we introduced two GGGP-based FL methods, DK-M3GP and DKA-M3GP, that incorporate domain knowledge and aggregation into the GGGP search process. As our methods are generalizable we applied them to different domains. We compared them for a set of benchmarks with M3GP across two metrics: performance and interpretability. For most use cases our methods improve the performance when compared to M3GP. Furthermore, we have shown three interpretability benefits of our methods over M3GP and traditional GP. Our methods (1) represent categorical features as categories, (2) bar nonsensical interactions of CFRNs, and (3) allow for representing building blocks in a tree-like format.

To evaluate the generalizability of our methods, we included them within ML pipelines and compared them with traditional GP, PCA, and no FL. We ran the FL methods and evaluated their performance on unseen data. Our methods did not outperform M3GP or no FL, but they did outperform PCA. Traditional GP performed best across use cases. The performance benefits of our methods at training are not translated to performance improvements at testing. As this is seen across FL methods, we suspect that either the training data does not represent the test data well, or we are overfitting the training data.

Still, our methods improve the interpretability of the feature set compared to M3GP, traditional GP, and PCA. As such, depending on the needs of interpretability, our methods can also be beneficial within ML pipelines.

Apart from giving the steps to implement our methods, we analyzed when our methods are most beneficial. The incorporation of domain knowledge in DK-M3GP was most beneficial in methods with many CFRNs do not have an intrinsic order. When order exists (consider seasons), it is less likely for our methods to generate explanations that consider this order (pattern matching on each one is another option), while when using scalars, it is more likely for them to be used in arithmetic decisions, arguably less interpretable, but more efficient. Therefore, when interpretability is important, we recommend considering incorporating domain knowledge, regardless.

Aggregation was beneficial more consistently, showing benefits for all but one use case. As the most significant benefits of aggregation are seen for the regression use cases, we hypothesize that regression problems especially benefit from aggregation. However, more benchmarks should be included to show this.

To conclude, domain knowledge and historical data can be beneficial for both performance and interpretability in supervised learning, requiring some, but minimal, effort from a domain expert.

Acknowledgements. This work was supported by Fundação para a Ciência e Tecnologia (FCT) in the LASIGE Research Unit under the ref. UIDB/00408/2020 and UIDP/00408/2020, by the CMU-Portugal project CAMELOT (LISBOA-01-0247-FEDER- 045915), the RAP project under the reference (EXPL/CCI-COM/1306/2021), and FCT Advanced Computing projects (2022.15800.CPCA.A1, CPCA/A1/402869/2021, CPCA/A2/6009/2020, and CPCA/A1/5613/2020). We thank Sara Silva for her feedback and José Eduardo Madeira for his help implementing the grammars.

References

1. Adadi, A., Berrada, M.: Peeking inside the black-box: a survey on explainable artificial intelligence (XAI). IEEE Access **6**, 52138–52160 (2018)
2. Aguiñaga, A.R., Delgado, L.M., López-López, V.R., Téllez, A.C.: EEG-based emotion recognition using deep learning and M3GP. Appl. Sci. **12**(5), 2527 (2022)
3. Amin, M., Ali, A.: Performance evaluation of supervised machine learning classifiers for predicting healthcare operational decisions. Wavy AI Research Foundation: Lahore, Pakistan, vol. 90 (2018)
4. Arnaldo, I., O'Reilly, U.M., Veeramachaneni, K.: Building predictive models via feature synthesis. In: Proceedings of the 2015 Annual Conference on Genetic and Evolutionary Computation, pp. 983–990. Association for Computing Machinery, New York (2015)
5. Arroba, P., Risco-Martín, J.L., Zapater, M., Moya, J.M., Ayala, J.L.: Enhancing regression models for complex systems using evolutionary techniques for feature engineering. J. Grid Comput. **13**(3), 409–423 (2015)
6. Azzali, I., Vanneschi, L., Silva, S., Bakurov, I., Giacobini, M.: A vectorial approach to genetic programming. In: Sekanina, L., Hu, T., Lourenço, N., Richter, H., García-Sánchez, P. (eds.) EuroGP 2019. LNCS, vol. 11451, pp. 213–227. Springer, Cham (2019). https://doi.org/10.1007/978-3-030-16670-0_14
7. Bacardit, J., Brownlee, A., Cagnoni, S., Iacca, G., McCall, J., Walker, D.: The intersection of evolutionary computation and explainable AI. In: Genetic and Evolutionary Computation Conference: GECCO 2022. ACM (2022)
8. Batista, J.E., Cabral, A.I., Vasconcelos, M.J., Vanneschi, L., Silva, S.: Improving land cover classification using genetic programming for feature construction. Remote Sens. **13**(9), 1623 (2021)

9. Batista, J.E., Silva, S.: Comparative study of classifier performance using automatic feature construction by M3GP (2022). https://doi.org/10.1109/CEC55065.2022.9870343

10. Boddu, J.: Boom bikes demand analysis (2022). https://www.kaggle.com/code/jayantb1019/boom-bikes-demand-analysis/data

11. Burlacu, B., Kronberger, G., Kommenda, M.: Operon C++: an efficient genetic programming framework for symbolic regression, pp. 1562–1570. Association for Computing Machinery, New York (2020). https://doi.org/10.1145/3377929.3398099

12. La Cava, W., Silva, S., Vanneschi, L., Spector, L., Moore, J.: Genetic programming representations for multi-dimensional feature learning in biomedical classification. In: Squillero, G., Sim, K. (eds.) EvoApplications 2017. LNCS, vol. 10199, pp. 158–173. Springer, Cham (2017). https://doi.org/10.1007/978-3-319-55849-3_11

13. Cherrier, N., Poli, J.P., Defurne, M., Sabatié, F.: Consistent feature construction with constrained genetic programming for experimental physics. In: 2019 IEEE Congress on Evolutionary Computation (CEC), Paris, France, pp. 1650–1658. IEEE (2019)

14. Detrano, R., et al.: International application of a new probability algorithm for the diagnosis of coronary artery disease. Am. J. Cardiol. **64**(5), 304–310 (1989)

15. Dong, G., Liu, H.: Feature Engineering for Machine Learning and Data Analytics. CRC Press, Boca Raton (2018)

16. Dua, D., Graff, C.: UCI machine learning repository (2017). http://archive.ics.uci.edu/ml

17. Espada, G., Ingelse, L., Canelas, P., Barbosa, P., Fonseca, A.: Data types as a more ergonomic frontend for grammar-guided genetic programming. In: Scholz, B., Kameyama, Y. (eds.) Proceedings of the 21st ACM SIGPLAN International Conference on Generative Programming: Concepts and Experiences, GPCE 2022, Auckland, New Zealand, 6–7 December 2022, pp. 86–94. ACM (2022). https://doi.org/10.1145/3564719.3568697

18. Horn, F., Pack, R., Rieger, M.: The `autofeat` Python library for automated feature engineering and selection. In: Cellier, P., Driessens, K. (eds.) ECML PKDD 2019. CCIS, vol. 1167, pp. 111–120. Springer, Cham (2020). https://doi.org/10.1007/978-3-030-43823-4_10

19. Hyndman, R.J., Athanasopoulos, G.: Forecasting: Principles and Practice. OTexts (2018)

20. Ingelse, L., Espada, G., Fonseca, A.: Benchmarking representations of individuals in grammar-guided genetic programming. Evo* 2022, p. 5 (2022)

21. Kanter, J.M., Veeramachaneni, K.: Deep feature synthesis: towards automating data science endeavors. In: 2015 IEEE International Conference on Data Science and Advanced Analytics, DSAA 2015, Paris, France, 19–21 October 2015, pp. 1–10. IEEE (2015)

22. Kohonen, T.: The self-organizing map. Proc. IEEE **78**(9), 1464–1480 (1990)

23. La Cava, W., Singh, T.R., Taggart, J., Suri, S., Moore, J.H.: Learning concise representations for regression by evolving networks of trees. arXiv preprint arXiv:1807.00981 (2018)

24. Li, Y., Yang, C.: Domain knowledge based explainable feature construction method and its application in ironmaking process. Eng. Appl. Artif. Intell. **100**, 104197 (2021). https://doi.org/10.1016/j.engappai.2021.104197

25. Lourenço, N., Pereira, F.B., Costa, E.: SGE: a structured representation for grammatical evolution. In: Bonnevay, S., Legrand, P., Monmarché, N., Lutton, E., Schoenauer, M. (eds.) EA 2015. LNCS, vol. 9554, pp. 136–148. Springer, Cham (2016). https://doi.org/10.1007/978-3-319-31471-6_11

26. Lourenço, N., Ferrer, J., Pereira, F.B., Costa, E.: A comparative study of different grammar-based genetic programming approaches. In: McDermott, J., Castelli, M., Sekanina, L., Haasdijk, E., García-Sánchez, P. (eds.) EuroGP 2017. LNCS, vol. 10196, pp. 311–325. Springer, Cham (2017). https://doi.org/10.1007/978-3-319-55696-3_20

27. Mégane, J., Lourenço, N., Machado, P.: Probabilistic grammatical evolution. In: Hu, T., Lourenço, N., Medvet, E. (eds.) EuroGP 2021. LNCS, vol. 12691, pp. 198–213. Springer, Cham (2021). https://doi.org/10.1007/978-3-030-72812-0_13

28. Muñoz, L., Silva, S., Trujillo, L.: M3GP – multiclass classification with GP. In: Machado, P., et al. (eds.) EuroGP 2015. LNCS, vol. 9025, pp. 78–91. Springer, Cham (2015). https://doi.org/10.1007/978-3-319-16501-1_7

29. Nau, B.: Daily website visitors (time series regression) (2022). https://www.kaggle.com/datasets/bobnau/daily-website-visitors/metadata

30. Olson, R.S., La Cava, W., Orzechowski, P., Urbanowicz, R.J., Moore, J.H.: PMLB: a large benchmark suite for machine learning evaluation and comparison. BioData Min. 10(1), 36 (2017). https://doi.org/10.1186/s13040-017-0154-4

31. Ryan, C., Collins, J.J., Neill, M.O.: Grammatical evolution: evolving programs for an arbitrary language. In: Banzhaf, W., Poli, R., Schoenauer, M., Fogarty, T.C. (eds.) EuroGP 1998. LNCS, vol. 1391, pp. 83–96. Springer, Heidelberg (1998). https://doi.org/10.1007/BFb0055930

32. Song, H.: AutoFE: efficient and robust automated feature engineering. Ph.D. thesis, Massachusetts Institute of Technology (2018)

33. Sovrano, F., Sapienza, S., Palmirani, M., Vitali, F.: Metrics, explainability and the European AI act proposal. J 5(1), 126–138 (2022)

34. Spector, L.: Assessment of problem modality by differential performance of lexicase selection in genetic programming: a preliminary report. In: Proceedings of the 14th Annual Conference Companion on Genetic and Evolutionary Computation, pp. 401–408 (2012)

35. Tran, B., Xue, B., Zhang, M.: Class dependent multiple feature construction using genetic programming for high-dimensional data. In: Peng, W., Alahakoon, D., Li, X. (eds.) AI 2017. LNCS (LNAI), vol. 10400, pp. 182–194. Springer, Cham (2017). https://doi.org/10.1007/978-3-319-63004-5_15

36. Uriot, T., Virgolin, M., Alderliesten, T., Bosman, P.A.: On genetic programming representations and fitness functions for interpretable dimensionality reduction. In: Proceedings of the Genetic and Evolutionary Computation Conference, pp. 458–466 (2022)

37. Virgolin, M., Alderliesten, T., Witteveen, C., Bosman, P.A.N.: Improving model-based genetic programming for symbolic regression of small expressions. Evol. Comput. 29(2), 211–237 (2021)

38. Whigham, P.A.: Search bias, language bias, and genetic programming. Genet. Program. 1996, 230–237 (1996)

39. Whigham, P.A., Dick, G., Maclaurin, J., Owen, C.A.: Examining the "best of both worlds" of grammatical evolution. In: Proceedings of the 2015 Annual Conference on Genetic and Evolutionary Computation, pp. 1111–1118 (2015)

40. Wilcoxon, F.: Individual comparisons by ranking methods. Biometrics Bull. **1**(6), 80–83 (1945). http://www.jstor.org/stable/3001968
41. Wold, S., Esbensen, K., Geladi, P.: Principal component analysis. Chemom. Intell. Lab. Syst. **2**(1–3), 37–52 (1987)
42. Zou, J., Schiebinger, L.: AI can be sexist and racist-it's time to make it fair (2018)
43. Zytek, A., Arnaldo, I., Liu, D., Berti-Equille, L., Veeramachaneni, K.: The need for interpretable features: motivation and taxonomy. arXiv preprint arXiv:2202.11748 (2022)

Genetic Improvement of LLVM Intermediate Representation

William B. Langdon(✉), Afnan Al-Subaihin, Aymeric Blot, and David Clark

CREST, Department of Computer Science, UCL,
Gower Street, London WC1E 6BT, UK
{W.Langdon,a.alsubaihin,david.clark}@ucl.ac.uk,
aymeric.blot@univ-littoral.fr
http://www.cs.ucl.ac.uk/staff/W.Langdon, https://afnan.ws/,
https://www-lisic.univ-littoral.fr/author/blot/,
http://www.cs.ucl.ac.uk/staff/D.Clark, http://crest.cs.ucl.ac.uk/

Abstract. Evolving LLVM IR is widely applicable, with LLVM Clang offering support for an increasing range of computer hardware and programming languages. Local search mutations are used to hill climb industry C code released to support geographic open standards: Open Location Code (OLC) from Google and Uber's Hexagonal Hierarchical Spatial Index (H3), giving up to two percent speed up on compiler optimised code.

Keywords: Genetic programming · GP · Linear representation · Clang · Static single assignment (SSA) · Mutational robustness · SBSE · Software resilience · Automatic code optimisation · World wide location · Plus codes · Zip code

1 Introduction

LLVM https://www.llvm.org/ is now a well established freely available open source software package containing the clang C/C++ language compiler and other tools to support human software engineers with maintaining and developing software. Clang converts program source code to LLVM's intermediate representation (IR). We speedup two programs (one from Google's OLC and the other from Uber's H3) by applying genetic improvement [1,2] directly to IR. LLVM IR is independent of both the source code language and the target hardware. The clean separation of the two has facilitated LLVM support for additional imperative and functional languages (e.g. Fortran, Rust, Haskell) and multiple processor types (e.g. Intel X86, ARM and nVidia). LLVM IR is like a typed hardware-independent assembly language, with a clean separation of code, memory and single assignment registers, meaning all registers are created with a fixed typed value which they keep until they deleted, e.g. on exiting the function containing them. The task of mapping the code, memory and this infinite set of registers into real hardware is left to the compiler backed. The compiler comes with many optimisation passes which transform the LLVM IR. Indeed it is possible to write in C++ additional LLVM IR passes. Although LLVM IR can be stored both in memory and binary files, we use the human-readable text files format.

G. Pappa et al. (Eds.): EuroGP 2023, LNCS 13986, pp. 244–259, 2023.
https://doi.org/10.1007/978-3-031-29573-7_16

Genetic Improvement [1,2] applies search-based software engineering [3] techniques, principally genetic programming [4], to existing human written software. Genetic Improvement has been applied to automatic porting [2], transplanting code [5,6] code optimisation [7] and automatic bugfixing. Indeed genetic programming [8] and other optimisation techniques are increasingly being used to automatically repair programs [9–16].

In Sect. 3 we describe our chromosome's representation: a variable length list of 3 different LLVM IR deletion mutations. (Delete is the most common way programmers speed up code [17].) Fitness (Sect. 4) is based on speed up whilst retaining each program's ability to process the locations of many thousands of zip codes. Our local search GI is detailed in Sect. 5 and the results given in Sect. 6 and Table 1. In three cases GI gives modest generalised speed-ups by specialising industrial C code for global locations to a definite application (postal delivery addresses in Great Britain, see Fig. 1). However one of the four cases also gives a speed up and passes 9 999 holdout tests but fails the very last holdout test. The GI code changes, generalisation and future work are discussed in Sect. 7 before Sect. 8 summarises. But first the next section describes the existing GI work on evolving LLVM IR.

2   Background

We have demonstrated genetic improvement of real world GPU applications [18–21], including BarraCUDA [22], the first GI code to be accepted into actual use [23]. At EuroGP'19 [24], we showed GI could also speed up parallel CPU code, this time Intel AVX vector instructions were optimised. The resulting GIed RNAfold [25] was accepted into production and like the GI version of BarraCUDA has been downloaded many thousands of times (for example [26]). We applied genetic programming to human written CUDA (or C) source code, whereas Tony Lewis showed GP could be used to evolve nVidia's PTX GPU assembler [27]. More recently Jhe-Yu Liou et al. [28–30] have applied grammatical evolution (GE) [31] to LLVM IR for CUDA applications running on nVidia parallel hardware and shown further real world examples where GI finds considerable improvement on hand optimised high level GPU code. GE runs were either for two or seven days. Shuyue Li and Hannah Peeler, et al. [32–34] applied linear genetic programming [35] to selecting and ordering existing LLVM optimisation passes (Sect. 1). Their GP automatically tailors the compiler pass sequence to examples from Thomas Stuetzle's ACOTSP and Parth Shirish Nandedkar's backtrack algorithm for the subset sum problem (SSP).

3   Mutating LLVM IR

3.1   Representation

The changes to the LLVM IR are stored as a list of line numbers and local registers to be mutated separated by semicolons ";". After all the changes have been made to the LLVM IR, Clang converts it to binary executable machine code. Due to neutrality [36,37], the changes may or may not alter the overall program's behaviour.

Where multiple changes to an individual line are possible, e.g. conditional branches, which branch is to be deleted is indicated by appending its number to the line number, separated by a colon ":". For example, 2046:2 means the second branch option on line 2046 is deleted. This is actually implemented by setting the one bit (i1) conditional <cond> to true. In the following LLVM IR code snippet, the local register variable %32 is replaced with 1 forcing the code to branch to label %37. Notice LLVM IR local label identifiers, such as %37, have the same format as local register identifiers such as %32.

```
br syntax             br i1 <cond>, label <iftrue>, label <iffalse>.
clang .ll line 2046   br i1 %32, label %37, label %33.
mutation 2046:2       br i1 1, label %37, label %33 ;deleted 2
```

3.2   LLVM IR define Functions

The LLVM IR call instruction is used to pass control to LLVM IR subroutines. These are delimited by define and } and contain local registers, whose names always start with a % character. Local registers names are reused by each subroutine. The closing } shows where local registers go out of scope.

3.3   Mutable LLVM IR

Our system is able to mutate the following lines of LLVM IRDash (unless we have already deleted them):

- store
- call
- conditional branches
- assignments to local registers (except from alloca). E.g. %25 = load i32, i32* %3, align 4

  There are at least 33 types of mutable assignments to local registers.

  We chose not to make alloca mutable since it declares the local register to be a pointer. Although we can delete a pointer by setting it to null, this will usually cause a run time exception. This means that there is usually a group of unmutable local registers at the start of a function. These correspond to the function's arguments and its variables. In the following example the C program entry point main(int argc, char *argv[]) {... in the human written source code is translated by the clang compiler into the define statement and the following alloca assignments for the local registers which correspond to main's arguments and some of its variables.

```
define dso_local i32 @main(i32 noundef %0, i8** noundef %1) local_unnamed_addr #1 {
    %3 = alloca %struct.LatLng, align 8
    %4 = alloca i64, align 8
    %5 = alloca %struct.LatLng, align 8
```

```
%6 = alloca i64, align 8
%7 = alloca i32, align 4
%8 = alloca double, align 8
%9 = alloca double, align 8
```

3.4 Compiling C/C++ etc. to Generate LLVM IR

The source code is compiled in the usual way, except instead of generating an object file the clang `-emit-llvm -S` command line option is used to direct the clang compiler to generate an .ll file holding LLVM IR. Similarly the linker is replaced by using the LLVM linker command `llvm-link -S` to create a single file containing all the LLVM IR.

3.5 Selecting Which LLVM IR to Optimise

The LLVM linker will generate LLVM IR for all the compiled code, including functions which are not called. Either the user can list the functions they wish to be optimised or we can recursively select all the functions which can be called by the program's **main** routine.

3.6 Deleting LLVM IR

LLVM IR **store** instructions and **call** of functions without a return value can be deleted by removing the line. (Actually to improve traceability they are commented out using the LLVM IR comment character ";".)

Assignment statements are *not* deleted. Instead all other occurrences of the left hand local register are replaced with zero. We use LLVM IR's **zeroinitializer** to ensure the zero matches the type of the "deleted" local register. Notice here the fact that LLVM is static single assignment (SSA) means that we are guaranteed that the register is only set once.

Functions which do return a value are actually assignment statements, with their return value being written into a local register. If the return value is not a pointer, to avoid disrupting the LLVM IR naming convention, the call instruction is replaced by a dummy **add** or **fadd** instruction. This adds two zeros together to generate a zero value of the same type as the removed function. To deal with integer, floating point and other types, we use the LLVM IR **zeroinitializer** to generate zero. Thus ensuring the local register has the same type as before but the function is not called. Note when the LLVM IR is compiled to executable binary code, the clang compiler may optimise the code and so remove unneeded instructions and memory.

If the function (which may be a system call) returns a pointer, then the call instruction is replaced with an **alloca** instruction, again ensuring the local register's type is unchanged. As with other assignments the local register is flagged as having been deleted and replaced by zero (i.e. null) everywhere else in the LLVM IR. Again for traceability, the original LLVM IR code is retained as a comment.

As mentioned above (Sect. 3.1), conditional branches are deleted by forcing the condition to be either true or false. Unconditional branches br and return ret instructions cannot be deleted.

By taking care of both syntax and types we ensure the mutated code remains valid LLVM IR and it compiles into executable binary code.

4   Fitness Function

There are multiple aspects of a mutation's fitness: 1) could we perform the mutation, 2) does the mutated LLVM IR compile without error, 3) is the mutated binary code different from the original version, 4) does the mutant program run ok on each test case, 5) does it produce output files, 6) how different are those outputs from the outputs of unmutated code, 7) how long does it take.

In these experiments all mutants pass (1) and (2). In a few cases, e.g. due to the clang compiler's optimisations, although the LLVM IR is changed, the binary machine code executable file is identical to that of the human written code (3). Since we already know their performance will be identical to that of the original code, such mutants are discarded without fitness testing[1]. Note, except for using -S -emit-llvm to generate the LLVM IR, we use the same compiler options as are normally used to compile the program.

In check (4), both the framework running the mutant on each of the test cases (see Sects. 4.3 to 4.5) and the mutant itself, can signal a problem via the usual unix exit status. In either case, the framework attempts to continue as usual. For (5) and (6) it will attempt to inspect the expected output files (one per test case) and compare them with those produced by the unmutated human written code, which (on the test cases) always successfully terminates. However due to the exit status error, fitness for that test case will be reduced. Finally (7) the GI framework will extract timing output generated by the unix perf command (see Sect. 4.2).

The GI framework will attempt to run the mutant program on all the test cases, summing the fitness for each test case. Section 4.2 describes how times for individual test cases are combined to lessen the impact of noisy outliers.

4.1   Test Cases for Google's OLC and Uber's H3: GB Post Codes

Both Google's Open Location Code (OLC)[2] and Uber's Hexagonal Hierarchical Geospatial Indexing System (H3)[3] are open industry standards. We obtained their human written sources from GitHub (total sizes OLC 14 024 and H3 15 015 lines of source code, LOC). Both OLC and H3 include C programs which convert global positions (i.e. pairs of latitude and longitude numbers) into their own

---

[1] In [38] we used a similar idea to test if mutated code is identical by inspecting X86 assembler generated by the GNU gcc compiler. Also Mike Papadakis et al. [39] compared compiler output to look for equivalent mutants.

[2] https://github.com/google/open-location-code downloaded 4 August 2022.

[3] https://github.com/uber/h3 downloaded 3 August 2022.

internal codes (see Table 1). For OLC we used their 16 character coding and for
H3 we used their highest resolution (-r 15) which uses 15 characters. Rather
than work on abstract locations, we use as test cases the actual locations of
homes and commercial premises.

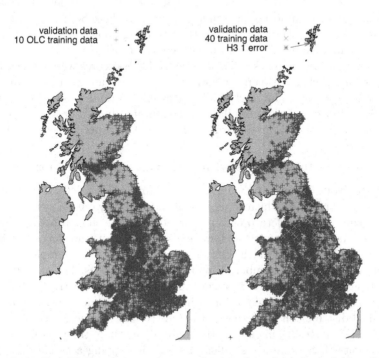

Fig. 1. Left: Ten OLC training points randomly selected in the neighbourhood of
Aberdeen (red). + holdout set (blue) GB post codes (zip codes). Right: Forty training
points randomly selected from ten H3 runtime classes (see also Figs. 2 and 3). + holdout
set (blue), locations of ten thousand random GB post codes (no overlap with H3
training or OLC (left) holdout data). Both OLC mutants and H3 -O3 pass all their
holdout tests. (Color figure online)

For Google's OLC, the location of the first ten thousand GB postcodes
(zip codes) were obtained[4]. For training (see next section) ten pairs of latitude
and longitude were selected uniformly at randomly (see Fig. 1). The unmutated
code was run on each pair and its output saved (16 bytes). For each test case
each mutant's output is compared with the original output.

Uber's H3 was treated similarly. However the H3 utility comprises about 13
times as much C code as the OLC utility does (see Table 1). Although significant
speed up could be obtained with the same 10 training data as OLC and for more

---

[4] https://www.getthedata.com/downloads/open_postcode_geo.csv.zip dated 16 March
2022. The data are alphabetically sorted starting with AB1 0AA, which is in
Aberdeen.

than 90% of post codes the mutated programs generalised, it was decided to increase the number of training data to 40 selected from a much wider pool of GB post codes (see red × right of Fig. 1). To get not only a geographic spread but also a spread of difficulty, the original code was timed on 10 000 uniformly chosen post codes and divided into ten classes see Fig. 2). Where possible, four points were chosen uniformly at random from each class. Some run time classes were empty (see Fig. 3), in which case their training points were allocated to the next slower non-empty class (shown with crosses in Fig. 2).

Although (see previous section) a series of fitness scores were defined to deal with partial matches between the correct and the mutant's output. For brevity they are omitted, since in practise all worthwhile mutants produced exactly the required output on all tests. Similarly if the mutant aborted or itself reported an error, on any test case, it was discarded. The secondary aspect of fitness is run time.

4.2   Counting Instructions with `perf stat -e instructions -x`,

In some previous GI work we had used actual run time, e.g. [40]. However in [40], we evolved subroutines which could be called directly by our GI system, whereas here we will test complete programs and so need the unix process time. Also runtime is notoriously noisy and we had previously found success using the unix perf tool, e.g. [41], which easily reports statistics for a complete program, including reporting the number of instructions actually used.

Although `perf stat -e instructions` is much less variable than elapse time, we run each mutant ten or forty times. Since run time typically has a noisy long tailed distribution [40], we sort the ten (H3 40) instruction counts and use the 3rd (11th) fastest. This means there are about three times as many (7, 29) larger counts than there are smaller (2, 10). Thus giving a stable average. We need not worry about a systematic bias, as the fitness function only ever compares the average count with other average counts obtained in the same way.

4.3   Sandboxing to Prevent Running Mutations Causing Harm

In software engineering mutation testing [42] there may be the possibility of rogue mutants doing unwanted things, such as writing to unprotected files. Therefore it may be necessary to protect system calls, such as fopen, or to run the mutants in a sandbox. In our experiments, the mutations are constrained and, for example, they cannot change file names but we still needed to guard against mutants consuming excessive resources, such as running into indefinite loops or the output file becoming too big (see next two sections).

4.4   Timeouts to Stop Poor Mutants Delaying Search

In these experiments each test case normally completes in well under a second. We used two Linux tcsh commands to impose a limit on mutants:

`limit cputime 2` The tcsh limit command can impose run time limits on many resources consumed by a unix process. `limit cputime 2` prevents a process using more than two seconds of CPU time.

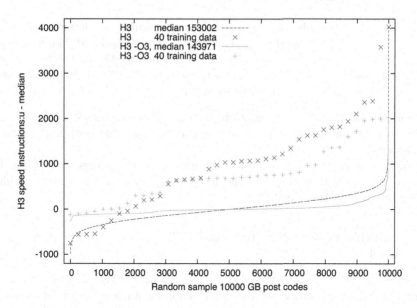

Fig. 2. Lines show distribution of H3 run times (perf instruction counts), when compiled without compiler optimisation (dashed blue) and with -O3 (red solid). GP training points are allocated to try to cover full range of H3 run times (see also Fig. 3). To plot -O3 data on the same vertical scale, run times have been adjusted by their median (5000) value (153 002 instructions, 143 971 with -O3). (Color figure online)

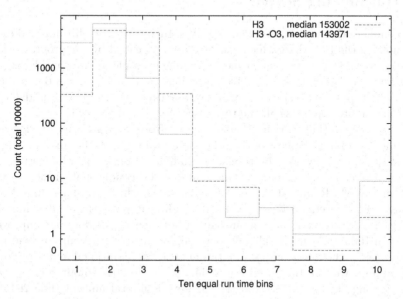

Fig. 3. H3 run time distributions (see lines in Fig. 2) separately split into ten equal run time ranges. Note post codes where H3 is slow (right hand side) are rare. To include a wide range of difficulty in the training data, where possible 4 training points are chosen randomly from each bin. The chosen data are plotted as crosses in Fig. 2. Note non-linear vertical scale.

Sadly this was not sufficient, as during development, a mutant managed to open an empty input (stdin) and then wait indefinitely (consuming no CPU time) for the first byte to arrive.

timeout 2 This timeout command aborts a process if it runs for more than a couple of seconds.

Although it should not be necessary to use both limit cputime 2 and timeout 2, it seemed safer to retain both and the overhead of using both appears to be negligible.

Where a system imposed limit is exceeded and the process terminated, the fitness function will detect the non-success exit status and give the mutant a zero score on that test case.

4.5   Limiting Output Size to Avoid Filling Disk or Exceeding Disk Quota

In our example, typical output size is 18 or 19 bytes. Nonetheless we use the linux tcsh limit filesize 1M command to ensure a rogue mutant does not fill the disk. The 1 megabyte limit is deliberately excessive, since it will avoid the disk filling problem and we found (with Centos 7) limits close to the expected output size could trigger the file size exceeded exception early.

5   Hillclimbing Search

The complete C source is compiled to LLVM IR but the search is focused on the functions in the LLVM IR which can be indirectly called by the program's main C entry point. Depending which geopositioning example we are considering, this gives between 219 and 2113 possible individual mutations (column 6 in Table 1). In the first pass we test them all one at a time. Between 37% and 63% of individual mutations pass all the test cases ($\frac{\text{column 7}}{\text{column 6}}$ in Table 1).

In the second (hill climbing) pass, we start from the fastest individual mutation and try adding the first of the other non-fatal mutants. For example, in one run of H3 compiled with -O3, the fastest individual mutation which still passes all the test cases was 10633:1. This mutates the conditional branch instruction on LLVM IR line 10633 (see Sect. 3.1). The first non-fatal mutation is on line 1972 (speed up 7 instructions). Adding it gives the double mutation 10633:1;1972;. However the combination of both mutations does not improve on mutation 10633:1 by itself. Therefore we do not include 1972, and instead move on to consider the next non-fatal mutation. The first additional mutation to give a speed up is that on LLVM IR line 2044. So we add 2044 to our current search point (giving 10633:1;2044;) and try the next non-fatal mutant (line 2045). In this way we work through all the non-fatal mutations in a single pass. This is fast, $O(n)$, but does not consider all possible combinations, $O(2^n)$. Nevertheless it does find a combined set of mutations which give still further speed up on pass one and continues to pass all the test cases.

6   Results

The results are summarised in Table 1. For both OLC and H3 we conducted two experiments. Firstly with default parameters for the clang C compiler and secondly using the -O3 optimisation flag.

Table 1. Size of C sources for Google's OLC and Uber's H3 code optimised. The rows labeled -O3 are for the same programs but when compiled with LLVM 14.0.0 clang's optimisation flag -O3. Columns 2 and 3 refer to the source files, including C .h header files. The (used) column gives the size of C code, excluding header files, to be optimised. Column 5 is the total size of the intermediate representation, whilst column 6 shows how many .ll lines we try to optimise. Column 7 gives the number of mutants which may run faster or slower, but do not change the program's output. The complete optimisation (columns 8–10) is assembled from these. Speedup is the average reduction in unix perf's instructions per test case. Average run times are for 1 core on a 3.6GHz Intel i7-4790 desktop.

C	files	LOC	(used)	LLVM IR		no output-	Mutant			GI duration
				total	mutable	change	size	speed up	holdout	
OLC	4	586	(127)	2546	294	141	2	698	682	5 min
-O3	4	586	(127)	2248	219	82	5	683	681	7 min
H3	43	5708	(1615)	19415	2113	955	51	2897	2631[a]	2.5 h
-O3	43	5708	(1615)	15680	1762	1108	46	3272	2985	3.25 h

[a]One holdout test failed

For OLC clang generates 2546 lines of LLVM IR (2248 with -O3). Considering only mutable LLVM IR in functions which are reachable from main' (see Sect. 3.5), there are 294 (-O3 219) possible individual mutations[5]. Of these 141 (-O3 82) can be individually applied without impacting OLC's output on the ten training cases. The hill climbing search described in the previous section, finds six $2323\%11; 2185; 2052\%168; 2329; 2356; 2323\%25;$ (-O3 15) which together give an average reduction of 698 (-O3 683) instructions. Of these four can be removed without changing OLC's performance, leaving just two $2323\%11; 2052\%168;$ (-O3 5), which together give a speedup of 698 (-O3 683). When tested on ten thousand uniformly chosen post codes (excluding those used to select the training data) the combined mutation gives an average reduction in number of instructions of 682 (-O3 681).

The results for H3 are given in the lower two lines of Table 1. The hill climbing search described in Sect. 5 finds 89 (-O3 113) individual changes which together give an average reduction of 2874 (-O3 3267) instructions. Again some (38, -O3 67) can be removed without changing H3's output and with little impact on its speed (see columns 8 and 9 in Table 1). When tested on ten thousand uniformly chosen post codes (excluding all those used to select the training data) the combined mutation gives an average reduction in number of instructions of

---

[5] Mutable conditional **br** instructions give rise to two mutations per line, Sect. 3.1.

2631 (-O3 2985). However, although the 46 changes to H3 -O3 LLVM IR pass all 10 000 holdout tests, the 51 changes to LLVM IR compiled without -O3 fail just the last holdout test ZEX XXX (see purple ∗ in Fig. 1). ZEX XXX is unusual in being almost the last post code (99.97%th). Also, H3's run time on ZEX XXX of 156 466 is 3845 above the average and so there is only one of the forty H3 training points which has a similar run time (see Fig. 2). Section 7.2 suggests ways to perhaps further increase the number or diversity of the training data, which might increase the mutant code's generalisation.

7   Discussion

7.1   Types of Improvement Found

OLC Two Deletions 2323%11;2052%168; Fortunately OLC compiled without -O3, gives us a simple example to start with. The two changes are independent.

2323%11 deletes local register %11 (from the scope defined by the main routine on line 2323). This has the effect of disabling a sanity check:

```
fprintf(stderr, "need two arguments latitude longitude\n");
  return 1;
```

which, as all the tests are well formed, is never invoked. Oddly this gives a greater speed up than the equivalent conditional branch mutation 2334:1 which disables the preceding if(argc != 1+2){. Although both mutations enter pass two (Sect. 5), 2323%11 is first in the list and as adding 2334:1 gives no additional speed up, only 2323%11 is retained.

2052%168 removes the line setting local register %168 (declared in the scope starting on LLVM IR line 2052, which is where function print_OLC_Encode is defined). In the LLVM IR produced by clang with no optimisation, this has the effect of removing the call to printf("\n") at the end of print_OLC_Encode. (With -O3 clang converts printf("\n") into a more efficient putchar(10) but gives a much more complicated mapping between C and LLVM IR.) Removing printf("\n") reduces the size of the output by one byte and so reduces the OLC's run time but makes no difference to its functionality. Again the mutated OLC code has been made slightly faster by removing non-essential code.

OLC -O3 five deletions 2148:2;1895%6;1905:2;1895%40;1895%178;
1895%178 deletes local register %178 (from scope starting on line 1895 print_OLC_Encode). This has the effect of deleting the call to putchar(10) mentioned in the previous section as having been generated by clang -O3 to replace printf("\n"). I.e. GI has found the same optimisation in more convoluted LLVM IR.
2148:2 br i1 1, label %7, label %4 ;deleted 2 disables the if(argc != 1+2){ mentioned in the previous section. So again skipping the number of arguments sanity check in main. With the -O3 optimised code, the mutation equivalent to 2323%11 is retained in the second pass (Sect. 5) but then is correctly

found to be redundant and eliminated in the cleanup pass (Sect. 6). The three other deletions are all in print_OLC_Encode.

print_OLC_Encode is only called by main and its third parameter is always 16, whereas kMaximumDigitCount is 15. Thus:

```
if (length > kMaximumDigitCount) {
  length = kMaximumDigitCount;
}
```

always sets length to 15. Mutation 1895%6 deletes local register %6 (i.e. sets it to zero). This has the knock on that the following %7 = select i1 %6, i64 %5, i64 15 is forced to set register %7 to 15 (i.e. kMaximumDigitCount). Effectively eliminating if (length > kMaximumDigitCount).

With -O3 clang inlines the calls to adjust_latitude(lat, length) and normalize_longitude(lon). However both lat and lon are always already normalised. Therefore the conditional branch mutation 1905:2 is able to force the branch on line 1905 to effectively skip over much inlined code.

1895%40 deletes local register %40 so removing the comparison lon_degrees < -kLonMaxDegrees of the first while loop in normalize_ longitude and forcing the following conditional branch to always jump to over the never needed adjustment lon_degrees += kLonMaxDegreesT2;

Thus again GI has sped up OLC (above that obtained by clang's -O3 optimisations) by specialising it to the training data and removing some internal checks and branches which either can never be taken or which must be taken. It seems that GI has been helped by clangs -O3 extensive inlining greatly expanding the LLVM IR in the print_OLC_Encode scope. Although the LLVM IR with and without -O3 are different, it is not yet clear why GI was unable to exploit the same opportunities when the LLVM IR was split into several smaller called functions with equivalent functionality, or if there are further similar but unexploited opportunities.

H3 51 Deletions Speedup 2631, H3 -O3 46 Deletions Speedup 2985. There are too many H3 improvements to describe them all in detail. Several follow the same ideas as OLC, with redundant operations being removed. Such as removing calls to normalise data which are always already normalised and simplifying H3 command line processing. (For example, the post code tests never invoke H3's "help" command line option.)

As an example consider 10508%74, which in one run gave the biggest individual saving (872 instructions). Again with -O3, clang inlines functions. In particular, doCoords (which converts the inputs, given in degrees, into radians and so must be called) is inlined into main's LLVM IR. Mutation 10508%74 forces local register %74 to be zero, so causing the immediately following conditional branch to always call doCoords. The direct mutation 10633:1 has the same effect, but due to noise in perf gets a speed up measurement of 871 and so 10508%74 is preferred. (In the run described in Sect. 5 mutation 10633:1 was the fastest.) Naturally 10633:1 gives no additional improvement in the hill climbing phase and so is dropped in favour of the conceptually slightly more complicated (but equivalent) 10508%74 mutation.

7.2   Discussion: Future Work, Co-evolution, Perf, Fitness Landscape

The H3 example is an order of magnitude bigger than the OLC. In retrospect, we should have been surprised if the simplistic choice of training data which works so well for OLC was sufficient for H3. Although the open source makes white box software engineering techniques (e.g. fuzzing) to target edge cases and branch coverage feasible, we have, so far tried to avoid in-depth analysis of the program's internal behaviour. Instead we used external measures, such as run time and geographic spread to increase the fraction of "difficult" cases in the training data, Fig. 2. It seems further improvements in the training data may be necessary, in which case an antagonistic co-evolutionary approach, perhaps where a population of training points is optimised to adversarially increase run time, might be beneficial.

As expected [41], when perf is used to measure whole program performance, it offers considerable noise reduction compared to the unix time command. However even perf's count of instructions executed is noisy. It is also subjected to systematic variation, e.g. between test cases, and also due to changes in the program's environment. For example, systematic changes in the harness running the test program, such as when more data are held in unix global environment variables as training progresses, can increase measured run time.

We have targeted only functions that can be called (see Sect. 3.5). In principle it should be possible to use LLVM profiling tools to target more finely individual LLVM IR instructions that are executed, possibily multiple times, during training. At present we use run time as a final pass to eliminate individual changes which appear to have no or little effect (see Sect. 6). This could be because they are never executed or because their beneficial effect is also obtained by other changes in the combined mutation. Although noisy, using run time potentially allows a (Pareto) tradeoff between size of the GI change and the benefit it gives [43][6].

We have considered only a few types of mutation. Many others, swaps and crossover could be included. It seems that in a few cases individual changes are independent and can all be applied to give each's own improvement. However some changes interfere, giving rise to an epistatic fitness landscape [36,44] for which genetic search may be suitable.

8   Conclusions

LLVM is a mature open collection of tools to support human programmers working with high level language comprised of compilers, linkers, profilers, debuggers and other tools. Although initially targeting C and C++ and Intel x86, the range of languages, supported hardware and analysis tools continues to grow. LLVM IR offers an intermediate target for genetic improvement (GI) which is independent of both the source code language and underlying hardware. Its

---

[6] Our LLVM IR representation allows ready calculation of how many lines of LLVM IR are impacted, as an alternative to counting the number of mutations.

simple line originated syntax offers a universal GI target without the need for specialised grammars. However the current black box training needs strengthening, perhaps using additional LLVM tools, or white box analysis. So far we have taken two examples from industry standard codes (Google's OLC and Uber's H3) and shown GI on IR can in a few minutes or hours (rather than days or weeks) give 0.5% (OLC) and 2% (H3) speed up even on compiler optimised code.

Acknowledgements. We are grateful for help from H.Wierstorf (gnuplot) and F.Pfenning (ϕ nodes). Funded by the Meta Oops project.

References

1. Petke, J., et al.: Genetic improvement of software: a comprehensive survey. IEEE TEVC **22**(3), 415–432 (2018). https://doi.org/10.1109/TEVC.2017.2693219
2. Langdon, W.B., Harman, M.: Evolving a CUDA kernel from an nVidia template. In: Sobrevilla, P. (ed.) WCCI, pp. 2376–2383 (2010). https://doi.org/10.1109/CEC.2010.5585922
3. Harman, M., Jones, B.F.: Search based software engineering. Inf. Softw. Technol. **43**(14), 833–839 (2001). https://doi.org/10.1016/S0950-5849(01)00189-6
4. Poli, R., Langdon, W.B., McPhee, N.F.: A field guide to genetic programming. Published via http://lulu.com and freely available at http://www.gp-field-guide.org.uk (2008). http://www.gp-field-guide.org.uk
5. Marginean, A., Barr, E.T., Harman, M., Jia, Y.: Automated transplantation of call graph and layout features into Kate. In: Barros, M., Labiche, Y. (eds.) SSBSE 2015. LNCS, vol. 9275, pp. 262–268. Springer, Cham (2015). https://doi.org/10.1007/978-3-319-22183-0_21
6. Marginean, A.: Automated software transplantation. Ph.D. thesis, University College London (2021). https://discovery.ucl.ac.uk/id/eprint/10137954/1/Marginean_10137954_thesis_redacted.pdf
7. Langdon, W.B., Harman, M.: Optimising existing software with genetic programming. IEEE TEVC **19**(1), 118–135 (2015). https://doi.org/10.1109/TEVC.2013.2281544
8. Weimer, W., Nguyen, T., Le Goues, C., Forrest, S.: Automatically finding patches using genetic programming. In: Fickas, S. (ed.) ICSE, pp. 364–374 (2009). https://doi.org/10.1109/ICSE.2009.5070536
9. Weimer, W., Forrest, S., Le Goues, C., Nguyen, T.: Automatic program repair with evolutionary computation. Commun. ACM **53**(5), 109–116 (2010). https://doi.org/10.1145/1735223.1735249
10. Haraldsson, S.O., Woodward, J.R., Brownlee, A.E.I., Siggeirsdottir, K.: Fixing bugs in your sleep: how genetic improvement became an overnight success. In: Petke, J., White, D.R., Langdon, W.B., Weimer, W. (eds.) GI-2017, pp. 1513–1520 (2017). https://doi.org/10.1145/3067695.3082517
11. Le Goues, C., Pradel, M., Roychoudhury, A.: Automated program repair. Commun. ACM **62**(12), 56–65 (2019). https://doi.org/10.1145/3318162
12. Monperrus, M.: Automatic software repair: a bibliography. ACM Comput. Surv. **51**(1), Article no. 17 (2018). https://doi.org/10.1145/3105906
13. Alshahwan, N.: Industrial experience of genetic improvement in Facebook. In: Petke, J., Tan, S.H., Langdon, W.B., Weimer, W. (eds.) GI-2019, ICSE Workshops Proceedings, p. 1 (2019). https://doi.org/10.1109/GI.2019.00010

14. Harman, M.: Scaling genetic improvement and automated program repair. In: Kechagia, M., Tan, S.H., Mechtaev, S., Tan, L. (eds.) International Workshop on Automated Program Repair (APR 2022) (2022). https://doi.org/10.1145/3524459.3527353

15. Kirbas, S., et al.: On the introduction of automatic program repair in Bloomberg. IEEE Softw. **38**(4), 43–51 (2021). https://doi.org/10.1109/MS.2021.3071086

16. Kechagia, M., Tan, S.H., Mechtaev, S., Tan, L. (eds.): 2022 IEEE/ACM International Workshop on Automated Program Repair (APR) (2022). https://ieeexplore.ieee.org/xpl/conhome/9474454/proceeding

17. Callan, J., Krauss, O., Petke, J., Sarro, F.: How do Android developers improve non-functional properties of software? Empr. Soft. Eng. **27**, Article no. 113 (2022). https://doi.org/10.1007/s10664-022-10137-2

18. Langdon, W.B., Harman, M.: Genetically improved CUDA C++ software. In: Nicolau, M., et al. (eds.) EuroGP 2014. LNCS, vol. 8599, pp. 87–99. Springer, Heidelberg (2014). https://doi.org/10.1007/978-3-662-44303-3_8

19. Langdon, W.B., Modat, M., Petke, J., Harman, M.: Improving 3D medical image registration CUDA software with genetic programming. In: Igel, C., et al. (eds.) GECCO, pp. 951–958 (2014). https://doi.org/10.1145/2576768.2598244

20. Langdon, W.B., Harman, M.: Grow and graft a better CUDA pknotsRG for RNA pseudoknot free energy calculation. In: Langdon, W.B., Petke, J., White, D.R. (eds.) GI, pp. 805–810 (2015). https://doi.org/10.1145/2739482.2768418

21. Langdon, W.B., Lam, B.Y.H., Modat, M., Petke, J., Harman, M.: Genetic improvement of GPU software. Genet. Program Evolvable Mach. **18**(1), 5–44 (2016). https://doi.org/10.1007/s10710-016-9273-9

22. Klus, P., et al.: BarraCUDA - a fast short read sequence aligner using graphics processing units. BMC Res. Notes **5**(27) (2012). https://doi.org/10.1186/1756-0500-5-27

23. Langdon, W.B., Lam, B.Y.H.: Genetically improved BarraCUDA. BioData Min. **20**(28) (2017). https://doi.org/10.1186/s13040-017-0149-1

24. Langdon, W.B., Lorenz, R.: Evolving AVX512 parallel C code using GP. In: Sekanina, L., Hu, T., Lourenço, N., Richter, H., García-Sánchez, P. (eds.) EuroGP 2019. LNCS, vol. 11451, pp. 245–261. Springer, Cham (2019). https://doi.org/10.1007/978-3-030-16670-0_16

25. Lorenz, R., et al.: ViennaRNA package 2.0. Algorithms Mol. Biol. **6**(1) (2011). https://doi.org/10.1186/1748-7188-6-26

26. Andrews, R.J., et al.: A map of the SARS-CoV-2 RNA structurome. NAR Genom. Bioinform. **3**(2), lqab043 (2021). https://doi.org/10.1093/nargab/lqab043

27. Lewis, T.E., Magoulas, G.D.: TMBL kernels for CUDA GPUs compile faster using PTX. In: Harding, S., et al. (eds.) GECCO, pp. 455–462 (2011). https://doi.org/10.1145/2001858.2002033

28. Liou, J.Y., Forrest, S., Wu, C.-J.: Genetic improvement of GPU code. In: Petke, J., Tan, S.H., Langdon, W.B., Weimer, W. (eds.) GI-2019, ICSE Workshops Proceedings, pp. 20–27 (2019). https://doi.org/10.1109/GI.2019.00014

29. Liou, J.Y., Wang, X., Forrest, S., Wu, C.J.: GEVO: GPU code optimization using evolutionary computation. ACM Trans. Archit. Code Optim. **17**(4), Article no. 33 (2020). https://doi.org/10.1145/3418055

30. Liou, J.Y., et al.: Understanding the power of evolutionary computation for GPU code optimization. arXiv (2022). https://doi.org/10.48550/ARXIV.2208.12350

31. Ryan, C., Collins, J.J., Neill, M.O.: Grammatical evolution: evolving programs for an arbitrary language. In: Banzhaf, W., Poli, R., Schoenauer, M., Fogarty, T.C. (eds.) EuroGP 1998. LNCS, vol. 1391, pp. 83–96. Springer, Heidelberg (1998). https://doi.org/10.1007/BFb0055930

32. Li, S.S., et al.: Genetic improvement in the Shackleton framework for optimizing LLVM pass sequences. In: Bruce, B.R., et al. (eds.) GECCO, pp. 1938–1939. Association for Computing Machinery (2022). https://doi.org/10.1145/3520304.3534000

33. Peeler, H., et al.: Optimizing LLVM pass sequences with Shackleton: a linear genetic programming framework. In: Trautmann, H., et al. (eds.) GECCO Comp, GECCO 2022, pp. 578–581. Association for Computing Machinery (2022). https://doi.org/10.1145/3520304.3528945

34. Peeler, H., et al.: Optimizing LLVM pass sequences with Shackleton: a linear genetic programming framework. arXiv (2022). https://arxiv.org/abs/2201.13305

35. Brameier, M., Banzhaf, W.: Linear Genetic Programming. No. XVI in Genetic and Evolutionary Computation (2007). https://doi.org/10.1007/978-0-387-31030-5

36. Petke, J., et al.: A survey of genetic improvement search spaces. In: Alexander, B., Haraldsson, S.O., Wagner, M., Woodward, J.R. (eds.) GECCO, pp. 1715–1721 (2019). https://doi.org/10.1145/3319619.3326870

37. Rainford, P., Porter, B.: Using phylogenetic analysis to enhance genetic improvement. In: Rahat, A., et al. (eds.) GECCO, GECCO 2022, pp. 849–857. Association for Computing Machinery (2022). https://doi.org/10.1145/3512290.3528789

38. Langdon, W.B.: Genetic improvement of genetic programming. In: Brownlee, A.S., Haraldsson, S.O., Petke, J., Woodward, J.R. (eds.) GI @ CEC 2020 Special Session (2020). https://doi.org/10.1109/CEC48606.2020.9185771

39. Papadakis, M., Jia, Y., Harman, M., Le Traon, Y.: Trivial compiler equivalence: a large scale empirical study of a simple fast and effective equivalent mutant detection technique. In: ICSE (2015). https://pages.cs.aueb.gr/~mpapad/papers/ICSE15B.pdf

40. Langdon, W.B., Petke, J., Bruce, B.R.: Optimising quantisation noise in energy measurement. In: Handl, J., Hart, E., Lewis, P.R., López-Ibáñez, M., Ochoa, G., Paechter, B. (eds.) PPSN 2016. LNCS, vol. 9921, pp. 249–259. Springer, Cham (2016). https://doi.org/10.1007/978-3-319-45823-6_23

41. Blot, A., Petke, J.: Using genetic improvement to optimise optimisation algorithm implementations. In: Hadj-Hamou, K. (ed.) ROADEF 2022. INSA Lyon (2022). https://www.cs.ucl.ac.uk/staff/a.blot/files/blot_roadef_2022.pdf

42. Harman, M., Jia, Y., Langdon, W.B.: A manifesto for higher order mutation testing. In: du Bousquet, L., Bradbury, J., Fraser, G. (eds.) Mutation 2010, pp. 80–89 (2010). https://doi.org/10.1109/ICSTW.2010.13

43. Sitthi-amorn, P., Modly, N., Weimer, W., Lawrence, J.: Genetic programming for shader simplification. ACM Trans. Graph. 30(6), Article no. 152 (2011). https://doi.org/10.1145/2070781.2024186

44. Langdon, W.B., Veerapen, N., Ochoa, G.: Visualising the search landscape of the triangle program. In: McDermott, J., Castelli, M., Sekanina, L., Haasdijk, E., García-Sánchez, P. (eds.) EuroGP 2017. LNCS, vol. 10196, pp. 96–113. Springer, Cham (2017). https://doi.org/10.1007/978-3-319-55696-3_7

Spatial Genetic Programming

Iliya Miralavy[1,2(✉)] and Wolfgang Banzhaf[1,2,3]

[1] Department of Computer Science and Engineering, Michigan State University,
East Lansing, MI, USA
miralavy@msu.edu
[2] BEACON Center of Evolution in Action, Michigan State University,
East Lansing, MI, USA
[3] Ecology, Evolution and Behavior Program, Michigan State University,
East Lansing, MI, USA

Abstract. An essential characteristic of brains in intelligent organisms is their spatial organization, in which different parts of the brain are responsible for solving different classes of problems. Inspired by this concept, we introduce Spatial Genetic Programming (SGP) - a new GP paradigm in which Linear Genetic Programming (LGP) programs, represented as graph nodes, are spread in a 2D space. Each individual model is represented as a graph and the execution order of these programs is determined by the network of interactions between them. SGP considers space as a first-order effect to optimize which aids with determining the suitable order of execution of LGP programs to solve given problems and causes spatial dynamics to appear in the system. RetCons are internal SGP operators which enhance the evolution of conditional pathways in SGP model structures. To demonstrate the effectiveness of SGP, we have compared its performance and internal dynamics with LGP and TreeGP for a diverse range of problems, most of which require decision making. Our results indicate that SGP, due to its unique spatial organization, outperforms the other methods and solves a wide range of problems. We also carry out an analysis of the spatial properties of SGP individuals.

Keywords: Genetic Programming · Spatial Computing · Evolutionary Computation

1 Introduction

Even though evolutionary algorithms have proven to be applicable for solving a wide range of computationally represented problems, they often are abstractions of their natural counterparts and do not account for the impactful dimensions of *time* and *space* in nature. Spatial Computing [7] is a relatively new field in computer science that states the distribution of computational elements in *space* can enhance the performance and the feasibility of computation. It further argues that it is more important to include *space* in our computational models as our understanding of natural computing systems and coupling of computational

© The Author(s), under exclusive license to Springer Nature Switzerland AG 2023
G. Pappa et al. (Eds.): EuroGP 2023, LNCS 13986, pp. 260–275, 2023.
https://doi.org/10.1007/978-3-031-29573-7_17

models and physical elements increases. Additionally, Spatial Computing offers a more natural approach to parallel computation. Parallelism is an essential part of natural systems where elements, be it particles in physics, molecules in chemistry, or individual agents in biology, are all bound by *space* and perform their functions in time simultaneously. The spatial properties of these elements play a critical role in the performance of these systems.

The primary contribution of this paper is introducing Spatial Genetic Programming (SGP), a Genetic Programming (GP) system controlled by a 2D space that evolves Linear Genetic Programming (LGP) [4] programs. An SGP model consists of one or more LGP programs which are represented as graph nodes located in a 2D space. In SGP, *space* plays an integral role in determining the order of execution of LGP programs. In each individual, these nodes form a network of interactions, responsible for regulating the order of execution of the LGP programs based on their spatial properties and the internal dynamics of the system. If necessary, the flexible representation of SGP allows for controlling the evolution of iterative behavior to develop more compact models. To show the effectiveness of the proposed system, we utilize SGP to solve different classes of problems and compare it to two common GP paradigms.

2 Related Literature

The evolution of the SGP models consists of two main parts: evolving the structural properties of the system (i.e., the graphs representing the network of interactions between LGP program nodes) and the instructions of the LGP programs.

Various works in the literature focus on evolving graphs capable of representing solutions to computational problems. Tree GP (TGP) [15] uses graph representation of tree data structures. As the most common type of GP, TGP has been previously used for various types of applications such as transportation [24], symbolic regression [1,3], image processing [22], classification [2] and others. Although the tree data structure is simple for understanding and evolving solutions, traversing these structures is not a computationally trivial task and often causes bloat problems.

Cartesian Genetic Programming (CGP) [16] is another mainstream graph-evolving GP that has shown good performance for solving computational problems. CGP uses integer values as genes representing nodes in a graph, their functions, links between the nodes, and how inputs and outputs are connected to these nodes. Compared to TGP, CGP is computationally less expensive, and therefore its evaluation time is faster and is less prone to bloat [17]. An interesting feature of CGP is its ability to encode and control computational systems similar to Artificial Neural Networks [14,23]. CGP also has various applications in agent control [10], image processing [9] and circuit design [12]. Similar to CGP, in SGP, the computational cost of creating network graphs are reduced by a mechanism that controls the system with a 2D grid.

It is possible to evolve GP models that do not rely directly on graph representations. LGP is among these types of GP. This paradigm is represented as a series of instructions, usually in the form of imperative programming language or machine language that execute sequentially. LGP supports branching operators, which allow the execution pointer to jump between instructions. One particular weakness of LGP is correctly determining the number of internal registers, which, if chosen wrong, will drastically undermine the performance of the solutions [19]. On the other hand, LGP programs are quite fast because they can be designed to run on the processor directly. This strength was the reason for choosing LGP programs to be a part of the SGP system. Another common GP variant that does not use graphs as their representation is Stack-based Genetic Programming. In such GP, fundamentally similar stack-based programming languages are responsible for obtaining operands for the program operators from a data stack and pushing the results of the operations to these stacks. Depending on the designed rules, multiple data stacks for different data types might exist. Generally, Stack-based Genetic Programming models are faster than tree structures, and it is possible to create bloat-free mutations and crossover mechanisms. Push GP [21] is one of the most famous stack-based systems and has been previously used for various applications such as automatic code simplification [11] and Python code synthesis [20].

Tangled Programming Graphs (TPG)s [13] are among the systems that evolve both computer programs and the relationship between them in the form of a graph. This system has been previously used for solving Visual Reinforcement Learning problems such as Atari games and has produced comparable results to deep learning algorithms. TPGs are one of the closest works in the literature to the idea of SGP since it is constructed based on mechanisms that control the execution flow of programs until a terminal state is reached; however, there are some key differences between the two systems. In SGP, the execution order is determined by minimizing a traverse cost value between the source program and every other program in the system. Furthermore, SGP supports iterative behaviors by allowing programs to execute more than once. In contrast, TPGs use a bidding system among teams of programs to determine the pathways taken to execute programs. SGP is controlled by a 2D space which makes the spatial properties of the nodes important for selecting the following programs to execute. Finally, unlike TPGs, and much like more conventional GP systems, in SGP a population of mutually exclusive individuals is used. In the next section, we'll be exploring the implementation of the SGP system in more detail.

3   Spatial Genetic Programming

SGP models are program nodes spread in a 2D coordinate system. The aim of the SGP interpreter is to choose the order of program executions until a termination condition is met while minimizing the traversing cost between programs. A cost function is used to calculate the cost of trajecting from the source coordinate (starting from $(0,0)$ as null program) to every other program node. In other words, in each step, a weighted network of interactions between all the program nodes is made in which the weights are the cost of traversing from a source

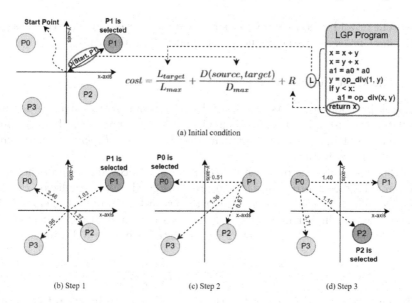

$$cost = \frac{L_{target}}{L_{max}} + \frac{D(source, target)}{D_{max}} + R$$

(a) Initial condition

(b) Step 1   (c) Step 2   (d) Step 3

Fig. 1. Model representation and interpretation steps for an SGP model with 4 programs. a) Different contributions to the cost function. b) Step 1: P1 is selected since it has the lowest traverse cost from the starting point. Red values indicate cost c) Step 2: P1 is the source point and P0 with the lowest traverse cost is selected. d) Step 3: P0 is the starting point and P2 is selected.

coordinate to a destination program node. These weights alter as the source coordinate and the internal state of the system change.

The program with the lowest traverse cost is then selected to execute prior to the others. If termination conditions are not met, the same process repeats. The position of the most recently executed program is then set to be the source coordinate to determine the next program to be executed. In Fig. 1a, an overview of an SGP model is illustrated in its initial conditions. Each node represents a program and is labeled with the program name. Each program contains instructions that manipulate internal memory registers shared between all programs and outputs a single value corresponding to an internal register, an input, or a constant value. In step 1 (Fig. 1b), the cost of traversing from $(0,0)$ to every other node is calculated (details of which can be found in the next section). Since P1 has the lowest cost, it is selected for execution. In the next step (Fig. 1c), P1 is the source point for calculating the costs to every other node, and therefore P0 is chosen for execution. The same principle continues until a termination condition is reached. Algorithm 1 (see supplementary material [18]) shares the details of how SGP selects the next program in line for execution. All of the individual programs are stored in a list. A loop on the program list is performed to calculate the cost of traveling to each program. Safeguards for protecting against infinite cost values and revisiting a node in case of loop-free configuration are in place to prevent invalid selections. The program with the lowest traverse cost is stored in the *next_program* variable and is the final program's output.

3.1   The Cost Function

The cost function considers the spatial and internal states of the system to calculate the cost of traversing to a given program node based on a source coordinate. SGP operates in two main modes: *Spatial* and *Programmatical*, which will be described in turn.

In *Spatial* mode, the distance between the source coordinate and the target node and the target program's length are calculated and normalized. *Program length* is defined as the number of instructions in that program, meaning that programs with a higher number of instructions have a slightly lower chance of being selected. The cost in this mode is calculated using the following equation:

$$cost = \frac{L_{target}}{L_{max}} + \frac{distance(source, target)}{D_{max}}$$

In which L denotes program length, $D()$ is an internal function that returns the Euclidean distance between two coordinates, and D_{max} denotes the maximum distance between two nodes of SGP. For an evolved SGP model in *Spatial* mode, normalized length and distance between every two nodes are constant, meaning that the order of execution of the programs does not change, forming a static solution graph that is not affected by input values.

In the *Programmatical* mode, other than the metrics considered in the *Spatial* mode, the output variable of the target program is also taken into account. Therefore, calculating the cost in this mode follows the following equation:

$$cost = \frac{L_{target}}{L_{max}} + \frac{D(source, target)}{D_{max}} + R$$

In which R denotes the current value of the parameter set to be the output of the target *LGP* program. Each *LGP* program terminates with a return statement that outputs a numerical value for R. The rest of the variables are the same as the ones in the *Spatial* mode. The current value of R cannot be pre-computed and at every step highly depends on the previously executed programs and how the internal registers have been manipulated prior to the cost calculation step. The impact of program length and the distance between nodes are normalized; however, the value of R is not bounded to any range and depends on the problem inputs. This design decision might increase the impact of R on selecting the next program significantly; however, the cost function is configurable and can be modified to normalize the scale of R. It is prevalent for models evolved in this mode to take different execution routes with different sets of given inputs, forming dynamic solution graphs. This feature enables the opportunity to evolve localization in the system so that different sections of an SGP model respond to different sets of stimuli, a known characteristic of the brain in natural organisms.

3.2   Outputs, Termination Conditions and Model Execution

An imperative SGP system has four means of producing outputs. First is the numeric value returned by the last executed SGP program. Second, SGP also outputs the system's internal state, which is all the register values (initially set to 0) manipulated during the run-time of a model. Third, SGP operators are allowed to manipulate an external file, a computational object, or a third-party environment. Finally, it is possible to associate terminal programs with discrete actions or outputs. In other words, if a terminal program is reached, the action associated with that program is performed in the problem environment ending the individual execution. Depending on the model inputs, a different final program might get selected and thus produce a different action.

Multiple conditions can end the execution of a model. Each model program has a chance to become a terminal node and end the execution. Suppose a model does not have any terminal program. In that case, a limit equal to the total number of programs in that model is set, breaking the execution if the count of executed programs exceeds that limit. Execution also ends if there are no more candidate programs or if the execution time exceeds a time threshold. Algorithm 2 (see supplementary material [18]) shows the details of executing an individual model.

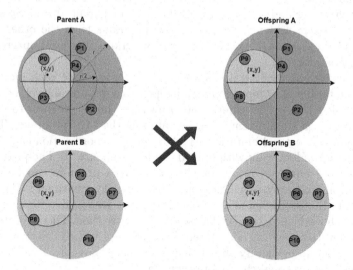

Fig. 2. Crossover between two SGP models. Suppose (x, y) is the randomly chosen point within $\frac{r}{2}$ distance of the center (0, 0). Programs within $\frac{r}{2}$ distance of (x, y) form S_i of parent A (red circle) and parent B (green circle), and the rest of the programs form S_o (blue circle for parent A and purple circle for parent B). Offspring A is a combination of the programs in S_i of parent A and S_o of parent B and Offspring B is a combination of the programs in S_i of parent B and S_o of parent A.

3.3   Evolution of Models and the Genetic Operators

Initially, a population of random SGP individual models is generated, in which the number of programs in each model, their length, and the initial coordination of the nodes within an allowed 2D space are randomly chosen. Next, an object pool of operators and operands is created from which the operator and operand(s) of each statement or instruction are randomly selected. If there is no suitable operand for an operator, it will be removed from the selection pool. Operator and operand objects are reusable and therefore do not add to the computational cost of the system.

After evaluating models in each generation of the evolution, tournament selection is applied to the population. The two best competitor models are selected and have a chance to crossover to produce two offsprings or directly make it to the next generation of models after mutation. If the crossover happens, a mutation with a chance is also applied to the two new offsprings.

The 2D space of the SGP models is bounded by a radius parameter r, meaning that the program coordinates must be within r distance from $(0,0)$. A random coordinate point within $\frac{r}{2}$ distance from $(0,0)$ is selected to be utilized while performing crossover between two individual models. Let us denote the set of programs within $\frac{r}{2}$ distance from the randomly chosen point of an individual with S_i and the set of programs outside that radius with S_o. Then, in the crossover between parent A and B, every program in S_i of parent A and S_o of parent B form one offspring while the rest of the programs form the other offspring (Fig. 2). There is no limit to the number of programs that are impacted by the crossover operator.

After crossover, there is a chance for every program of each individual to undergo mutation. SGP mutations can happen on a structural level, i.e., altering a program location or switching a program type (input program to output or vice versa), or on a statement level, i.e., altering the LGP programs. There are three types of structural mutations. 1) A program's coordination can change by performing a random walk with a fixed random step size. 2) The program type can alter from input to output or vice versa. 3) A program can be added or removed from/to the system. These modifications, along with an LGP mutation that targets the return value of the programs, are responsible for changing the behavior of how the programs will be selected for execution. There are three types of LGP mutations, which add statements to the program, delete a statement from the program or modify an existing statement if possible. By default, these mutations have an equal chance of occurring.

3.4   Conditional Return Statements

One of the abilities of SGP is to evolve rational pathways that change in response to the problem inputs. Conditional operators such as the basic *if* statements can help build a logic behind the return values of each program, forcing a different order of execution when different input values are given to the system. By default, however, SGP requires each program to have a final single return statement that cannot be connected to any other operators, such as being tied to a conditional *if* statement. Since allowing evolution to use a combination of conditional operators and internal state values to evolve conditional pathways is not trivial, we come up with the idea of replacing the normal return statements of the program with a custom conditional operator called RetCon (stands for Return Conditions). This operator forces a condition on the return statement in a way that if the condition is true, an internal state value or a constant value will be returned. Otherwise, another return value will be selected. The two return values could be the same.

4   Experiments and Results

In this section, we apply SGP to a set of problems classified into two case studies to analyze the behavior of the system by comparing different modes of otherwise identical SGP setups with classical TGP and LGP. The TGP included in the DEAP framework [8] and the same LGP system used for the SGP programs were used to conduct the experiments. The use of RetCon in SGP facilitates the evolution of conditional pathways, making SGP models well-suited for addressing problems that require decision-making. The specific problem set for each case study was selected due to the presence of a decision-making component.

4.1   Case Study: Classic Control Problems

OpenAI Gym [5] is a library of Reinforcement Learning problems in Python which helps with the development and comparison of problem-solving algorithms by providing a straightforward environment-to-algorithm API. In particular, we tackled Cart Pole, Mountain Car, Pendulum and the Acrobat problems from the Gym library; details of which can be found in [5].

Figure 3 shows the results for tackling the OpenAI Gym classic control problems. 50 replicate experiments with different random seed values were conducted for each of the four problems. The median Fitness values over generations for the best-evolved models of each replicate are illustrated. The shaded areas represent the 25 and the 75 quantiles, while the solid lines represent the median. Three configurations of SGP are tested against these problems and are compared with classical TGP and LGP. *Prog* refers to the Programmatical mode of the system; *Prog RetCon* indicates the usage of conditional return statements in the Programmatical mode, and *Spatial* refers to the Spatial mode. In the spatial mode, the usage of RetCon operators does not make a difference since the return statements of the programs do not change the execution order. All of the experiments are run for 1000 generations; however, depending on the problem, after a certain number of generations, the fitness values cease to change, and therefore, a portion of the generations are selected to ease the analysis of the results. Finally, even though all the classic control problems are deterministic, the starting conditions are slightly randomized (e.g., the position of the car in the mountain car problem) to help the problem solvers find a generalized solution. For all of the experiments, *if*, *assign*, and basic math operators are used as the function/operator set of LGP and SGP.

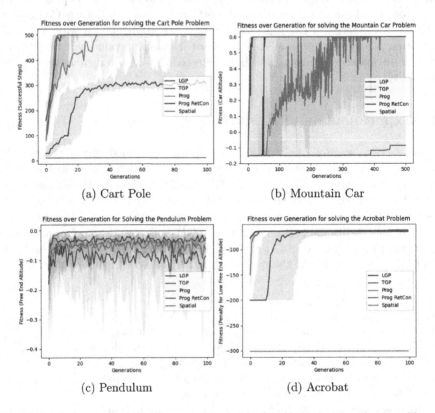

(a) Cart Pole

(b) Mountain Car

(c) Pendulum

(d) Acrobat

Fig. 3. The fitness over generations plot for solving the four classic control problems. a) Fitness equals to the number of steps in which the pole is held in an upright position. b) Fitness represents the car altitude at the end of each evaluation. c) Fitness equals to the altitude of the free end of the Pendulum at the end of each evaluation. d) Fitness indicates a -1 penalty for each step in which the free end has not passed the threshold line.

To solve the Cart Pole problem, SGP is configured to use discrete outputs in which each individual must consist of two terminal nodes, each associated with an action of either accelerating the cart towards left or right. 3a shows the results for solving the Cart Pole problem. SGP with RetCon and TGP solve this problem in less than 20 generations. Programmatical settings without RetCon also solve the problem. However, it takes more generations to solve, and the shaded green area shows that it takes more time for all the individuals in all the replicates to be able to solve this problem while all the individuals of the replicates for the RetCon settings solve the problem in less than 30 generations.

LGP also solves the problem but its performance is not as good as the rest of the approaches The Spatial setting fails to solve the task over all generations since, in this setting, the network inputs do not change the execution order of the graph. In other words, the same discrete action is always taken; therefore, constant fitness is achieved over generations.

Same as the Cart Pole problem, for the Mountain Car problem, SGP is configured to use discrete outputs. As illustrated in Fig. 3b the RetCon outperforms the other two configurations by solving the problem for all the replicates in less than 50 generations. LGP performs slightly worse, solving the problem in approximately 60 generations. The Programmatical setting without RetCon has a cold start, but the replicates mostly solve the problem at around 450 generations. However, the difference between the fitness of the best models among all the replicates varies greatly. The shaded orange area shows that there are individuals in the TGP approach that solve the problem but the median results are worse than the other approaches in 500 generations. Once again, the spatial mode fails to solve the problem while producing a constant fitness.

The nature of the Pendulum problem is slightly different from the other problems since it requires a continuous input indicating the amount of torque applied. Unlike the other three problems, the Spatial configuration performs comparably to the other approaches. TGP outperforms all other approaches; however as shown in Fig. 3c fitness values of approximately 10^{-4} were achieved by the best individuals of all the approaches showing almost an upright position of the pendulum. The high fluctuation of the median line is due to the high impact of the random starting position of the pendulum on the outcome of the evaluation.

The final classic control problem tackled in this paper is the Acrobat problem. As illustrated in Fig. 3d, SGP manages to solve the problem in both Programmatical modes with or without RetCon in less than 5 generations and improves its performance until 10 generations managing to reach the specified line in all of the best models in about 60 steps. Like the other discrete output problems, the Spatial mode drastically fails by only producing a constant output. TGP has a slightly worse performance while LGP solves the problem in 40 generations.

4.2   Case Study: Custom Toy Problems

The custom Toy Problems is a custom library of three Reinforcement Learning problems included with the SGP source code that can be briefly described as the following (Fig. 4):

- **The Adventure Problem:** This problem is inspired by an Atari 2600 game called *Adventure* [6]
- **The Foraging Problem:** This is a famous classic Artificial Life problem in which an agent has to gather all the food spread in a 2D grid. The tiles are often blocked by obstacles or walls (Fig. 4b).

- **The Obstacle Avoidance Problem:** As illustrated in Fig. 4c a car agent is driving on a road that is occasionally blocked by randomly appearing road-blocks. The car agent has to avoid hitting the roadblocks for a specified number of time steps.

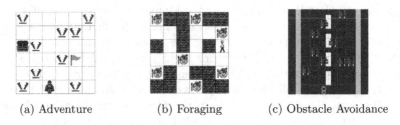

(a) Adventure   (b) Foraging   (c) Obstacle Avoidance

Fig. 4. Three different toy problems. *Icons used in the images are from:* https://www. flaticon.com/

Figure 5 depicts the results produced for solving the three Toy Problems for 50 replicates. The only non-deterministic problem is Obstacle Avoidance since the roadblocks spawn randomly.

In the adventure problem, the observation consists of 6 integer inputs corresponding to the agent's vision cone and a single bit corresponding to whether the agent has picked the treasure or not. The agent's vision cone shows two three-tile rows in front of the agent. The problem's action space consists of three discrete actions: moving one tile ahead, turning left, and turning right. All the entities in the problem grid and the empty tiles are coded with unique integer values and are visible to the agent. The agent can move to the treasure tile to automatically pick up the treasure. A small reward of 0.01 is given to the agents that move. The computational models are responsible for giving an agent instructions to solve the task through actions, and the individual's fitness equals the score the controlled agent achieves. A significant score of 10 is given to the agents that manage to grab the treasure, and a very significant score of 20 is given to the agents that reach the final destination while carrying the treasure. The simulation ends after 100 time steps or when the agent reaches the final destination or falls into a trap. The score is returned to the SGP evolver module as the controlling model's fitness value. Figure 5a depicts the results produced for solving the Adventure problem over 500 generations. As expected, the Spatial configuration fails to solve a task with discrete output. The programmatical settings manage to evolve agents capable of picking up the treasure; however, they fail to reach the final destination. The RetCon settings, however, solve the problem entirely in less than 250 generations. TGP slightly outperforms the Prog settings since in later generations, the best TGP individuals fully solve the problem. LGP on the other hand, only manages to find the treasure in the later generations but fails to completely solve the problem and the median line always stays low.

The Foraging problem has an observation space consisting of 6 inputs corresponding to the agent's vision cone. Like the adventure problem, the vision cone includes the two three-tile rows in front of the agent. The action space of the

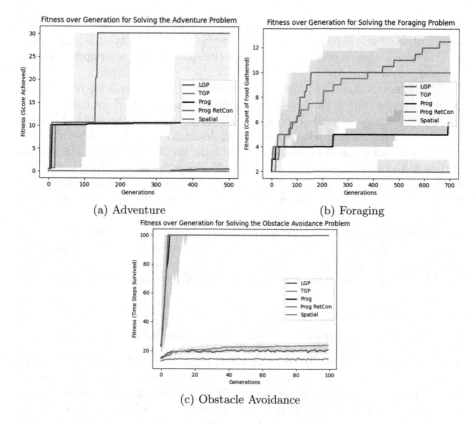

(a) Adventure   (b) Foraging

(c) Obstacle Avoidance

Fig. 5. The fitness over generations plot for solving the three toy problems. a) Fitness indicates the score of the agent at the end of evaluation. b) Fitness is equal to the number of food gathered by the agent c) Fitness is the total number of time steps that the agent has survived the environment

problem is the same as the Adventure problem consisting of three actions: moving, turning left, and right. A total of 20 food tiles are available for the agents to take, which are often placed at the end of a maze-like pattern in which the agent will have to return to the path taken to reach the food to get out (e.g., top left food in Fig. 4b). The simulation is run for 200 time steps while no reward is considered for moving. Compared to other problems, this is a more challenging task to solve since the maze-like patterns make it quite difficult to gather all the food in the allowed time steps. Figure 5b shows the result for solving this problem. Same as most cases, the Programmatical settings with RetCon outperforms the other two modes while being able to gather as much as 14 food at best among all the replicates. The changes in the median line seem to show evolution after 700 generations. Perhaps, running this task for a more extended period would help the system completely solve the problem. The programmatical SGP needs quite more time to evolve conditional logic to solve these types of problems only using a basic *if* statement, and the spatial mode fails to solve the task. The performance of TGP on this scenario is slightly worse than the RetCon settings while LGP only manages to perform better than the spatial mode after approximately 400 generations.

The observation space in the obstacle avoidance problem consists of 12 integer values corresponding to the vision cone of the car agent. This vision cone includes four three-tile rows in front of the agent. The action space of the problem is three discrete actions: moving left and right and doing nothing. To achieve a perfect score, the car agent must avoid all the roadblocks for 100 time steps. The number of time steps before the car agent hits a roadblock is the fitness of the controlling SGP model. Figure 5c shows the results produced for solving this task. Both SGP Programmatical configurations with or without RetCon manage to avoid all the obstacles in less than 20 generations. At the same time, it takes a bit longer for all the replicates to completely solve the problem for the setting without RetCon. The fluctuations in the case of Spatial mode are due to the randomness of the roadblock patterns selected to appear in the far front of the car. TGP and LGP do not achieve good fitness levels on this problem but outperform the spatial mode.

4.3   Impact of a Spatial Crossover on the Evolution of Programs

To check whether the mechanisms in the system impact the spatial properties of the SGP models, an experiment was conducted with a different crossover algorithm called the Spatial Crossover. This crossover is quite similar to the normal crossover used in SGP however, instead of choosing a random circular area to form S_i, programs that are located in the top right quadrant of the 2D space (x-coord and y-coord greater than or equal to 0) are selected to form S_i of the parent individuals. In this approach, always the same spatial portion of the individuals swap to form offspring. We tracked the position of all the individuals' programs (not just the best) in all the 50 replicates. We summarized the results in Table 1. SC stands for Spatial Crossover, NC stands for Normal Crossover and P1 and P2 refer to two test arbitrary problems. The 2D space of each individual is divided into four quadrants starting from the top right (Q1) and going clockwise to the top left (Q4). Results show that in the case of using the Spatial Crossover, programs tend to move out of the Q1 area in which the crossover is happening. This behavior is reflected by the significantly lower percentage of appearance of programs in Q1 compared to the case where the normal crossover is being applied.

Table 1. Position of the final programs of all individuals in the latest generation using Normal Crossover (NC) and Spatial Crossover (SC) for two test problems.

Quadrant	RetCon P1 (NC)	RetCon P1 (SC)	RetCon P2 (NC)	RetCon P2 (SC)	Spatial P1 (NC)	Spatial P1 (SC)	Spatial P2 (NC)	Spatial P2 (SC)
Q1	25.69%	**12.63%**	26.23%	**10.98%**	27.51%	**8.98%**	34.17%	**6.8%**
Q2	22.09%	30.21%	22.84%	27.07%	21.68%	29.78%	26.48%	28.58%
Q3	27.87%	27.72%	23.79%	29.32%	21.66%	30.38%	17.68%	33.52%
Q4	24.34%	29.64%	27.14%	32.63%	29.14%	30.87%	21.67%	31.1%
Total	38310	40030	27.14%	38493	38566	41742	38059	41597

5 Conclusion

This paper introduced a new GP paradigm which accounts for the dimension of space as a first-order effect to optimize. SGP works in two modes of Spatial and Programmatical, which bring unique characteristics to the system, allowing it to evolve static and dynamic graphs, respectively. The impact of these two operation modes was tested against two classes of problems while introducing conditional return statements. SGP was tested against four classic Control Problems of OpenAI Gym library. RetCon's ability to quickly evolve conditional statements to choose the right pathway of the graph by manipulating the weights of the underlying regulatory network was shown during these experiments. For all the cases except the Pendulum problem, RetCon quickly solved the control tasks. The Pendulum problem required less decision-making and more accuracy on the produced continuous outputs (amount of torque). The Programmatical mode without RetCon was able to solve the control problems as well. However, it takes more time for evolution to evolve the factual conditional statements in the LGP programs to reflect the same decision-making structures. The Spatial mode fails in producing discrete outputs while showing promise in the Pendulum problem that requires continuous outputs. This is because of the ability of the Spatial mode to refrain from using too many conditional statements and rely more on the power of LGP to produce continuous outputs. SGP was compared to two other approaches of TGP and LGP. Except for the Pendulum task which had more of a continuous nature, SGP outperformed the other two approaches in all cases.

Three custom Toy Problems were introduced in this paper, on which SGP was tested. These problems had a larger observation space compared to the classic control problems. Comparing the three tested configurations, SGP produced a similar result to the Control Problems, with RetCon outperforming the two other configurations. The more complex observation space did not significantly impact the performance of SGP showing better performance than TGP and LGP. The Foraging problem was not completely solved; however, improvements in the fitness values showed the possibility of solving this problem if run for an extended period.

A shortcoming of SGP is not having enough control to create a balance in evolving structural elements and LGP programs simultaneously. Perhaps, the utilization of parallel island models that decouple focusing on the evolution of the structural elements and the SGP programs from the main population while interacting with it now and then could be helpful in this scenario to achieve better results. Furthermore, a method for optimizing the system hyper-parameters during evolution could also be among the future directions of this work. As of now, the cost function used in the system adds the normalized distance and length to the return value of the SGP programs. This reduces the impact of distance and length compared to the possible return values. Perhaps a different cost function can result in more exciting results. Finally, to show the effectiveness of SGP, it is necessary to apply it to more realistic and complex problems and to compare it with state of the art problem solvers. Currently, we are working on including more evaluations to the proposed work to prepare it for a future journal submission. We aim to perform systematic analysis on the spatial properties of SGP and how the spatial properties are responsible for achieving the system's performance level and to compare it with TPGs and CGPs.

Code Availability and Supplemental Materials

SGP code in Python, description of main SGP algorithms, practical examples and tutorials for applying SGP and replicating the results of this article can be found in:

https://github.com/elemenohpi/EuroGP-SGP

References

1. Amir Haeri, M., Ebadzadeh, M.M., Folino, G.: Statistical genetic programming for symbolic regression. Appl. Soft Comput. **60**, 447–469 (2017)
2. Aoki, S., Nagao, T.: Automatic construction of tree-structural image transformations using genetic programming. In: Proceedings of the 10th International Conference on Image Analysis and Processing, pp. 136–141. IEEE (1999)
3. Augusto, D.A., Barbosa, H.J.: Symbolic regression via genetic programming. In: Proceedings, vol. 1. Sixth Brazilian Symposium on Neural Networks, pp. 173–178. IEEE (2000)
4. Brameier, M., Banzhaf, W.: Linear Genetic Programming. Springer, New York (2007). https://doi.org/10.1007/978-0-387-31030-5
5. Brockman, G., et al.: OpenAI Gym (2016)
6. Chance, G.: Adventure - atari - atari 2600. https://atariage.com/manual_html_page.php?SoftwareLabelID=1. Accessed 07 Aug 2022
7. DeHon, A., Giavitto, J.L., Gruau, F.: 06361 Executive report - Computing media languages for space-oriented computation. In: Computing Media and Languages for Space-Oriented Computation. Dagstuhl Seminar Proceedings (DagSemProc), vol. 6361, pp. 1–5. Schloss Dagstuhl - Leibniz-Zentrum für Informatik, Dagstuhl, Germany (2007)
8. Fortin, F.A., De Rainville, F.M., Gardner, M.A.G., Parizeau, M., Gagné, C.: DEAP: evolutionary algorithms made easy. J. Mach. Learn. Res. **13**(1), 2171–2175 (2012)

9. Harding, S., Leitner, J., Schmidhuber, J.: Cartesian genetic programming for image processing. In: Riolo, R., Vladislavleva, E., Ritchie, M., Moore, J. (eds.) Genetic Programming Theory and Practice X, pp. 31–44. Genetic and Evolutionary Computation. Springer, New York (2013). https://doi.org/10.1007/978-1-4614-6846-2_3

10. Harding, S., Miller, J.F.: Evolution of robot controller using cartesian genetic programming. In: Keijzer, M., Tettamanzi, A., Collet, P., van Hemert, J., Tomassini, M. (eds.) EuroGP 2005. LNCS, vol. 3447, pp. 62–73. Springer, Heidelberg (2005). https://doi.org/10.1007/978-3-540-31989-4_6

11. Helmuth, T., McPhee, N.F., Pantridge, E., Spector, L.: Improving generalization of evolved programs through automatic simplification. In: Proceedings of the Genetic and Evolutionary Computation Conference, pp. 937–944 (2017)

12. Hodan, D., Mrazek, V., Vasicek, Z.: Semantically-oriented mutation operator in cartesian genetic programming for evolutionary circuit design. Genet. Program. Evolvable Mach. 22(4), 539–572 (2021). https://doi.org/10.1007/s10710-021-09416-6

13. Kelly, S., Heywood, M.I.: Emergent tangled graph representations for Atari game playing agents. In: McDermott, J., Castelli, M., Sekanina, L., Haasdijk, E., García-Sánchez, P. (eds.) EuroGP 2017. LNCS, vol. 10196, pp. 64–79. Springer, Cham (2017). https://doi.org/10.1007/978-3-319-55696-3_5

14. Khan, M.M., Ahmad, A.M., Khan, G.M., Miller, J.F.: Fast learning neural networks using cartesian genetic programming. Neurocomputing 121, 274–289 (2013)

15. Koza, J.R.: Genetic Programming: On the Programming of Computer by Means of Natural Selection. MIT Press, Cambridge (1992)

16. Miller, J.F.: An empirical study of the efficiency of learning Boolean functions using a cartesian genetic programming approach. In: Proceedings of the Genetic and Evolutionary Computation Conference, vol. 2, pp. 1135–1142 (1999)

17. Miller, J.F.: Cartesian genetic programming: its status and future. Genet. Program. Evolvable Mach. 21(1), 129–168 (2020)

18. Miralavy, I., Banzhaf, W.: SGP supplementary materials and code (2023). https://github.com/elemenohpi/EuroGP-SGP

19. Oltean, M., Grosan, C.: A comparison of several linear genetic programming techniques. Complex Syst. 14(4), 285–314 (2003)

20. Pantridge, E., Spector, L.: Code building genetic programming. In: Proceedings of the Genetic and Evolutionary Computation Conference (GECCO-2020), pp. 994–1002 (2020)

21. Spector, L.: Autoconstructive evolution: push, pushgp, and pushpop. In: Proceedings of the Genetic and Evolutionary Computation Conference (GECCO-2001), vol. 137 (2001)

22. Tran, B., Zhang, M., Xue, B.: Multiple feature construction in classification on high-dimensional data using GP. In: 2016 IEEE Symposium Series on Computational Intelligence (SSCI), pp. 1–8. IEEE (2016)

23. Turner, A.J., Miller, J.F.: Cartesian genetic programming encoded artificial neural networks: a comparison using three benchmarks. In: Proceedings of the 15th Annual Conference on Genetic and Evolutionary Computation, pp. 1005–1012 (2013)

24. Yao, M.J., Hsu, H.W.: A new spanning tree-based genetic algorithm for the design of multi-stage supply chain networks with nonlinear transportation costs. Optim. Eng. 10(2), 219–237 (2009)

All You Need is Sex for Diversity

José Maria Simões(✉) , Nuno Lourenço , and Penousal Machado

CISUC, Department of Informatics Engineering, University of Coimbra,
Polo II - Pinhal de Marrocos, 3030 Coimbra, Portugal
josecs@student.dei.uc.pt, {naml,machado}@dei.uc.pt

Abstract. Maintaining genetic diversity as a means to avoid premature convergence is critical in Genetic Programming. Several approaches have been proposed to achieve this, with some focusing on the mating phase from coupling dissimilar solutions to some form of self-adaptive selection mechanism. In nature, genetic diversity can be the consequence of many different factors, but when considering reproduction Sexual Selection can have an impact on promoting variety within a species. Specifically, Mate Choice often results in different selective pressures between sexes, which in turn may trigger evolutionary differences among them. Although some mechanisms of Sexual Selection have been applied to Genetic Programming in the past, the literature is scarce when it comes to mate choice. Recently, a way of modelling mating preferences by ideal mate representations was proposed, achieving good results when compared to a standard approach. These mating preferences evolve freely in a self-adaptive fashion, creating an evolutionary driving force of its own alongside fitness pressure. The inner mechanisms of this approach operate from personal choice, as each individual has its own representation of a perfect mate which affects the mate to be selected.

In this paper, we compare this method against a random mate choice to assess whether there are advantages in evolving personal preferences. We conducted experiments using three symbolic regression problems and different mutation rates. The results show that self-adaptive mating preferences are able to create a more diverse set of solutions when compared to the traditional approach and a random mate approach (with statistically significant differences) and have a higher success rate in three of the six instances tested.

Keywords: Diversity · Sexual Selection · Mate Choice · Mating Preferences

1 Introduction

Natural Selection – or survival of the fittest, as described by Charles Darwin [1] – operates as the foundation for Genetic Programming (GP) as it does for other classes of Evolutionary Algorithms (EA). The adaptation of basic evolutionary principles has made nature-inspired algorithms suitable for several optimization tasks, as encoded solutions in a population evolve under selective pressure to

© The Author(s), under exclusive license to Springer Nature Switzerland AG 2023
G. Pappa et al. (Eds.): EuroGP 2023, LNCS 13986, pp. 276–291, 2023.
https://doi.org/10.1007/978-3-031-29573-7_18

solve specific problems [2–4]. By combining solutions into new ones (crossover) and/or changing existing ones (mutation), we aim at improving results in a given area (exploitation) while promoting a broad search for new regions (exploration) [5,6].

One of the main issues regarding EAs is that of premature convergence, where the population loses general diversity by focusing on one single promising area, which eventually leads to an inescapable set of similar solutions [5]. As such, the balance between exploration and exploitation is important in the search for the global optima, meaning that maintaining diversity throughout the evolutionary process is paramount [2,5,7,9].

Drawing inspiration from natural phenomena, some authors have studied the potential benefits of Sexual Selection as a means for diversity maintenance. Firstly proposed by Darwin as a way to support the idea that Natural Selection alone could not justify certain traits that seemed to hinder survival, Sexual Selection has since then evolved to be recognized as an important (and often complex) evolutionary force [10–13]. Whether acting on the same sex (e.g., battles for mates) or on the opposite sex (e.g., expressing preferences for certain attributes), Sexual Selection can promote the development of different and diverse traits [14].

Despite the seeming potential of Sexual Selection mechanisms in EAs, literature on this evolutionary force is scarce when it comes to its application in GP. Although not extensively, studies that incorporate Sexual Selection mechanisms in Genetic Algorithms are more common, where we can find gender separation [15–17], multiple genders [18] or even dissimilar Mate Choice [19,20], to name a few. Mate Choice based on dissimilarity has also been studied in GP [21] as well as self-adapting mate selection functions [22], yet the array of works in the field appears to become even narrower when we consider mainly individual preferences as a means to avoid early convergence. Recently, Leitão [23] proposed the PIMP method: Ideal partner representations as a way of modelling mate choice in GP. In PIMP, each individual has two chromosomes: one that represents the solution to the problem being tackled, and another that encodes the ideal solution (or partner) against which potential mates are compared. The proposed framework pointed towards performance improvements in several symbolic regression experiments when compared to a standard approach. Furthermore, the author presents a thorough study of the resultant dynamics, which also seems to be promoting exploratory gains. Throughout the reported analysis, some other interesting factors are worth noting, such as the emergence of different roles (i.e., female (or chooser) and male (courter)) merely resulting from distinct selective pressures.

Given the intricate nature of Sexual Selection itself (especially Mate Choice) and the potential benefits of transposing such mechanisms to GP, we set out to expand the existing literature on the topic by shedding light on what we consider to be a relevant question: Are there benefits in encoding ideal mates as opposed to a random mate choice? For that, we took advantage of the good results provided by the PIMP method and tested it against a simpler method in which the courter is chosen at random. In this work, a Standard Approach is also included for reference, and results are measured through performance and diversity metrics. Our results suggest that Mate Choice via the PIMP method is able to promote and maintain more diversity without compromising performance. Furthermore, the experiments suggest that the dynamics behind this mate choice mechanism behave differently from a simpler random mate choice.

The remaining of the article is organized as follows: We first describe the PIMP approach as presented by Leitão [23] and proceed to clarify our motivation behind this experiment. We then describe the methodology (such as the metrics used to validate the comparisons) and present the results followed by an analysis and discussion of the findings.

2   The PIMP Approach

Proposed by Leitão [23], PIMP (Mating Preferences as Ideal Mating Partners in the Phenotype Space) incorporates Mate Choice in GP by encoding a representation of an ideal partner. This approach is built upon a phenomenon often observed in nature regarding the preliminary stages of reproduction: a candidate (courter) has to get access to the opposite sex (chooser), which can set different selection pressures on each [14]. Of the mechanisms within Sexual Selection, this approach models preferences, meaning that mates are chosen based on attractiveness regardless of their fitness. When applied to GP, these different pressures act in parallel and independently, having the potential of driving the evolutionary process in different directions. These dynamics are intended to avoid early convergence in a self-adaptive fashion, as preferences evolve unpredictably, whereas fitness pressure is established at the beginning of the simulation.

Essentially, PIMP differs from a Standard Approach at two levels: the individuals and the selection phase, as described below.

Individuals

Each individual has two chromosomes: the solution and the encoded preference (see Fig. 1). Preferences are representations of the ideal mates, where the trees can be structurally identical to solutions, created from the same building blocks.

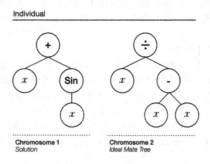

Fig. 1. Illustration of the composition of an individual under the PIMP approach

This means that these can be generated from the same set of terminals and functions used to solve the problem at hand. The preference chromosome has no direct impact on fitness, being computed only during selection.

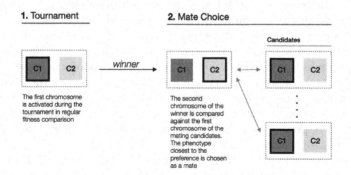

Fig. 2. Illustration of the selection scheme in PIMP

Selection

During selection, an individual is chosen via a standard tournament, becoming the chooser. The candidates (i.e., potential partners) are randomly selected from the population. Each candidate is then compared to the second chromosome of the chooser, which is done in the same way as comparing a solution to the fitness function. In a way, one could say that the encoded preference of the chooser becomes the fitness function for that set of candidates. After going through all candidates, the one closest to the preference is chosen. A simple illustration of the selection phase is presented in Fig. 2. Finally, it is important to note that both chromosomes are subject to recombination and mutation.

3   Motivation

In order to establish a comparison between random mate selection and Mate Choice, we decided to implement the aforementioned PIMP method by Leitão [23]. This choice was based on two major factors. First, the author provided an extensive analysis of the method and specifically targeted Mate Choice as the primary research subject, which is the main focal point of this article. Secondly, the method needs few implementation adjustments compared to a Standard Approach as encoded preferences are built upon the same arguments used to build the solutions.

Furthermore, the work suggests that modelling Mate Choice as in PIMP is advantageous in GP. The method showed performance gains when compared to a Standard Approach on 52 symbolic regression instances, reaching statistical significance on more than half when mutation was introduced. By having mates being selected through ideal representations, the potential mates are subject to a secondary force besides Natural Selection (via fitness function) which led to a role separation that evolved organically. As explained by the author, the self-adaptive nature of mating preferences coupled with Natural Selection promotes a more exploratory search, which in turn seems to be related to the observed performance gains. Also, it was stated that preferences do not always evolve in the same direction as the fitness function, meaning that in some test-cases divergence was observable and evolved independently of fitness pressure.

Hence, we hypothesize that testing this mechanism against a random mate choice might help us understand even better to what extent this approach is favourable. Under the PIMP framework, only the first parent is subject to fitness pressure, while the mate is subject to a direct comparison with the ideal solution encoded in the chooser (i.e., the first parent). By avoiding assembling couples merely through fitness metrics we allow more diversity among the population, which in turn improves performance. As such, we aim at exploring to what extent can a random mate choice lead to similar outcomes.

Moreover, considering that there is strong evidence that in PIMP preferences evolve in a self-reinforced fashion rather than arbitrarily – thus having the potential of evolving towards its own goal –, we will also be able to test this against a mate choice with no sustainable direction and access potential benefits. In the next section, we address the criteria used for this experiment.

4   Methodology

This study compares two existing approaches to an experimental one: PIMP, Random Mate Choice (first parent chosen via tournament selection and a mate chosen at random) and a Standard Approach acting as a reference where both parents are chosen through tournaments (see Table 1). The setup used for all these methods is shown in Table 2. Two different mutation probabilities are considered (of 5% and 10%). Regarding the genetic operators, Subtree mutation is applied and One Point Crossover is used for recombination. Three symbolic regression instances were used: Koza-1 (one variable), Nguyen-6 (one variable) and Pagie-1 (two variables) (a summary of these instances can be found in [24]). These functions were tested in the original work, thus providing us with some knowledge beforehand, particularly regarding Mean Best Fitness. While tested under Koza-1 (with a mutation rate of 1%) PIMP showed no statistically significant differences when compared to a Standard Approach, under Nguyen-6 PIMP performed better with significant differences. We also included Pagie-1 (where PIMP also performed better with significant results) given that it can be more difficult to solve than the last two [24].

All testing instances were performed in sets of 30 runs, and each run had its exclusive seed shared with all three approaches. Although the main setup is similar to the original work on PIMP, some changes were performed, such as the number of generations (500 originally) and the mutation rates (1% in the original work). By increasing the number of generations, we can have a better understanding of how the population behaves in the long run, while by having higher mutation rates we simultaneously promote more genetic diversity (especially in the Standard Approach).

Table 1. Approaches used

PIMP	Parent 1 (Chooser)	Tournament Selection (size = 5)
	Parent 2 (Courter)	Random Set (size = 5)
Random Mate	Parent 1	Tournament Selection (size = 5)
	Parent 2	Random
Standard	Both Parents	Tournament Selection (size = 5)

4.1 Measures

We compare the mentioned approaches at two main levels: performance and diversity. The metrics used for each are described below.

Performance Measures

To assess performance differences we compute the Mean Best Fitness (MBF) measured through Mean Squared Error for all algorithms under study. Considering that we are testing them under symbolic regression instances with known solutions, we defined a success rate regarding the quality of the solutions. Therefore, we have:

- Mean Best Fitness (Best individual of the final population in each run)
- Success Rate (Number of runs (out of 30) in which at least one solution is better than 1E–4)

Diversity Measures

While performance metrics are quite straightforward to establish, diversity metrics require a bit more discussion. When studying diversity, different techniques can be applied to examine its dynamics and effects within the population [7]. Knowing that in GP different trees can have the same fitness, we decided not to measure population diversity by fitness variety. Instead, we performed a comparison at the level of the genotype to identify differences between trees (e.g., through Tree Edit Distance [8]). Nevertheless, this choice came with a cost that proved to be too expensive computation-wise. As such, it was decided to record only the number of unique solutions in the population every one hundred generations. This metric is not sufficient to tell us how different solutions are, but it can nonetheless be useful at a higher level: if an algorithm is not able to maintain unique solutions, then convergence is more likely to occur.

Another measure used to assess diversity is strictly linked to the original work on PIMP. In the experiments, the author opted for operation restrictions at the root node – which was never chosen as a crossover point or for mutation.

Table 2. General Set up

General Parameters	Population Size	100
	Generations	1500
	Elitism	None
Individual Builder	Ramped half-and-half	random(2,6)
Breeding	Crossover Prob.	0.9
	Mutation Prob.	[0.05, 0.1]
	Max Depth	17

This means that if these particular alleles managed to avoid extinction, they were forced to remain the same throughout the evolutionary process. Furthermore, under these conditions, if a particular root node is lost it is then lost forever. Thus, it was decided to keep this particular feature in implementation as we saw it as a way of studying which approach was better at maintaining diversity at the root node. To achieve this, we kept track of each individual's root node throughout evolution. This was a straightforward process as all functions used for testing share the same function set (see Table 3). A run where there the root node is the same in every individual on the last generation was considered a convergence instance. With this, we can list the two metrics used for diversity analysis:

– Number of unique solutions among the population
– Runs where root node convergence was avoided

Table 3. Function Set (same for all instances)

Functions	Constants
$+, -, \times, \%, \sin, \cos, e^n, \ln(|n|)$	None

4.2   Statistical Tests

After running the experiments and gathering the required data for analysis, a set of statistical tests was performed to better understand the significance of the observed differences. We highlight that in this experiment we are dealing with dependent samples, given that each run had a specific initial population.

For tests consisting of quantitative data (Mean Best Fitness (MBF) and Number of Unique Solutions (NUS)), a Shapiro-Wilk test was performed to assess the likelihood of having normally distributed data, while Bartlett's test was performed to test the variance of the data. For an alpha value of 5%, MBF data failed the normality assumption, and while NUS data only failed to meet the normality assumption under the function Pagie-1, the Bartlett's test showed

no signs of equal variance between groups (also for an alpha of 5%). As such, a
Friedman test was performed to test the differences between groups. Whenever
differences were found, a Wilcoxon signed rank was performed as a post hoc with
the correspondent Bonferroni correction [25].

For qualitative data (Success Rate and Root Node convergence), a Cochran's
Q test was used to find significant differences between approaches, proceeding
with a McNemar's Test whenever that was the case. Again, an alpha value of
5% was used when comparing the three approaches, and a Bonferroni correction
was applied for paired comparisons [25].

5   Results

In this section we present the results for each symbolic regression instance and
mutation variation. For the sake of readability, Mean Best Fitness results are
presented in two different tables (one for each mutation variant) as the sample
standard deviation is also included. Although the raw data regarding the success
rates is nominal, results are presented in the percentage form merely on a stylistic
choice. Whenever there is statistical significance between PIMP and Random
Mate approach these will be highlighted (**bold**). Furthermore, an asterisk sign
(*) next to a result indicates that particular approach is statistically different
from the Standard Approach.

Mean Best Fitness
Table 4 and Table 5 depict the results regarding the MBF for different mutation
rates as well as the sample standard deviation. From a general overview, PIMP
does not always outperform the Random Mate approach, which in turn shows
slightly better results in some instances. Similarly, whenever PIMP shows better
results, the differences between these two approaches do not seem to be critical.
Furthermore, including the Standard Approach as a benchmark, PIMP seems to
do better also with a seemingly small margin. Using a Random Mate approach
seems to provide competitive results to a Standard one, which might suggest
that under these specific conditions having both parents under fitness pressure
is not necessarily advantageous. This might be reinforced by the fact that no
statistically significant differences were found. However, it must be mentioned
that on the particular instance of the function Pagie-1 with a mutation rate of
10%, we obtained a *p-value* of ≈ 0.048. Although this meant that differences
were significant under the defined criteria, this *p-value* was considered to be
too close to the chosen alpha value. This was later confirmed when performing a
Wilcoxon signed rank, where no differences were found. As such, we have decided
to follow a conservative approach and conclude that no relevant differences were
found.

Table 4. Results for the MBF and StDev with a mutation rate of 5% over 30 runs

		PIMP	Random Mate	Standard
Koza-1	MBF	3.44E-4	1.75E-3	9.68E-4
	StDev	4.86E-4	6.29E-3	2.20E-3
Nguyen-6	MBF	6.54E-3	1.00E-2	5.82E-3
	StDev	1.72E-2	2.33E-2	1.52E-2
Pagie-1	MBF	9.11E-3	1.29E-2	1.57E-2
	StDev	1.06E-2	1.23E-2	1.46E-2

Table 5. Results for the MBF and StDev with a mutation rate of 10% over 30 runs

		PIMP	Random Mate	Standard
Koza-1	MBF	5.56E-4	1.28E-3	3.9E-4
	StDev	1.01E-3	4.75E-3	8.59E-4
Nguyen-6	MBF	5.63E-3	2.74E-3	6.23e-3
	StDev	1.5E-2	8.57E-3	1.76E-2
Pagie-1	MBF	9.62E-3	1.08E-2	1.50E-2
	StDev	1.08E-2	1.21E-2	1.44E-2

Success Rate

Just as observed in the MBF results, the data presented in Table 6 also does not seem to provide any consistent advantage for any approach when it comes to success rate. The main differences can be found on Nguyen-6 with a mutation rate of 5%, where PIMP and Random Mate promote a gain of 26.7% and 23.3%, respectively, over the Standard Approach. Conversely, by increasing the mutation rate to 10% in the same function the Standard Approach performs better, with a success rate 26.7% better than PIMP and only 3.4% over a Random Mate choice. On both Pagie-1 instances (which is considered to be more difficult to tackle [24]), a Random Mate approach fails to find any good solution under the defined terms, while PIMP manages to do so in 2 runs out of 30. Performing a Cochran's Q test on the data results in no significance for any instance.

Table 6. Success Rate (average of 30 runs)

	Mutation %	PIMP	Random Mate	Standard
Koza-1	5%	56.6%	56.6%	60%
	10%	60%	53.3%	43.3%
Nguyen-6	5%	60%	56.6%	33.3%
	10%	43.3%	66.6%	70%
Pagie-1	5%	6.6%	0%	3.3%
	10%	0%	0%	3.3%

Diversity Analysis

While there seems to be no performance gains in using any of the studied approaches over the others, the same can not be said regarding diversity maintenance. Table 7 depicts the results on the average percentage of unique trees after 1500 generations, where there is a consistent pattern throughout all instances. It can be observed that under the Standard Approach the final population tended to contain roughly 35% of repeated solutions on average, which might be the consequence of submitting both parents to fitness pressure. In terms of genetic diversity, Random Mate provides an improvement over this, which is further improved when using PIMP, where on average it produced 15% of repeated solutions. Statistic results from Friedman tests show significant differences in all instances, which was later confirmed through a paired Wilcoxon Signed Rank holding two exceptions: in Pagie-1 with mutation rates of 5% and 10%, PIMP and Random Mate have no significant differences between them. Figures 3a and 3b illustrate the percentage of unique solutions throughout the evolutionary process each 100 generations for Koza-1 (mutation rate of 5%) and for Pagie-1 (mutation rate of 5%).

Table 7. Unique Solutions After 1500 Generations (Average of 30 runs)

	Mutation %	PIMP	Random Mate	Standard
Koza-1	5%	84.8%*	77.2%*	64.26%
	10%	86.4%*	77.4%*	66.6%
Nguyen-6	5%	84.2%*	79%*	63.2%
	10%	86.8%*	78.3%*	62.2%
Pagie-1	5%	86%*	83.8%*	61.5%
	10%	88.4%*	84.6%*	67.6%

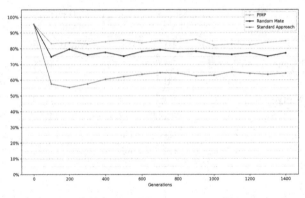

(a) Koza-1 Mutation rate 5%

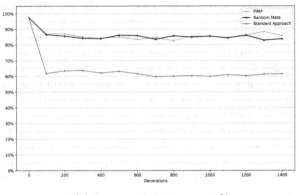

(b) Pagie-1 Mutation rate 5%

Fig. 3. Percentage of unique solutions each 100 generations (average).

Avoiding Root Node Convergence

Finally, Table 8 provides an overview of the ability of each approach to prevent convergence to a single function at the root node. Similarly, there is a gradual improvement from the Standard Approach to Random Mate, followed by PIMP. These results are also an interesting aspect of this study considering that, as discussed earlier, preserving root node diversity under the current conditions is a difficult task. Although there seems to be an advantage in using PIMP to achieve this (mainly in Koza-1 and Nguyen-6), the only instances where statistical significance was reached was in Koza-1 with a mutation rate of 5%, where PIMP was statistically different from Random Mate and Standard Approach (p-$values$ ≈ 0.004 and ≈ 0.0009 respectively) and in Pagie-1 with a mutation rate of 10% – both PIMP and Random Mate held no differences (p-$value$ ≈ 0.54) but both were statistically different from the Standard Approach (p-$values$ of ≈ 0.0009 and ≈ 0.002).

6   Discussion and Additional Remarks

Within the limits of the current study, the results presented throughout the last section allow us to draw some conclusions that should be discussed in more detail. Although the central point of this work is not to establish a direct comparison between a Mate Choice mechanism (PIMP or Random Mate) and a Standard Approach, it is nonetheless relevant to make some notes on performance and diversity. It seems that in the long run (i.e., after 1500 generations), reducing fitness selective pressure does not affect significantly the performance, whether it is achieved by choosing a partner via preferences or at random. Comparing our results to that of the original work on PIMP, it seems that elongating the running time improves performance by a quite small margin in some symbolic regression instances. Arguably, mutation might also have something to do with this. As the original work provides a wider range of testing instances, it might be useful to extend the current experiment in the future.

Comparing PIMP directly against a random mate choice, the former seems to provide slightly better results regarding MBF, but we found these differences to be insufficient to reach statistical significance. Again, testing a more diverse set of symbolic regression instances could be beneficial in this regard. As such, under the conditions applied in this experiment, it seems that PIMP and Random Mate are similar performance-wise.

On the other hand, in general, results regarding diversity favour PIMP. We believe this to be a particularly interesting – and arguably counterintuitive – finding: the dynamic evolution of preferences in PIMP seems to have a higher potential for promoting unique solutions rather than a simple random mate choice. This is also a topic to be addressed with more detail in the future, that is, a more focused analysis on how the evolvable preferences tend to promote more diversity than a random approach. Arguably, this might be a result of the directional Sexual Selection force that emerges throughout the evolutionary process, which in turn might be doing a better job at racing against fitness pressure than a non-directional random selection.

Table 8. Instances where root node convergence was avoided after 1500 Generations (Out of 30 runs)

	Mutation %	PIMP	Random Mate	Standard
Koza-1	5%	**11/30***	**3/30**	0/30
	10%	12/30	7/30	5/30
Nguyen-6	5%	7/30	3/30	2/30
	10%	10/30	4/30	3/30
Pagie-1	5%	5/30	4/30	2/30
	10%	11/30*	9/30*	0/30

Although this analysis requires an in-depth review, an answer to this difference may be laying precisely in how Sexual Selection via preferences shapes the search space of the ideal mates. In the original work of PIMP, a clear role separation throughout the evolutionary process was observed: the population tended to evolve into three main types (choosers, courters and both). To understand whether the same effect could be achieved via a Random Mate selection, we kept track of the selected individuals to establish a direct comparison. As such, we used the same terminology: Choosers (individuals that were only selected via tournament), Courters (individuals that were exclusively selected by a Chooser) and Both (individuals that in the same generation acted as a Chooser and a Courter). The evolutionary lineages of these can be seen in Fig. 4a for PIMP and in Fig. 4b for Random Mate selection. Note that only one instance is shown (Koza-1 with a mutation rate of 5%), but the behaviour was similar in all symbolic regression instances. The small shifts observed in the first generations using the PIMP approach suggest an adaptation rather than a predefined rule (as seen in Random Mate, where on average all roles emerge in the same range in which they terminate). It is interesting to note that although a Random Mate choice does not provide a specific direction regarding choice, it manages to produce role segregation anyway. Another difference between these two figures is that Random Mate seems to be producing more individuals that act as Both roles, while in PIMP there's a larger percentage of Choosers.

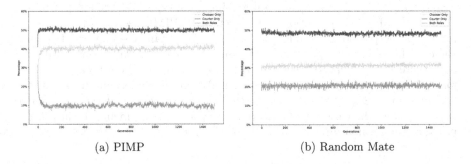

(a) PIMP   (b) Random Mate

Fig. 4. Role Seggregation - Koza-1 Mutation rate 5%.

We set out to analyse whether this stronger segregation had any impact on the MBF of each role (Figs. 5a and 5b).

As expected, in both approaches the Choosers follow a better fitness path as they are subject to fitness pressure. Yet interestingly, a Random Mate choice seems more prone to preserve individuals with good fitness values while in PIMP these seem to be performing some sort of exploration. Furthermore, the MBF of individuals assuming both roles also seems to be evolving differently: In Random Mate, it looks like these behave as Choosers fitness-wise, while in PIMP this type of role assumes a more unpredictable behaviour, and these differences might be contributing to a more diverse population overall. By choosing a mate randomly, it means that the individuals that made it through the tournament have the same

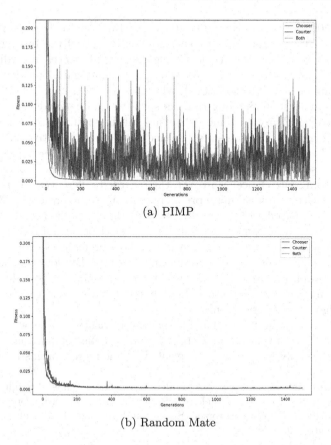

(a) PIMP

(b) Random Mate

Fig. 5. MBF evolution for different roles - Koza-1 Mutation rate 5%.

chance of being selected as a partner, while PIMP might be forcing a particular portion of the population to adapt to the existing range of chooser's preferences. Considering that even the chromosomes that encode the ideal partner are subject to mutations and crossover, this might also be contributing to the diversity of the whole population. As such, a detailed study on how PIMP shapes these separated roles might be valuable to understand the inner mechanisms of mating preferences in GP.

7   Conclusion

Diversity maintenance is known to be an important factor in EAs. Sexual Selection, particularly Mate Choice, has shown promising results when transposed to algorithmic practice, yet its dynamics and potential benefits are poorly understood, especially when it comes to GP. In this article, we took advantage of PIMP – mate choice based on ideal mates – to address whether there were tangible differences between guiding the evolutionary process via a random mate choice and by having a set of individual mating preferences.

After 1500 generations (in sets of 30 runs), PIMP revealed competitive results regarding MBF and had a better success rate in half of the testing instances. Although there were no statistically significant differences regarding performance, these were achieved when considering diversity gains (particularly in the average number of unique solutions), where PIMP has shown a better capacity of doing so at least in the long run. As such, we can conclude that within our experiments PIMP promoted more diversity within populations without compromising performance. These results, coupled with other indicative factors such as root node diversity (although in this regard fewer differences were statistically significant), seem to suggest that there are in fact differences between PIMP and a Random Mate choice. Moreover, the recorded differences are interesting: a mate choice based on ideal mate preferences seems to be promoting more diversity within the population (even fitness-wise) than a random mate search. We believe these to be meaningful results in the field of Sexual Selection applied to GP, which deserves an in-depth study on how preferences evolve, mainly on the evolutionary dynamics of choosers and courters. By doing so, we may not only understand in more detail how this force is promoting more diversity than a random mate choice but also if (and how) this approach can be shaped to improve algorithmic performance.

Finally, it must be noted that for such an intricate task, it would be beneficial in the future to have a more broad study on the comparison of these two approaches, mainly with a broader set of symbolic regression instances.

Acknowledgements. This work is funded by national funds through the FCT - Foundation for Science and Technology, I.P., within the scope of the project CISUC - UI/BD/151046/2021 and by European Social Fund, through the Regional Operational Program Centro 2020.

References

1. Darwin, C.: On the Origins of Species by Means of Natural Selection. Murray, London (1859)
2. Poli, R., Langdon, W.B., McPhee, N.F.: A Field Guide to Genetic Programming. Lulu Enterprises, London (2008)
3. Eiben, A., Smith, J.: From evolutionary computation to the evolution of things. Nature **521**, 476–482 (2015)
4. Back, T., Hammel, U., Schwefel, H.-P.: Evolutionary computation: comments on the history and current state. IEEE Trans. Evol. Comput. **1**(1), 3–17 (1997)
5. Eiben, A.E. and Smith, J.E.: Introduction to Evolutionary Computing. In: Natural Computing Series, 2nd edn. Springer, Heidelberg (2015). https://doi.org/10.1007/978-3-662-44874-8
6. Eiben, A.E., Schippers, C.A.: On evolutionary exploration and exploitation. Fundamenta Informaticae **35**(1), 35–50 (1998)
7. Burke, E., Gustafson, S., Kendall, G.: Diversity in genetic programming: an analysis of measures and correlation with fitness. IEEE Trans. Evol. Comput. **8**, 47–62 (2004)
8. Zhang, K., Shasha, D.: Simple fast algorithms for the editing distance between trees and related problems. SIAM J. Comput **18**, 1245–1262 (1989)

9. Hien, N.T., Hoai, N.X.: A brief overview of population diversity measures in genetic programming. In Proceedings 3rd Asian-Pacific Workshop on Genetic Programming, Hanoi, Vietnam, pp. 128–139 (2006)

10. Gayon, J.: Sexual selection: another darwinian process. Comptes Rendus Biologies **333**(2), 134–144 (2010)

11. Alonzo, S. H., Servedio, M.R.: Grey zones of sexual selection: why is finding a modern definition so hard? Proc. Roy. Soc. B: Biol. Sci. **286** (2019)

12. Clutton-Brock, T.: Sexual selection in males and females. Science **318**, 1882–1885 (2007)

13. Ralls, K., Mesnick, S.: Sexual dimorphism. In: Perrin, W.F., Würsig, B., Thewissen, J.G.M. (eds), Encyclopedia of Marine Mammals, 2nd edn., pp. 1005–1011. Academic Press, London (2009)

14. Jones, A., Ratterman, N.: Mate choice and sexual selection: what have we learned since Darwin? Proc. Natl. Acad. Sci. **106**(Supplement 1), 10001–10008 (2009)

15. Drezner, T., Drezner, Z.: Gender-specific genetic algorithms. INFOR Inf. Syst. Oper. Res. **44**(2), 117–127 (2006)

16. Bandyopadhyay, S., Pal, S.K., Maulik, U.: Incorporating chromosome differentiation in genetic algorithms. Inf. Sci. **104**(3–4), 293–319 (1998)

17. Zhu, Y., Yang, Z., Song, J.: A genetic algorithm with age and sexual features. In: Huang, D.-S., Li, K., Irwin, G.W. (eds.) Intelligent Computing. ICIC 2006. LNCS, vol. 4113. Springer, Heidelberg (2006). https://doi.org/10.1007/11816157

18. Vrajitoru, D.: Natural selection and mating constraints with genetic algorithms. Int. J. Model. Simul. **28**(2), 188–194 (2008)

19. Varnamkhasti, M.J.: Sexual selection and evolution of male and female choice in genetic algorithm. Sci. Res. Essays **7**(31), 2788–2804 (2012)

20. Varnamkhasti, M.J., Lee, L.S.: A genetic algorithm based on sexual selection for the multidimensional 0/1 knapsack problems. Int. J. Mod. Phys.: Conf. Ser. **9**, 422–431 (2012)

21. Fry, R., Smith, S.L., Tyrrell, A.M.: A self-adaptive mate selection model for genetic programming. In: IEEE Congress on Evolutionary Computation, vol. 3, pp. 2707–2714 (2005)

22. Smorodkina, E., Tauritz D.: Toward automating EA configuration: the parent selection stage. In IEEE Congress on Evolutionary Computation, pp. 63–70 (2007)

23. Leitão, A.: Sexual Selection through Mate Choice in Evolutionary Computation. PhD thesis, Universidade de Coimbra (2019)

24. McDermott, J., et al.: Genetic programming needs better benchmarks. In: Proceedings of the 14th Annual Conference on Genetic and Evolutionary Computation, pp. 791–798. Association for Computing Machinery, Philadelphia (2012)

25. Sheskin, D.J.: Handbook of Parametric and Nonparametric Statistical Procedures, 2nd edn. Chapman & Hall/CRC, Boca Raton (2000)

On the Effects of Collaborators Selection and Aggregation in Cooperative Coevolution: An Experimental Analysis

Giorgia Nadizar[1] and Eric Medvet[2]

[1] Department of Mathematics and Geosciences, University of Trieste, Trieste, Italy
[2] Department of Engineering and Architecture, University of Trieste, Trieste, Italy
emedvet@units.it

Abstract. Cooperative Coevolution is a way to solve complex optimization problems by dividing them in smaller, simpler sub-problems. Those sub-problems are then tackled concurrently by evolving one population of solutions—actually, *components* of a larger solution—for each of them. However, components cannot be evaluated in isolation: in the common case of two concurrently evolving populations, each solution of one population must be coupled with another solution of the other population (the *collaborator*) in order to compute the fitness of the pair. Previous studies have already shown that the way collaborators are chosen and, if more than one is chosen, the way the resulting fitness measures are aggregated, play a key role in determining the success of coevolution. In this paper we perform an experimental analysis aimed at shedding new light on the effects of collaborators selection and aggregation. We first propose a general scheme for cooperative coevolution of two populations that allows to (a) use different EAs and solution representations on the two sub-problems and to (b) set different collaborators selection and aggregation strategies. Second, we instantiate this general scheme in a few variants and apply it to four optimization problems with different degrees of separability: two toy problems and two real prediction problems tackled with different kinds of model (symbolic regression and neural networks). We analyze the outcomes in terms of (a) effectiveness and efficiency of the optimization and (b) complexity and generalization power of the solutions. We find that the degree to which selection and aggregation schemes differ strongly depends on the interaction between the components of the solution.

Keywords: Cooperative coevolution · Collaborator selection · Fitness aggregation · Symbolic Regression · Neuroevolution · Genetic Programming · Evolutionary Strategies

1 Introduction and Related Works

Cooperative coevolutionary algorithms (CCEAs) are optimization techniques leveraging a divide-and-conquer approach for addressing complex problems

© The Author(s), under exclusive license to Springer Nature Switzerland AG 2023
G. Pappa et al. (Eds.): EuroGP 2023, LNCS 13986, pp. 292–307, 2023.
https://doi.org/10.1007/978-3-031-29573-7_19

where the solution can be decomposed into simpler *components* [16,17]. The main idea behind Cooperative Coevolution (CC) is that of evolving, i.e., optimizing, components independently, and than proceeding to aggregate them into the final, complex, solution to the problem. The rationale being that it might be easier to navigate smaller search spaces—those of components—separately, rather than the larger search space of complete solutions, making it simpler to converge to satisfying solutions.

The components identification and/or emergence is itself an interesting research topic, where continuous coadaptation plays a fundamental role [16]. Here, without loss of generality, we consider problems where the solutions have only two hand-designed components. In this case, evolution happens in an ecosystem with 2 species, which, as in the biosphere, are independent, since they are, in principle, related to completely different aspects of the problem. Therefore, each species is evolved in its own population, and the two evolutions are independent from each other.

However, in most practical cases, the components of the solution are completely meaningless on their own, making their independent fitness evaluation practically unfeasible. In fact, oftentimes the components acquire a meaning only when combined to make a complete solution to the problem at hand. Hence, the need for forming *collaborations* to enable the fitness evaluation. To make a clarifying example consider the scenario of a robot being optimized to perform a task, where CCEAs could be used to separately evolve the "body" and the "brain" of an artificial agent: it is immediately clear how neither a body without a brain, nor a brain without a body, can be assessed against the proposed task.

The process of forming collaborations introduces additional degrees of freedom: (1) on how to select the collaborators from the opposite component population (how many and which ones), and (2) on how to define the fitness of a component which has been evaluated in combination with different collaborators, i.e., how to *aggregate* multiple fitness values. Several approaches have been proposed in the literature for both: Ma et al. [4] provide a neat categorization of the existing ones (we refer the reader to the cited survey for further details). Collaborators selection methods mainly differ in terms of greediness: ranging from best collaborator selection [16], the greediest, to random or worst collaborator selection [13,21], which maintain higher diversity in the population [10,18]. Such methods can also be generalized or combined in a hybrid manner [14], in order to sample multiple collaborators. For the fitness aggregation schemes, the rationale is similar: the approaches range from the most optimistic ones, considering the fitness of the best individual [5], to pessimistic ones, using the worst fitness found [21], going through more moderate approaches, employing the average or the median fitness in the sample [7].

Clearly, the choice of the collaborator selection and the fitness aggregation have strong implications on the outcome of CC, affecting efficiency, computational costs, generalization ability, to only mention a few. However, the recipe for choosing the right scheme for a given problem remains somewhat unclear. Some studies have indeed tried to tackle this issue, as [3,12,15,21], which have all proposed experimental comparisons of some existing methods. Yet, all these studies focused on synthetic problems, which generally have different characteristics from the real-world ones. Moreover, to the best of our knowledge, all the existing works have only considered a particular combination of EAs, lacking the generality which CC enables in that direction.

In this context, our goal is that of filling the gap, by first formalizing a general and fully modular scheme of CCEA, and by then employing it for an experimental evaluation. We tailor the CCEA scheme to be completely agnostic with respect to the EAs chosen for evolving the two sub-populations, as long as they follow an iterative structure. Moreover, we devise it to be able to employ a variety of collaborator selection strategies, in combination with different fitness aggregation criteria. Concerning the experimental analysis, we aim at (a) providing a proof of concept of the generality of the formalized schema, and at (b) exploiting it for measuring the effects of different combinations in the modules (EAs, collaborator selection, and fitness aggregation). To this extent, we focus on two toy problems first, and then on two real-world problems: symbolic regression and neuroevolution for data classification.

Our results suggest the following insights. First, the number of collaborators has a clear impact on efficiency and a fuzzier impact on effectiveness of the optimization: in general, few collaborators are enough for co-evolving effective solutions without affecting efficiency too severely. Second, optimistic aggregation of fitness values, i.e., assigning to a collaborator the fintess of the best collaboration pair, is often better than other schemes (namely, median and worst). Third, depending on the interaction of solution components, fitness aggregation may have a strong impact on the structure of evolved solutions, by relying more on one of the two populations for exploring the conjoint search space. Finally, we do not observe any evidence of increased generalization power by using more collaborators or using unfit collaborators.

2   A General Scheme for CC

We here describe a CCEA that instantiates a general scheme for CC. We consider the case with only two solution components and we assume that the way they are assembled together, in order to form a complete solution for the problem at hand, is fixed in advance. We deem our results conceptually portable to cases with more components, where emergence is allowed [16].

Our CCEA works by internally exploiting two other EAs. The only requirements we impose on the two EAs is that they are (a) iterative and (b) population-based. We assume that they comply to the structure displayed in Fig. 1, according to which an EA corresponds to a function $\texttt{solve}(f)$ that, given a fitness function $f : P \rightarrow \mathbb{R}$ to be minimized[1], should return a solution (see Footnote 1) $p^\star$, such that $q^\star = f(p^\star)$ is minimal. Note that a given EA might internally

---

[1] Our CCEA does not constrain the fitness to be minimized, nor to be a single number; similarly the inner EAs may return many solutions, not just one; we pose here these limitations just for clarity.

search a space G different than P and use a genotype-phenotype mapping function $\phi : G \to P$ for obtaining solutions to be evaluated with f. Our CCEA does not pose any constraint concerning this possibility; in particular, the two search spaces G_1 and G_2 of the inner EAs may be different, as well as the two mapping functions ϕ_1 and ϕ_2—that is, the two EAs may employ different solution representations.

An iterative, population-based EA can be described by providing the bodies for the `init()`, `stop()`, `update()`, and `extractSolution()` functions used by `solve()`. The progress of the EA is stored in its state s, which hosts the population, together with additional information (specific to each EA) about the ongoing optimization. The state is initialized by the `init()` function, and is then iteratively updated within `update()`: these functions respectively perform the initialization of the population, which is then stored in the state, and its update via some forms of genetic operators. The optimization process then iteratively proceeds until a termination condition is met, tested in `stop()`, e.g., an evaluation budget has been exceeded, and finally the solution is extracted from the state s via `extractSolution()` and returned.

```
function solve(f):
    s ← init(f)
    while !stop(s) do
    |   s ← update(s, f)
    end
    return extractSolution(s, f)
end
```

Fig. 1. General structure of an iterative EA. s is the generic state of the EA. $f : P \to \mathbb{R}$ is the fitness function.

Our CCEA is itself an iterative, population-based EA, and hence we can define it by describing its `init()`, `stop()`, `update()`, and `extractSolution()` functions, which are shown in detail in Fig. 2.

The working principle of our CCEA is that it delegates the evolution of the two components of the solution to two inner EAs, that we denote with EA_1 and EA_2, feeding them with *disposable* fitness functions created on-the-fly. Such disposable fitness functions are needed because the inner EAs evolve components of the solution, rather than full solutions. Hence, they cannot be assessed with f, which is defined on P, but need a $f_1 : P_1 \to \mathbb{R}$ and $f_2 : P_2 \to \mathbb{R}$, where P_1 and P_2 are the set of possible components for EA_1 and EA_2, i.e., their solution spaces. Moreover, by building a disposable f_1 (or f_2) on-the-fly, we can enclose the collaborator selection and aggregation steps inside the function, thus making it dependent on the current population of candidate collaborators.

Each disposable fitness function evaluates an individual of the corresponding population (i.e., a component of the solution) by performing these three steps: (1) select a portion of the other population (*collaborator selection*); (2) use f to compute the fitness of each pair resulting by combining the current individual with a collaborator of the selected subset; (3) aggregate the resulting fitness

```
function init(f):                          function update(s,f):
    f₁ ← p₁ ↦ 0                                P′₁ ← cSel(s.s1.pop)
    f₂ ← p₂ ↦ 0                                P′₂ ← cSel(s.s2.pop)
    s₁ ← init₁(f₁)                             f₁ ← p₁ ↦ cAggr([f(p₁ ⊕ p₂)]_{p₂∈P′₂})
    s₂ ← init₂(f₂)                             f₂ ← p₂ ↦ cAggr([f(p₁ ⊕ p₂)]_{p₁∈P′₁})
    P′₁ ← cSel(s₁.pop)                         s.s1 ← update₁(s.s1, f₁)
    P′₂ ← cSel(s₂.pop)                         s.s2 ← update₂(s.s2, f₂)
    f₁ ← p₁ ↦ cAggr([f(p₁ ⊕ p₂)]_{p₂∈P′₂})     return s
    f₂ ← p₂ ↦ cAggr([f(p₁ ⊕ p₂)]_{p₁∈P′₁})  end
    s.n ← 0
    s.s1 ← init₁(f₁)                       function extractSolution(s,f):
    s.s2 ← init₂(f₂)                           P′₁ ← [p₁ : (p₁,q) ∈ s.s1.pop]
    return s                                   P′₂ ← [p₂ : (p₂,q) ∈ s.s2.pop]
end                                            (p*₁,p*₂) ←   arg max    f(p₁ ⊕ p₂)
                                                          (p₁,p₂)∈P′₁×P′₂
function stop(s):                              return p₁ ⊕ p₂
    return stop₁(s.s1) ∨ stop₂(s.s2)      end
end
```

Fig. 2. The `init()`, `update()`, `stop()`, and `extractSolution()` functions for our CCEA. The $\oplus$ operator composes a solution $p = p_1 \oplus p_2 \in P$ from two components $p_1 \in P_1$ and $p_2 \in P_2$. $\mathrm{init}_1()$, $\mathrm{init}_2()$, $\mathrm{update}_1()$, $\mathrm{update}_2()$, $\mathrm{stop}_1()$, and $\mathrm{stop}_2()$ are the corresponding functions for the first and second EA, searching respectively in P_1 and P_2. $x \mapsto y$ represents a literal for a function that maps an x to an y: hence, $f_1 \leftarrow p_1 \mapsto 0$ means that f_1 becomes a function that maps any p_1 to $f_1(p_1) = 0$.

values (*fitness aggregation*). In the algorithms of Fig. 2, the collaborator selection step is performed by the `cSel()` function, which receives a population, i.e., a multiset of pairs (p, q), p being the solution and q being its fitness, and returns a multiset of solutions p, extracted from the population passed as argument. The fitness aggregation step is performed by the `cAggr()` function, which takes a multiset of real values in input, and outputs a single real value.

Going more into details about each part of the CCEA, the core of the CCEA lies in the `update()` function, which is responsible for performing each evolutionary step. It starts by selecting the collaborators P'_1 and P'_2, using `cSel()`, to evaluate the solutions components, from the first and the second current subpopulation, respectively. Then, `update()` proceeds by defining the disposable fitness functions f_1 and f_2 to evaluate the components: internally, both rely on f to compute the fitness of a pair of solution components. To complete the evolutionary step, `update()` invokes $\mathrm{update}_1(s.\mathrm{s1}, f_1)$ and $\mathrm{update}_2(s.\mathrm{s2}, f_2)$ to have the inner EAs perform an evolutionary step and update their states (and populations).

The function `init()`, responsible for starting the CCEA, has a similar structure to the `update()` function. However, since the collaborator selection `cSel()` takes a population $[(p^{(i)}, q^{(i)})]_i$ of individuals, rather than just a population $[p^{(i)}]_i$ of components, there is an additional preliminary step where we define two dummy fitness functions, which map all solutions to a 0 value of fitness. At this point, the `cSel()` can effectively be applied to select the individuals from each population, even though here the selection is driven only by randomness.

Then, the two fitness functions are defined, and the initialization is concluded with the invocation of the inner initializers, $\text{init}_1(f_1)$ and $\text{init}_2(f_2)$.

The last two building blocks of the CCEA are the termination condition check and the extraction of the solution to be returned. The first is performed within the stop() function, which checks if any of the two inner EAs has achieved its termination condition. The second, performed by the extractSolution() function, computes all the possible combinations among the components, and returns the best one according to the fitness function f. This should increase the likelihood of achieving a good solution in the end, regardless of the criterion used during evolution to sample the collaborators. In fact, some criteria might be helpful at steering the evolutionary search, but they might not be suitable for selecting the final combination of components.

Following the provided schema, we highlight that there is great freedom in selecting (a) the inner EAs, as long as they comply to the structure of Fig. 1, (b) the collaborator selection, which can be freely chosen implementing cSel() as desired, and (c) the fitness aggregation function, which can be customized in the cAggr() function. The latter two are the main focus of this study.

Concerning the collaborator selection, we consider two *sorting* variants and different *proportion rates*. For the sorting we opt for *First* and *Last*, whereas for the proportion rate we experiment with different rates in $r \in \{0.1, 0.25, 0.5, 0.75, 1\}$. Namely, we first order the population in ascending or descending order (depending on first/last), and then we take the top $r|s.\text{pop}|$ individuals, corresponding to the desired proportion of the population $s.\text{pop}$.

Regarding the fitness *aggregator*, we consider three cases: the *Best*, the *Worst*, and the *Median*. We prefer the median over the mean, as it can be generalized to non-numerical fitness values, and it is also less sensitive to outliers.

In the following, we denote an instance of our CCEA as $\text{EA}_1 + \text{EA}_2/s/r/a$, where s is the sorting (F for First and L for Last), r is the proportion rate, a is the aggregator (B for Best, M for Median, and W for Worst). For instance, ES+GA/F/0.1/M is the CCEA that uses ES and GA, selects the best 10% of the other population as collaborators, and takes the median fitness of the resulting solutions.

3   Case Studies

We consider four case studies to evaluate the practicability and the generality of the proposed CCEA scheme. Moreover, we use them as a test bed for assessing the impact of different pairs of inner EAs, collaborator selection schemes, and fitness aggregation functions. To this end, we start from two simple toy problems, which serve mostly as a proof-of-concept, and then we move towards two real-world problems, namely Symbolic Regression (SR) and Neuroevolution (NE).

For each of the examined case studies, we split the search of a solution $p \in P$, into the search of two components $p_1 \in P_1$ and $p_2 \in P_2$, such that $\forall p_1, p_2 : p_1 \oplus p_2 \in P$, and we measure the quality of p with a fitness function $f : P \to \mathbb{R}$. We delegate the optimization of p_1 and p_2 to two inner EAs, which search two genotype spaces, G_1 and G_2 respectively, and make use of two genotype-phenotype mapping functions, $\phi_1 : G_1 \rightarrow P_1$ and $\phi_2 : G_2 \rightarrow P_2$, to obtain the solution from the genotype. Clearly, the spaces G_1, G_2, P_1, P_2, P and the functions $\phi_1, \phi_2, \oplus, f$ strongly depend on the problem at hand. We hence describe them in detail in the following sections.

3.1   Toy Problems

Point Aiming. The goal of the point aiming problem is to find a point in $\mathbb{R}^n$ as close as possible, according to the Euclidean distance, to a target point $x^* \in \mathbb{R}^n$. The solution space P is hence $\mathbb{R}^n$, and the fitness of a solution $p = x$ is the Euclidean distance between x and x^*, $f(x) = \|x - x^*\|_2$. We split the search into the search of the first $\lfloor \frac{n}{2} \rfloor$ components ($P_1 = \mathbb{R}^{\lfloor \frac{n}{2} \rfloor}$), and the last $\lceil \frac{n}{2} \rceil$ components ($P_2 = \mathbb{R}^{\lceil \frac{n}{2} \rceil}$), and we rely on a direct encoding for both, i.e., $G_1 = P_1$, $G_2 = P_2$. We compose $p = x$ from $p_1 = x_1$ and $p_2 = x_2$ through simple concatenation.

We note that this problem is fully-separable [11], meaning that it can be solved by optimizing each decision variable independently. These types of problems are, in general, easily tackled by CCEAs [4].

Bimodal Point Aiming. The bimodal point aiming problem increases the difficulty of the point aiming problem, by introducing a bimodal fitness landscape. Namely, two possible target points, $x^*, x^{**} \in \mathbb{R}^n$, are considered, and the fitness of a candidate solution x, is measured as $f(x) = \min(\|x - x^*\|_2, \|x - x^{**}\|_2)$, i.e., as its Euclidean distance to the closest among x^* and x^{**}. This causes the problem to not be fully-separable anymore, as the solution components need to agree on the direction in which they are moving, in order to both get closer to either x^* and x^{**}.

3.2   Symbolic Regression

We consider SR as our first real-world problem, where the goal is to find a symbolic formula $h : \mathbb{R}^d \rightarrow \mathbb{R}$ which best approximates the relation between a data point $x \in \mathbb{R}^d$ and a dependent variable $y \in \mathbb{R}$ expressed in a dataset $\{(x^{(i)}, y^{(i)})\}_{i=1}^{i=n}$. Hence, the solution consists of a symbolic formula $h : \mathbb{R}^d \rightarrow \mathbb{R}$, and we measure its fitness $f(h)$ in terms of mean squared error (MSE) over the training data: $f(h) = \text{MSE}\left(h, \{(x^{(i)}, y^{(i)})\}_{i=1}^{i=n}\right)$.

We split the problem into the search of the skeleton of the formula (p_1) and the optimization of the numerical constants which appear therein (p_2), as inspired by [19]. More in details, we define P_1 as the space of parametric symbolic formulae $p_1 : (\mathbb{R}^d, \mathbb{R}^m) \rightarrow \mathbb{R}$, and P_2 as $\mathbb{R}^m$. We rely on a tree-based encoding for the formula, as in standard Genetic Programming (GP), hence, G_1 becomes the space of trees where the inner nodes are mathematical operators (namely, $\bullet + \bullet, \bullet - \bullet, \bullet \times \bullet, \bullet \div \bullet, \bullet \div^* \bullet, \log \bullet, \log^* \bullet, \exp \bullet, \sin \bullet, \cos \bullet, \frac{1}{\bullet}, -\bullet, \sqrt{\bullet}, \bullet^2, \bullet^3, \max(\bullet, \bullet), \min(\bullet, \bullet)$, where $\bullet$ represents an operand, and operators marked with * are protected), and the leaf nodes are either problem features $x_1, \ldots, x_d$ or parameters $c_1, \ldots, c_m$. Conversely, we use a direct encoding for the second component. For composing the solutions, we substitute each parameter c_i in the formula p_1 with the corresponding element of p_2, $c_i = p_{2,i}, \forall i = 1, \ldots, m$.

We refrain from using linear scaling, although it can dramatically increase the performance of SR on real-world problems [20], as it could pollute the results achieved through CC. In fact, our goal here is not that of achieving state-of-the-art performance, but we aim mainly at investigating the effects of different aspects within CC.

We highlight that this problem is far from being separable, as there is a strong intertwining between the two components. This poses additional hurdles on the CCEA, and enables us to observe its behavior under more difficult circumstances.

3.3   Neuroevolution

The second real-world problem we analyze is NE, where the aim is that of finding a suitable Artificial Neural Network (ANN) to solve a task. We here consider binary classification, hence we search the ANN $h : \mathbb{R}^d \to \{y^+, y^-\}$ which best captures the relation between a data point $x \in \mathbb{R}^d$ and its class $y \in \{y^+, y^-\}$ expressed in a dataset $\{(x^{(i)}, y^{(i)})\}_{i=1}^{i=n}$. The solution thus consists of an ANN h with d input neurons (one per feature), and 2 output neurons (one per class): $h(x)$ is y^+ if the first neuron activation is larger than the second neuron activation, or y^- otherwise. We measure fitness with the Balanced Error Rate (BER) over the training data: $f(h) = \text{BER}\left(h, \{(x^{(i)}, c^{(i)})\}_{i=1}^{i=n}\right)$.

Motivated by the work of Gaier and Ha [1], we split the NE task in the search of a suitable architecture for the ANN and in the optimization of the ANN weights. However, to constrain and ease the optimization—we remark our goal is not that of excelling at NE—we take inspiration from the practice of pruning [22], which has been shown to yield to well performing ANNs even in presence of high sparsity [8,9]. Hence, we fix the architecture of the ANN in terms of number and size of the hidden layers and of activation functions (we always use tanh), but allow for synapses to be present or not, ranging from a completely disconnected ANN to a fully connected one. Therefore, p_1 consists of the weights of the ANN, as if it was fully connected, whereas p_2 consists of a binary mask for the ANN, indicating which synapses are pruned. For composing the final solution p, we simply mask the weights of p_1 using p_2. Concerning the evolution of the components, we encode the weights as real-valued vectors, and the mask as a bit-string. Hence, let w be the maximum number of weights of the ANN, $G_1 = \mathbb{R}^w$ and $G_2 = \{0, 1\}^w$.

4   Experimental Analysis

We hereon present the details and the results of our experimental evaluation. We implemented the proposed CCEA in JGEA [6]. For all the problems described in the following sections we performed 10 independent evaluations for each considered scenario, unless otherwise specified.

4.1   Toy Problems

For both toy problems we considered the same settings. We set the problem size to $n = 25$, resulting in $P = \mathbb{R}^{25}$, $P_1 = \mathbb{R}^{12}$, $P_2 = \mathbb{R}^{13}$. For the point aiming problem we set $\boldsymbol{x}^* = 2 \cdot \mathbf{1}_{25}$, while for the bimodal point aiming problem we set $\boldsymbol{x}^* = 2 \cdot \mathbf{1}_{25}$, $\boldsymbol{x}^{**} = -1 \cdot \mathbf{1}_{25}$. We considered two EAs: a simple form of Evolutionary Strategy (ES) and a Genetic Algorithm (GA). We experimented with both EAs on their own (i.e., without CC) as a baseline, and as inner EAs in the CCEA. For the latter case, we had ES+ES/s/r/a, ES+GA/s/r/a, and GA+GA/s/r/a, with s in {F,L}, r in {0.1, 0.25, 0.5, 0.75, 1}, and a in {B,M,W}. For both EAs, we evolved a population of size $n_{\text{pop}} = 50$ for $n_{\text{gen}} = 70$ generations. We initialized individuals sampling each component uniformly in $[0, 1]$. Then for the ES, we used a $(1 + \lambda)$ model, where we generate the offspring applying a Gaussian mutation ($\sigma = 0.1$) to the point-wise mean of the parents, selected as the top fourth of the current population. For the GA, we used a $(\mu + \lambda)$ scheme, with $\mu = \lambda$. We relied on tournament selection ($n_{\text{tour}} = 5$) for sampling the two parents, which we combined with geometric crossover followed by a Gaussian mutation ($\sigma = 0.1$) to produce the offspring.

We report the results the experimental evaluation in Figs. 3, 4 and 5 for point aiming, and in Figs. 6 and 7 for bimodal point aiming. These results constitute a proof of concept for the practicability of our CCEA, as we successfully combined different EAs, and experimented with various collaborator selection methods and fitness aggregation criteria. Focusing more on the single figures, we can gain different insights. First, in Fig. 3, we report the median across the 10 runs of the fitness of the best evaluated individual vs. the amount of fitness evaluations performed during evolution. Clearly, since we used the number of generations as termination criterion, some lines stop earlier than others. This highlights the trade-off between the amount of collaborators chosen and the computational effort required: a higher rate r, requires more evaluations to compute an evolutionary step. From the figure we can also note that at the end of evolution all combinations reach comparable fitness levels, meaning that evolution was always able to achieve reasonable solutions regardless of the employed components. These findings are also confirmed by Fig. 4, where we plot, for each generation, the median across the 10 runs of the fitness of the best evaluated individual. In fact, we can see the plots almost perfectly overlapping at the end of evolution. Given the insights gained from these two figures, we limit ourselves to showing the progression of fitness along generations for the other problems, in order to save space.

Moving on to Fig. 5, we can reason further on the impact of collaborator selection and fitness aggregation on the solutions found by the CCEA. We here report the median fitness of the best individual found at the end of evolution. Concerning the fitness aggregation (one per column), Best generally yields to better results, although comparable to both Median and Last. Focusing on the rate r, we observe larger values to be more effective with the Best aggregation scheme, whereas we note a fuzzier impact for Median and Large. Last, regarding the collaborator selection, First performs better than Last, especially for ES+ES.

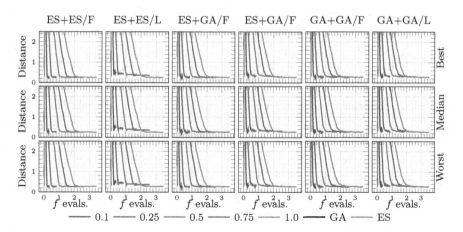

Fig. 3. Fitness (distance from the target) vs. number of fitness evaluations [$\times 10^5$] for the point aiming problem, one line for each rate or baseline EA. The ES line overlies the GA line.

We speculate this could be because both ES populations focus on a specific point in the search space, the mean, and the combination with the another components tends to move it causing instability.

For the bimodal problems we obtained similar results, see Figs. 6 and 7. In fact, we notice higher differences between plots during evolution, which were smoothed out at the end of evolution. This leads us to conclude that both toy problems considered were easy enough for evolution to eventually converge to a reasonable solution, regardless of the components involved.

4.2 Symbolic Regression

For the SR problem we considered the Boston Housing dataset [2], which consists of 506 examples and 13 features. We used 5 different 80%/20% splits of the dataset, which we respectively used as training and test sets, thus resulting in $5 \cdot 10 = 50$ independent evolutionary runs. We set the number of numerical constants to $m = 10$, hence $P_2 = \mathbb{R}^{10}$. We employed GP as EA_1, and ES as EA_2, but we also considered, for reference, a variant with GP on its own, where the terminal nodes could be the problem features or a constant among $\{0.1, 1, 10\}$. We set $n_{pop} = 100$ and $n_{gen} = 500$. For GP we used a $(\mu + \lambda)$ scheme, with *ramped half-and-half* initialization and tournament selection ($n_{tour} = 10$). We used either subtree crossover or subtree mutation to compute the offspring, chosen with $p_{xo} = 0.8$ and $p_{mut} = 1 - p_{xo} = 0.2$. For the ES we used the same configuration as described in Sect. 4.1.

We report the results for the SR problem in Figs. 8, 9 and 10. In Fig. 8, we show the median fitness of the best evaluated individual along generations. Concerning the fitness aggregation, we see an ordering: Best is better than Median, which is better than Worst. Regarding the collaborator selection, instead, the

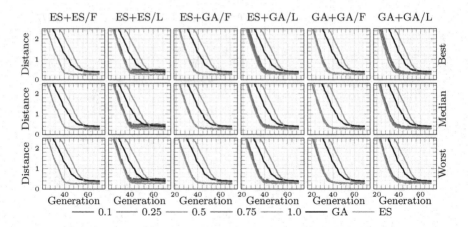

Fig. 4. Fitness (distance from the target) vs. generation for the point aiming problem, one line for each rate or baseline EA.

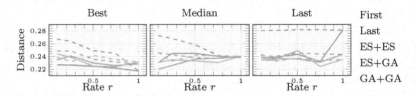

Fig. 5. Fitness at the end of evolution vs. rate r for the point aiming problem.

results are not too different, and, in fact, different rates and First and Last achieve similar fitness results at the end of evolution.

More insights can be gained from Fig. 9, where we report the fitness (MSE on the training set) of the best solutions found at the end of evolution, together with the MSE of said solutions on the test set. From what we see in this figure, the previous findings appear confirmed, also in terms of generalization: the collaborator selection plays a secondary role compared to the fitness aggregation, where the Best is clearly better than Median or Worst. We can also note, once again, that the amount of collaborators, i.e., the rate r, does not influence the performance of the CCEA, hence we deem convenient to use a lower rate to constrain the amount of computational resources used. From another point of view, these results seem to suggest that the optimistic expectation that coupling each component with unfit components of the other population results in a more robust solution is not met.

Last, we investigated the structure of the final solutions found by the CCEA for SR. In Fig. 10 we plot the median amount of constants in the best evaluated individual along generations. From here, the differences between columns, i.e., between fitness aggregation schemes, are evident: the more pessimistic the fitness aggregation, the less cooperation between populations is preferred. In fact,

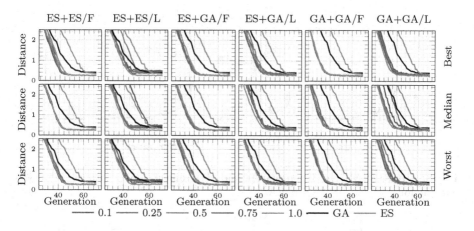

Fig. 6. Fitness (distance from the target) vs. generation for the bimodal point aiming problem, one line for each rate or baseline EA.

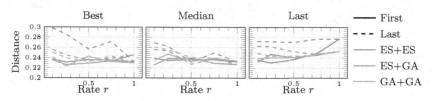

Fig. 7. Fitness at the end of evol. vs. rate r for the bimodal point aiming problem.

evolution tends to try to stabilize the fitness of the tree component by avoiding the combination with constants, as few poor performing ones could hinder the final fitness of a tree, good though it might be in combination with other values.

4.3 Neuroevolution

For the NE problem we used the German Credit dataset [2], consisting of 1000 examples and 20 features. As in SR, we considered 5 splits for the dataset. However, here, we performed 5 experiments on each split, rather than 10, totaling $5 \cdot 5 = 25$ independent evaluations. We chose an ANN with 3 hidden layers, each of size 26, to match the size of the input. In fact, some of the 20 features were categorical, hence we transformed them with one-hot encoding. This architecture resulted in a maximum number of weights $w = 2160$, hence $P_1 = R^{2160}$ and $P_2 = \{0, 1\}^{2160}$. We relied on ES as EA_1 and on a GA as EA_2. As a baseline, we also considered ES on its own, leaving all the weights unpruned. We set $n_{pop} = 100$ and $n_{gen} = 700$. For the ES we used the configuration described in Sect. 4.1. For the GA we used a similar scheme as for the toy problems, but given the different search space we had some key differences. Namely, we initialized the individuals sampling each component from $\{0, 1\}$ with equal probability, and we computed the offspring with either uniform crossover or bitflip mutation,

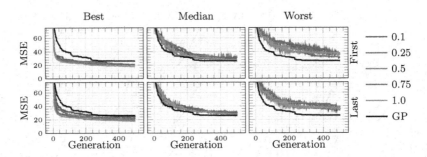

Fig. 8. Fitness (MSE) vs. generation for the SR problem (Boston Housing dataset), one line for each rate or baseline EA.

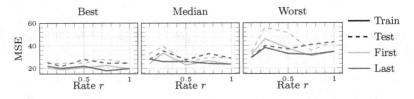

Fig. 9. Fitness (MSE on training set) at the end of evolution and MSE on the test set vs. rate r for the SR problem.

chosen with equal probability ($p_{xo} = p_{mut} = 0.5$). Again, we used tournament selection, with $n_{tour} = 10$.

We report the results in Figs. 11 and 12. In Fig. 11 we display the median of the fitness of the best evaluated individual along generations. Here we can clearly notice a difference for the third column: with the Worst aggregation, evolution is not able to converge for rates $r > 0.1$. In fact, instead of achieving high selective pressure, the pessimist fitness aggregation is steering evolution in the wrong direction. We speculate this does not happen for a low rate of collaborators as this corresponds to a smaller pool of individuals among which the worst fitness is extracted, meaning that it is more unlikely to have a low value if the evaluated component is promising.

The results at the end of evolution, reported in Fig. 12 together with the BER resulting from re-evaluation on the test set, are in line with the previous findings, and confirm the poor performance of the CCEA with the Worst fitness aggregation scheme. Conversely, we do not note significant differences neither between the Best and the Median aggregation schemes, nor between the First and Last collaborator selection criteria. Anyway, it appears that a smaller rate of collaborators is not only sufficient, but also beneficial in most cases.

Regarding the solutions structure, for space reasons we avoid reporting any plots. However, the results showed a constant amount of weights pruned, around 50%, meaning that there was no evolutionary pressure in that sense.

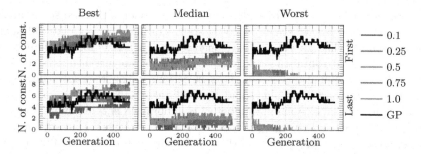

Fig. 10. Amount of constants in the best individual vs. generation for the SR problem, one line for each rate or baseline EA.

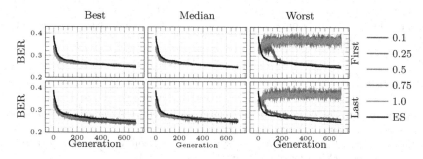

Fig. 11. Fitness (BER) vs. generations for the NE classification problem.

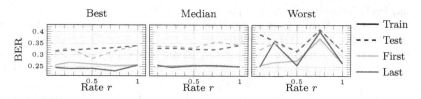

Fig. 12. Fitness (BER on training set) at the end of evolution and BER on the test set vs. rate r for the NN classification problem.

5 Concluding Remarks

We considered Cooperative Coevolution (CC), that aims at solving a problem by dividing it in simpler sub-problems, and we conducted an experimental campaign for investigating the impact of key CC components on the effectiveness and efficiency of optimization and on the complexity and generalization power of optimized solutions.

We first proposed a general scheme for a CC evolutionary algorithm (CCEA) for the significant case where the problem can be split in two sub-problems. Our CCEA formulation is general enough to use, as inner EAs for the two sub-problems, any population-based iterative EA with any representation. At the same time, it neatly modularizes the key CC components that we study in this paper, namely (a) how to *select* the collaborators in the other population (including how many of them) and (b) how to *aggregate* the fitness of many collaborator-collaborator pairs to obtain a single fitness value. Then, we instantiated our CCEA with different combinations of inner EAs, selection, and aggregation schemes and we applied it to two toy problems and two real prediction problems.

We found that: (a) a small number of collaborators is often enough to have good effectiveness and efficiency; (b) optimistic fitness aggregation schemes (i.e., choosing the best fitness among all the collaborator pairs) often results in better effectiveness; (c) depending on the way the problem is split, fitness aggregation may have a dramatic impact on the structure of evolved solutions. Moreover, we found no evidence of increased generalization power when more or unfit collaborators are selected. We believe that our study may help gaining insights in a sub-field of evolutionary computation, i.e., CC, that is promising, yet not completely characterized.

References

1. Gaier, A., Ha, D.: Weight agnostic neural networks. In: Advances in Neural Information Processing Systems, vol. 32 (2019)
2. La Cava, W., et al.: Contemporary symbolic regression methods and their relative performance. arXiv preprint arXiv:2107.14351 (2021)
3. Luke, S., Sullivan, K., Abidi, F.: Large scale empirical analysis of cooperative coevolution. In: Proceedings of the 13th Annual Conference Companion on Genetic and Evolutionary Computation, pp. 151–152 (2011)
4. Ma, X., et al.: A survey on cooperative co-evolutionary algorithms. IEEE Trans. Evol. Comput. **23**(3), 421–441 (2018)
5. Maniadakis, M., Trahanias, P.: Assessing hierarchical cooperative coevolution. In: 19th IEEE International Conference on Tools with Artificial Intelligence (ICTAI 2007), vol. 1, pp. 391–398. IEEE (2007)
6. Medvet, E., Nadizar, G., Manzoni, L.: JGEA: a modular Java framework for experimenting with evolutionary computation. In: Proceedings of the Genetic and Evolutionary Computation Conference Companion, pp. 2009–2018 (2022)
7. Moriarty, D.E., Miikkulainen, R.: Forming neural networks through efficient and adaptive coevolution. Evol. Comput. **5**(4), 373–399 (1997)
8. Nadizar, G., Medvet, E., Pellegrino, F.A., Zullich, M., Nichele, S.: On the effects of pruning on evolved neural controllers for soft robots. In: Proceedings of the Genetic and Evolutionary Computation Conference Companion, pp. 1744–1752 (2021)
9. Nadizar, G., Medvet, E., Ramstad, H.H., Nichele, S., Pellegrino, F.A., Zullich, M.: Merging pruning and neuroevolution: towards robust and efficient controllers for modular soft robots. Knowl. Eng. Rev. **37** (2022)
10. de Oliveira, F.B., Enayatifar, R., Sadaei, H.J., Guimarães, F.G., Potvin, J.Y.: A cooperative coevolutionary algorithm for the multi-depot vehicle routing problem. Expert Syst. Appl. **43**, 117–130 (2016)
11. Omidvar, M.N., Li, X., Mei, Y., Yao, X.: Cooperative co-evolution with differential grouping for large scale optimization. IEEE Trans. Evol. Comput. **18**(3), 378–393 (2013)

12. Panait, L., Luke, S.: A comparative study of two competitive fitness functions. In: Proceedings of the Genetic and Evolutionary Computation Conference (GECCO 2002), pp. 503–511. Citeseer (2002)
13. Panait, L., Luke, S., Harrison, J.F.: Archive-based cooperative coevolutionary algorithms. In: Proceedings of the 8th Annual Conference on Genetic and Evolutionary Computation, pp. 345–352 (2006)
14. Peng, X., Liu, K., Jin, Y.: A dynamic optimization approach to the design of cooperative co-evolutionary algorithms. Knowl.-Based Syst. **109**, 174–186 (2016)
15. Popovici, E., De Jong, K.A., et al.: A dynamical systems analysis of collaboration methods in cooperative co-evolution. In: AAAI Fall Symposium: Coevolutionary and Coadaptive Systems, pp. 26–34 (2005)
16. Potter, M.A., De Jong, K.A.: A cooperative coevolutionary approach to function optimization. In: Davidor, Y., Schwefel, H.-P., Männer, R. (eds.) PPSN 1994. LNCS, vol. 866, pp. 249–257. Springer, Heidelberg (1994). https://doi.org/10.1007/3-540-58484-6_269
17. Potter, M.A., Jong, K.A.D.: Cooperative coevolution: an architecture for evolving coadapted subcomponents. Evol. Comput. **8**(1), 1–29 (2000)
18. Tan, K.C., Yang, Y., Goh, C.K.: A distributed cooperative coevolutionary algorithm for multiobjective optimization. IEEE Trans. Evol. Comput. **10**(5), 527–549 (2006)
19. Vanneschi, L., Mauri, G., Valsecchi, A., Cagnoni, S.: Heterogeneous cooperative coevolution: strategies of integration between GP and GA. In: Proceedings of the 8th Annual Conference on Genetic and Evolutionary Computation, pp. 361–368 (2006)
20. Virgolin, M., Alderliesten, T., Bosman, P.A.: Linear scaling with and within semantic backpropagation-based genetic programming for symbolic regression. In: Proceedings of the Genetic and Evolutionary Computation Conference, pp. 1084–1092 (2019)
21. Wiegand, R.P., Liles, W.C., De Jong, K.A., et al.: An empirical analysis of collaboration methods in cooperative coevolutionary algorithms. In: Proceedings of the Genetic and Evolutionary Computation Conference (GECCO), vol. 2611, pp. 1235–1245. Morgan Kaufmann, San Francisco (2001)
22. Zullich, M., Medvet, E., Pellegrino, F.A., Ansuini, A.: Speeding-up pruning for artificial neural networks: introducing accelerated iterative magnitude pruning. In: 2020 25th International Conference on Pattern Recognition (ICPR), pp. 3868–3875. IEEE (2021)

To Bias or Not to Bias: Probabilistic Initialisation for Evolving Dispatching Rules

Marko Đurasević[1(✉)], Francisco Javier Gil-Gala[2], and Domagoj Jakobović[1]

[1] Faculty of Electrical Engineering and Computing, University of Zagreb, Zagreb, Croatia
{marko.durasevic,domagoj.jakobovic}@fer.hr
[2] Department of Computer Science, University of Oviedo, Campus de Viesques s/n, 33271 Gijón, Spain
giljavier@uniovi.es

Abstract. The automatic generation of dispatching rules (DRs) for various scheduling problems using genetic programming (GP) has become an increasingly researched topic in recent years. Creating DRs in this way relieves domain experts of the tedious task of manually designing new rules, but also often leads to the discovery of better rules than those already available. However, developing new DRs is a computationally intensive process that takes time to converge to good solutions. One possible way to improve the convergence of evolutionary algorithms is to use a more sophisticated method to generate the initial population of individuals. In this paper, we propose a simple method for initialising individuals that uses probabilistic information from previously evolved DRs. The method extracts the information on how many times each node occurs at each level of the tree and in each context. This information is then used to introduce bias in the selection of the node to be selected at a particular position during the construction of the expression tree. The experiments show that with the proposed method it is possible to improve the convergence of GP when generating new DRs, so that GP can obtain high-quality DRs in a much shorter time.

Keywords: Genetic programming · Dispatching rules · Unrelated machines environment · Scheduling · Individual initialisation

1 Introduction

The unrelated machines environment represents an important combinatorial optimisation problem with application in many real-world domains, such as in manufacturing [19], multiprocessor scheduling [18], and many others [15]. The goal in this environment is to schedule a set of jobs on a scarce set of machines to minimise one or more user defined scheduling objectives [8]. Often, these problems need to be solved under dynamic conditions, meaning that unexpected events can occur during the execution of the system, such as the arrival of new

G. Pappa et al. (Eds.): EuroGP 2023, LNCS 13986, pp. 308–323, 2023.
https://doi.org/10.1007/978-3-031-29573-7_20

jobs or machines breaking down. Because of this, it is not possible to create the entire schedule before the system starts execution. Therefore, alternative methods called dispatching rules (DRs), which can deal with such dynamic situations, are proposed and applied in the literature [12].

DRs incrementally construct the solution by determining the next job that should be scheduled on an available machine at each decision moment. This is usually done by ranking all jobs available at each decision moment using certain system parameters to determine the most appropriate job to be scheduled next. Therefore, these methods do not build the entire schedule up front, but only determine the next scheduling decision and immediately execute it. Since DRs use only the current state of the system to perform their decisions, they can quickly determine which job should be executed next. As a result, they can react to unexpected events that occur in the system, and can thus be applied for solving dynamic scheduling problems. However, DRs are not without their own issues, a significant one being the difficulty in their manual design. This motivated a great deal of research to focus on automated design of DRs using various machine learning and evolutionary computation methods [1].

Among the many methods used to automatically design DRs, genetic programming (GP) [10] has garnered the largest attention from the research community [1,6]. In the context of evolving new DRs, GP has been applied to generate expressions that, based on the current state of the system, assign a priority to each available job. By using the assigned priorities, all jobs are ranked and the job with the best rank is selected and scheduled on the available machine. Using such a process relieves domain experts of the arduous task of designing new dispatching rules manually, since this task can be automatised.

Although the process of generating new DRs is performed offline, prior to solving a certain scheduling problem, the evolution process is still computationally expensive, especially if the goal is to evolve DRs of high quality. Throughout the years, a plethora of methods has been proposed to improve the convergence speed of GP when designing DRs, such as using guided subtree selection in genetic operators [20,21], local search operators [2], surrogate models [7,23], using alternative selection operators [9], and many others [5,22]. One of the easiest ways to improve the convergence of evolutionary algorithms is to use customised population initialisation methods [3]. For example, in the context of solving scheduling problems, various DRs can be used to initialise a certain part of the population of an evolutionary algorithm, giving it a much better start [17]. With such an approach it was possible to significantly improve both the convergence and performance of a genetic algorithm. However, there is usually no heuristic by which individuals in GP could be initialised, especially since these individuals themselves represent scheduling heuristics. One possibility would be to introduce existing DRs in the initial population; however, there is only a limited number of existing DRs for a small number of scheduling criteria, and sometimes it is difficult to represent them as expressions. The most common way of generating the initial population in GP is still to randomly construct the expressions using the standard ramped-half-and-half initialisation method

[10], although several studies investigated the possibility of additionally using probabilistic information during the initialisation process [4].

The goal of this study is to investigate whether the convergence of GP during the evolution of new DRs can be improved by exploiting probabilistic information extracted from previously generated DRs, and then using it during the construction of individuals for the initial population. The idea of the proposed probabilistic initialisation approach is to calculate probabilities of each node appearing in a certain level of the tree or context from previously evolved rules, and use that information to introduce a bias when generating the individuals of the initial population. In that way, the initialisation procedure does not select nodes completely randomly, but is inclined to generate subexpressions that have appeared in good DRs. The contributions of this study can be summarised as:

- A novel probabilistic based initialisation procedure for generating expression trees of the initial population of GP for the evolution of DRs,
- Comparison of the proposed initialisation procedure with standard GP with randomly generated individuals for automated generation of DRs,
- An extensive analysis of the proposed probabilistic initialisation method.

The rest of the paper is structured as follows. The next section defines the problem and describes how to automatically design dispatching rules using GP. Section 3 provides details about the probabilistic individual initialisation methods. Sections 4 and 5 presents the experimental results and further analysis. Finally, in Sect. 6, we outline the conclusions and future work.

2   Background

2.1   Unrelated Machines Environment

In the unrelated machines environment, a set of n jobs needs to be scheduled on one of the m available machines. Each job j is defined with several properties, which include its processing time on machine i p_{ij}, the release time r_j, the due date d_j, and the job weight w_j. The release time of each job defines the moment when the job becomes available for scheduling, the due date the time until which the job should be completed, and finally the weight defines the job's importance. The objective of the considered problem is to construct a schedule which minimises the total weighted tardiness criterion, defined as $Twt = \sum_j \max(C_j - d_j, 0)$, where C_j represents the time when job j finished. As the described problem is considered under dynamic conditions, no information about the jobs is known before they are released into the system. Thus, at each decision point, the DRs can only perform their decision based on the information of currently released jobs in the system.

2.2   Designing Dispatching Rules with Genetic Programming

As previously indicated, DRs are simple heuristics that iteratively construct the schedule, and consist of a schedule generation scheme (SGS) and a priority function (PF). The SGS represents the outline of the DR, which determines when a scheduling decision needs to be performed, which jobs and machines can be considered at that moment, etc. In the SGS applied in this study, a scheduling decision needs to be performed each time there is at least one available machine and one unscheduled released job in the system [14]. At that moment, the SGS needs to determine on which of the machines the job should be scheduled. It should be stressed out that in this SGS all machines are considered at each scheduling decision, also the ones that are occupied at the current point. This is done since a better machine for executing the considered job could become available soon, and it would be better to delay the execution of the job until that machine becomes available. Therefore, if the SGS determines that the best choice would be to schedule a job on a machine that is currently occupied, the decision to schedule it will be postponed to later moment in time when the job can be scheduled on an available machine. It was demonstrated that using such delays in the schedule is beneficial and leads to significantly better results [14].

At each decision moment, the SGS needs to determine which of the jobs needs to be scheduled on which machine. To do that, the SGS uses a PF to rank the assignments of each job to each of the machines. Based on these ranks it selects the best decision and executes the corresponding job on the selected machine.

Designing a good PF, i.e. the strategy by which jobs should be scheduled, is a difficult task, and thus this design is often delegated to genetic programming (GP). To be able to use GP to design new PFs, it is required to specify the primitive set of nodes used to construct the expression for assigning priorities to scheduling decisions. Table 1 presents the set of terminals used for this purpose. This set contains some basic information about the system, like the processing times of jobs or their due dates, but also some composite information like job slack times (time remaining until a job becomes late). On the other hand, the set of function nodes consists of the addition, subtraction, multiplication, protected division (returns 1 if division by a number close to 0 is detected), and the unary positive operator (defined as $POS(x) = \max(x, 0)$). All terminal and function nodes were selected based on preliminary experiments [16].

Table 1. The terminal set

Node	Description
pt	processing time of job j on machine i
$pmin$	minimum processing time (MPT) of job j
$pavg$	average processing time of job j across all machines
PAT	time until machine with the MPT for job j becomes available
MR	time until machine i becomes available
age	time which job j spent in the system
dd	time before job j has to finish with its execution
w	weight of job j (w_j)
SL	slack of job j, $max(d_j - p_{ij} - t, 0)$

3   Probabilistic Individual Initialisation

The standard individual initialisation technique used in GP constructs the individual by randomly selecting a primitive that should be generated at each place in the expression. In this case, each primitive of the same type has the same probability of being selected, regardless of the level at which it is generated, or with regards to its parent or sibling. There are cases when not all primitives can be selected (to satisfy the constraints placed on the size of the expression), which is manifested in a way that only terminal or function nodes can be generated at a certain place, with all primitives having the same probability of being generated.

The goal of the proposed probabilistic individual initialisation procedure is to enhance this process by introducing a bias to select certain primitives with a higher probability than others at certain positions in the individual. Thus, instead of having the same probability, each primitive will have a dedicated probability with which it can be generated at different positions in the individual. To calculate these probabilities, a certain set of previously evolved DRs is required to extract information about the occurrence of different primitive nodes in them. In this study we calculate two probability types for each primitive: *level based probability* and *conditional probability*. To illustrate the calculation of both probabilities, a set of three individuals shown in Fig. 1 will be used.

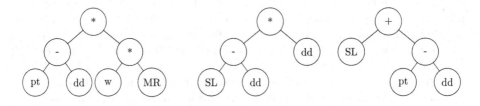

Fig. 1. Example individuals to calculate the probability values

The *level based probability* defines the probability of each primitive appearing at a certain level in the individual. This probability is calculated by determining the frequency of each primitive in the set appearing at each level of the tree. Therefore, each primitive will have a different probability of being generated depending on the level at which it needs to be generated. For example, based on the individuals in Fig. 1, only the * and + primitives appear at the first level, with * appearing two times, and + appearing only once. Therefore, the probability of generating the * root node would be 0.66, whereas the probability of generating the + node would be 0.33. All other primitives would have 0 probability of being generated at this level. On the second level there are six nodes, with the − primitive appearing 3 times thus assigning it a probability of 0.5 of being selected, and the other three *, *dd*, *SL* appearing only once, thus each obtaining a probability of 0.16. The same calculation would then also be performed for the final level.

The *conditional probability* represents the probability of a primitive being selected within a certain context, with the context representing the parent and sibling nodes of the position where a new node needs to be generated. Since a node can appear as a left or right child of a parent, for each of those two cases a set of separate probabilities will be calculated. The set of probabilities when generating the left child of a parent is calculated by counting the frequencies of all primitives appearing as the left child with the given parent node. Thus, only the information about the parent is used when generating the left child as it is the first child generated and has no siblings yet. Using the example from Fig. 1, this would mean that when the left child for the $*$ node needs to be generated, primitive $-$ will have a 0.66 probability of being selected, and w a probability of 0.33. On the other hand, the probabilities for generating the right child are calculated based on the frequencies of each primitive appearing with a given parent and sibling, since the left sibling is already generated at that point, and the information about it can be used as contextual information when generating the right child. For example, when generating the right child in the situation where the parent node is equal to $*$ and the left child is equal to $-$, primitives $*$ and dd would each have a 0.5 probability of being selected, as they each appear once in such a context in the set of sample DRs. From the previous description we see that these probabilities cannot be calculated for the root node, since it has no context (parent or sibling) based on which they could be calculated. As such all primitives have an equal probability of being selected as the root.

The level based and conditional probabilities can be used completely independently, but also in conjunction with each other. In that case, both probability values can simply be multiplied to calculate the joint probability that takes into account both the information about the level of the node and its context. Additionally, the probabilities are also normalised to ensure that their sum still totals to 1. Basically, with this combination we obtain an approximation of how likely a certain context is to appear at a certain level of the tree. For example, if a new individual is to be generated and the $*$ would be generated as the root node, then as previously denoted the $-$ and w nodes would have a probability different than 0 to be generated within this context. However, by additionally using the level based probabilities, only the $-$ node could be generated within this context, since the primitive w does not appear at all in this level of the tree. This can help to place context at the right positions in the individual.

From the previous descriptions of calculating the probabilities, one significant issue becomes immediately apparent. The problem is that it is possible that certain nodes will have zero probability of being generated at certain places in the individual, if they were not present in the set of the sample DRs at a certain level or within a certain context. Although this problem can be mitigated with genetic operators, as they can generate new individuals of various shapes, it could still create too strong a bias and have a negative effect. Therefore, an optional smoothing can be carried out when calculating the probabilities for each node, to ensure that any primitive can appear at any place in the individual. The applied smoothing procedure sets the frequency of each node appearing at each level or context to 1 instead of 0 at the beginning (before the probability estimation), which gives each node a small probability of being generated at each level and context. This probability is quite dependent of the size of the samples, but for the initial tests it was decided to keep the procedure simple.

4   Experimental Analysis

4.1   Benchmark Setup

In order to evaluate the proposed method, a benchmark set of 120 problem instances, divided by half into the training set and test set, was used. The training set was used by GP to evaluate individuals during the evolution process, whereas the test set was used to evaluate the best obtained individuals on unseen problems to determine their generalisation capability. Each set consisted of instances of different sizes, ranging from 12 to 100 jobs, and 3 to 10 machines.

In order to evolve new DRs the standard 3-tournament steady state GP was used, where in each iteration three individuals were randomly selected. The better two individuals were crossed over; the obtained child individual was mutated with a certain probability and used to replace the worst individual from the tournament. The GP used a population of 1 000 individuals, individual mutation probability of 0.3, and 80 000 function evaluations as a termination criterion. For crossover the subtree, uniform, context-preserving, and size-fair operators were used, whereas for mutation the subtree, hoist, node complement, node replacement, permutation, and shrink operators were used [10]. Tree depths of 3, 5, and 7 were used when evolving DRs to analyse the influence of the maximum allowed tree size on the individual initialisation procedure. Since several genetic operators were used, each time an operator needs to be performed one of the operators was randomly selected with an equal probability. These parameter settings were used for all individual initialisation methods. The population was initialised in a ramped-half-and-half fashion, with each individual generated using the random or probabilistic individual initialisation method.

The GP algorithm was executed with several individual initialisation strategies, to evaluate the performance of the proposed probabilistic initialisation techniques. First, GP was executed with the standard random individual initialisation method, denoted as GP-rand. Regarding the probabilistic initialisation method, several scenarios were tested. First, we tested the use the different probabilistic information types during the generation of the individual. Thus, the suffix "-lvl" denotes that only level based probabilities were used, the suffix "-ctx" that only conditional probabilities were used, and the suffix "-bth" that both probabilities were used in combination. Secondly, the influence of smoothing was also investigated, thus experiments in which GP used smoothed probability values are denoted with the suffix "-s". This resulted in six combinations for the probabilistic initialisation method. To calculate the level and conditional probabilities, a set of 50 sample rules was used, which contained the best individuals from previous experiments [16]. These rules were collected from 50 previously executed GP runs, from which the best individual in each run was stored.

Each experiment was executed 30 times to obtain a good sample of each method's performance. From each execution the best individual in the final population (on the training set) was stored and evaluated on the test set. Since for each experiment 30 individuals were obtained, the tables in the results report the minimum, median and maximum values from these executions. Furthermore, the Kruskal Wallis statistical test with the Bonferroni correction method and Dunn's post hoc test was used to determine whether there is any significant difference between the obtained results. The results are considered significantly different if a p-value smaller than 0.05 was obtained.

4.2   Results

Table 2 shows the results obtained by each of the initialisation methods after the given maximum number of evaluations. The values in the table denote the total weighted tardiness criteria as defined in Sect. 2, which is subject to minimisation. As can be seen from the table, GP obtains similar results with all tested individual initialisation methods after the given number of function evaluations, especially for the lowest tree depth. Although in some cases the probabilistic individual method does achieve slightly better median values, the statistical tests show that there is no significant difference between any of the methods. This suggests that given enough time, GP will obtain equally good solutions regardless of which individual initialisation method is used. However, this also suggests that using the probabilistic initialisation method does not result in GP getting trapped in a local optimum due to a higher bias towards generating certain individuals in the starting population.

Table 2. Results for the tested individual initialisation methods

Method	Depth 3			Depth 5			Depth 7		
	min	med	max	min	med	max	min	med	max
GP-rand	13.50	14.21	15.39	12.80	13.71	14.90	12.68	13.90	15.99
GP-lvl	13.87	14.21	14.24	12.89	13.66	15.55	12.70	13.88	15.44
GP-ctx	13.50	14.21	14.27	13.00	13.88	14.79	12.56	13.62	15.69
GP-bth	14.19	14.21	14.27	12.79	13.71	14.52	13.29	14.08	15.64
GP-s-lvl	14.19	14.21	14.24	12.68	13.78	14.67	12.81	14.17	15.99
GP-s-ctx	13.50	14.21	14.32	12.90	13.57	14.41	13.02	13.93	15.98
GP-s-bth	13.52	14.21	14.24	12.90	13.60	15.16	13.11	13.95	15.11

The differences between the various initialisation methods can better be observed when investigating the convergence patterns of GP throughout the evolution process. The convergence patterns of the methods for all three tree depths are shown in Fig. 2. The figure outlines the average fitness on the test set calculated over the 30 executions of GP for each individual initialisation method.

First of all, we can observe that GP-rand has the slowest convergence rate among all the tested methods. This is especially evident at the start of the evolution process, where its best solutions are much worse than those obtained by probabilistic initialisation. Using the probabilistic method, GP is able to converge much faster and obtain good DRs in a smaller number of evaluations. However, given enough time, in the end GP-rand can reach solutions of the same quality as the probabilistic initialisation methods. It is interesting to observe how the probabilistic initialisation method generates a quite good individual in the starting population, providing GP with a much better start at the beginning.

Among the tested probabilistic methods, the ones using both level based and conditional probabilities usually achieve the best convergence. This shows that combining both information types results in the best initialisation procedure, which generates the individuals containing the best genetic material. On the other hand, conditional initialisation seems to result in the worst initial solution being generated, although GP converges faster using those initial solutions than those generated using the level based initialisation (except for the lowest tree depth). This suggests that using conditional probabilities populates the initial population with good building blocks which allows GP to obtain better solutions faster after applying genetic operators to them. Due to these building blocks being distributed all over the expression tree in the initial solutions, the quality of the initial population is not as good as when using the level based information. On the other hand, the level based initialisation leads to better initial solutions as it seems to allocate good nodes to the right places in the tree, but requires more time to create good building blocks using genetic operators. As such, the methods which use both types of information perform the best as it introduces good building blocks, but also try to allocate them at the right place in the tree, thus reducing the effort that GP would have to invest.

To get a better notion of the convergence speed, Table 3 reports the number of function evaluations required for each method to obtain an individual of a fitness smaller than the target fitness values denoted at the top of the table. If for the considered initialisation method certain values are not reached, the cells are left empty. This table demonstrates that when using tree depth 3, the initialisation methods quite quickly converge to the best individuals, only after a few thousand evaluations, and some already generate quite good solutions in the initial population. However, due to the quite limited search space, all methods converge quite quickly and do not improve their results any further. When observing the convergence speed for depth 5, it can again be seen that using probabilistic initialisation allow GP to obtain good results faster. It is also evident that the level based initialisation method has the slowest convergence rate, as at one point GP-rand can match its result in the same number of evaluations. The table also outlines that with certain types of probabilistic initialisation, GP can reach the same results as GP-rand but requires only half of the function evaluations. For depth 7, the probabilistic method also allows GP to reach good solutions in less time, but also reach better solutions in the end as well. As the

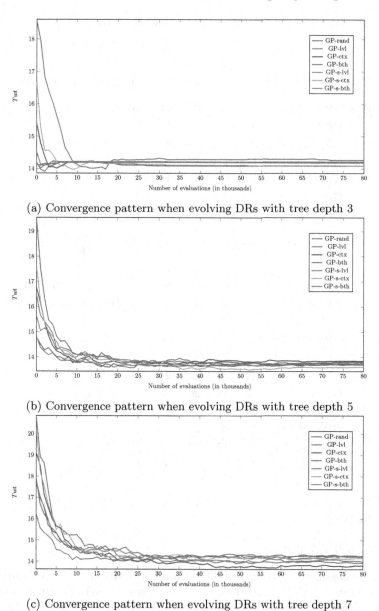

(a) Convergence pattern when evolving DRs with tree depth 3

(b) Convergence pattern when evolving DRs with tree depth 5

(c) Convergence pattern when evolving DRs with tree depth 7

Fig. 2. Convergence patterns when evolving DRs with various initialisation methods

depth increased we also witness that more evaluations were required to reach
equally good solutions in comparison when depth 5 was used, thus also outlining
the importance of selecting the appropriate tree depth.

Table 3. Number of function evaluations (in thousands) required to reach a solution of a better fitness

		Target fitness value										
		14.5	14.4	14.3	14.2	14.1	14	13.9	13.8	13.7	13.6	13.5
Depth 3	GP-rand	8	9	9	9	11	17					
	GP-lvl	0	0	0	0	1	1					
	GP-ctx	2	2	3	3	3						
	GP-bth	1	1	1	1							
	GP-s-lvl	0	0	0	0	1						
	GP-s-ctx	4	5	5	6	6	9					
	GP-s-bth	0	0	0	0							
Depth 5	GP-rand	8	9	9	9	10	15	19	22			
	GP-lvl	5	5	7	7	9	9	19	23	48		
	GP-ctx	4	5	5	6	7	9	9	9			
	GP-bth	6	6	6	9	9	10	10	22	26	35	52
	GP-s-lvl	5	5	5	9	10	15	21	24	33		
	GP-s-ctx	2	2	2	4	5	6	7	12	16	18	
	GP-s-bth	1	1	2	3	3	5	6	10	15	19	
Depth 7	GP-rand	20	26	27	33							
	GP-lvl	12	20	20	24	24	27					
	GP-ctx	12	12	16	20	20	25	32	43			
	GP-bth	11	11	17	19	19						
	GP-s-lvl	17	22	25	29							
	GP-s-ctx	11	17	17	17	21						
	GP-s-bth	7	7	9	9	9	33					

5   Analysis

5.1   Node Probabilities

Table 4 outlines level based probabilities for all primitives calculated from expressions generated with tree depth 5, as the best results were obtained using this depth. By analysing the table, several interesting observations can be made. First of all, + and − are the most common operators that appear in the expression. This is especially true for the root node, but also for several lower levels of the tree. As the level increases, the probability that other function nodes appear increases, especially with / appearing more frequently. This seems to suggest that the final expressions most commonly represent additions of smaller expressions, which are probably specialised for certain situations.

On the other hand, terminal nodes never appear in the first two levels, and only rarely in the third one. It is slightly surprising to see that terminals appear so rarely on lower levels, which suggests the expressions evolved by GP are more complex. However, with higher levels their probability of appearing increases, which is expected as they are required to complete the expression. The table shows that the most commonly used terminal node is the *pmin* terminal, whereas the least commonly used nodes are *age*, *pavg*, and *PAT*. Interestingly, the level of the tree does not seem to influence the probability of generating certain nodes, which means that terminals that generally appear more commonly will do so regardless of the level at which they are generated.

Table 4. Probabilities of selecting the primitives at each level of the tree

Level	+	−	*	/	pos	pt	SL	MR	pmin	pavg	dd	age	w	PAT
0	54	40	6	0	0	0	0	0	0	0	0	0	0	0
1	44	38	8	7	3	0	0	0	0	0	0	0	0	0
2	31	32	9	11	10	1	0	0	1	0	1	0	2	1
3	22	22	12	21	9	3	1	2	3	1	2	0	1	1
4	17	15	11	17	8	4	5	7	7	1	4	2	4	1
5	0	0	0	0	0	12	16	12	19	6	13	4	11	7

Figure 3 outlines the total frequency of all node occurrences calculated from the set of sample rules. The figure just confirms that function nodes appear more commonly than terminals, and that there is a high discrepancy in the occurrence of individual terminals, thus outlining which seem to be more important when calculating the priorities of DRs.

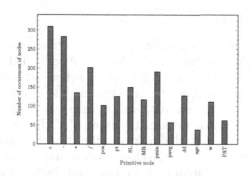

Fig. 3. Histogram of primitive node occurrences

Table 5 contains the conditional probabilities for the 10 contexts appearing most often in the case when only function nodes appear in them, and when both function and terminal nodes appear. By investigating function nodes, we see that the most common ones are those consisting out of the + and − operators, especially appearing as the root or the left child. As the right child the division operator tends to appear quite often, probably signalling that these contexts appear in the higher levels of the tree. On the other hand, for contexts consisting out of terminal nodes we see that certain terminal nodes like the due date or minimum processing time prevail in them. The context also shows some patterns appearing commonly, such as dividing certain job properties with w in order to scale them with the importance of jobs. These probabilities could potentially be used to introduce new candidate terminal nodes that could represent these contexts and thus relieve GP of having to construct them.

Table 5. Selected contexts with the largest occurrence probabilities

Contexts containing functions				Contexts containing termials			
parent	left child	right child	prob. (%)	parent	left child	right child	prob (%)
+	+	−	3.44	/	pmin	w	2.37
+	+	+	2.37	+	SL	pmin	2.04
−	+	−	2.26	+	MR	pt	1.51
+	+	/	2.15	+	dd	pmin	1.40
+	−	−	1.94	*	dd	pmin	1.29
−	−	−	1.61	/	dd	w	1.08
+	−	/	1.61	*	dd	dd	0.86
−	+	+	1.40	−	pmin	pt	0.75
*	+	/	1.18	−	pt	SL	0.75
+	*	/	1.18	/	pmin	pt	0.65

5.2   Method Ranking

To better outline the difference between the results obtained by random and probabilistic initialisation based methods, several additional metrics are calculated and shown in Table 6. During the evolution after each 1 000 evaluations the best individual on the training set was stored and later evaluated on the test set to obtain its performance, thus obtaining 80 values for each experiment. Since each experiment is executed 30 times, the means of all these 80 values were determined and used to calculate the metrics outlined in the table. First, the rank of each method for each of the 80 values was calculated, and the average rank (column "Avg. rank") and total rank (column "Tot. rank") of each method were calculated. Furthermore, the number of times each probabilistic method performed better than the random method for these 80 values is given in column "# better", while the column "Avg. diff" denotes the average relative difference between each probabilistic and the random initialisation method.

The table indicates that the GP-rand method always achieved the worst rank, thus clearly demonstrating to be inferior to probabilistic initialisation methods.

Table 6. Metrics outlining the performance of the initialisation methods

		GP-rand	GP-lvl	GP-ctx	GP-bth	GP-s-lvl	GP-s-ctx	GP-s-bth
Depth 3	Avg. rank	6.354	4.402	1.976	5.683	4.585	1.939	3.061
	Tot. rank	7	4	2	6	5	1	3
	# better	-	73	73	72	72	75	74
	Avg. diff. (%)	-	1.879	2.334	1.540	1.771	2.122	2.008
Depth 5	Avg. rank	5.988	5.049	5.622	2.085	4.524	2.134	2.598
	Tot. rank	7	5	6	1	4	2	3
	# better	-	66	35	80	65	80	80
	Avg. diff. (%)	-	0.824	0.758	1.997	1.038	1.727	2.061
Depth 7	Avg. rank	6.378	3.122	1.927	4.939	4.976	4.585	2.073
	Tot. rank	7	3	1	5	6	4	2
	# better	-	79	80	57	72	70	80
	Avg. diff. (%)	-	2.247	3.047	1.456	1.354	1.187	3.458

This is also demonstrated by the number of times that probabilistic initialisation methods achieve a better result than GP-rand, as they usually achieve better results for most of the 80 points, and in some cases in each of the 80 measured points GP using probabilistic initialisation achieved a better result than GP-rand. Regarding the probabilistic initialisation methods, it is evident that no method is constantly superior. However, the results still suggest that the methods using conditional or both types of probabilities perform better. This is also illustrated by the fact that using that method GP usually achieves the largest average difference in the results over GP-rand. Furthermore, the results also suggest that slightly better results are obtained when smoothing is used, however, the differences are not significant.

6   Conclusion

In this study a simple probabilistic based individual initialisation method was proposed. The methods use information from previously evolved DRs to introduce a bias towards generating certain nodes with a larger probability at certain positions in the individual. For that purpose, two types of probabilistic information were defined, namely the level based probability (probability of primitives appearing at each level of the tree) and conditional probability (probability of a primitive appearing within a given context). Even though the method by itself is quite simple and does not introduce any overhead in the execution time of GP (given existing DRs are available for extracting the information about the terminal distributions), it results in a much faster convergence of GP in comparison to using completely randomly generated individuals in the initial population. Although both types of probabilistic information resulted in an improved convergence, using both at the same time usually resulted in the fastest convergence rate. This demonstrates that using information from previously evolved DRs can be beneficial and used to improve the performance of GP when evolving new DRs, especially given the simplicity and low overhead of the method.

There are several possible future directions in which the work presented in this study can be extended. One possibility is to define new ways of calculating the probabilities of generating the nodes during the initialisation of a tree. The information used is quite local to the position of the node being generated, but it could be possible that using a wider context could potentially improve the performance of the method. Currently the calculated probabilistic information is used only during the initialisation of the trees. Therefore, another extension would be to adapt genetic operators to also use this probabilistic information. The obtained probabilities could also serve to detect patterns that commonly occur in the evolved expressions, which could be used for feature selection and generation. Another potential research avenue would be to apply Markov chains to define a better probabilistic model. Finally, the concept described in this paper can be extended to the problem of selecting DRs for constructing ensembles, where DRs were usually selected completely by random [11,13].

Acknowledgements. This research has been supported by the Croatian Science Foundation under the project IP-2019-04-4333 and by the Spanish State Agency for Research (AEI) under research project PID2019-106263RB-I00.

References

1. Branke, J., Nguyen, S., Pickardt, C.W., Zhang, M.: Automated design of production scheduling heuristics: a review. IEEE Trans. Evol. Comput. **20**(1), 110–124 (2016). https://doi.org/10.1109/TEVC.2015.2429314
2. Gil-Gala, F.J., Sierra, M.R., Mencía, C., Varela, R.: Genetic programming with local search to evolve priority rules for scheduling jobs on a machine with time-varying capacity. Swarm Evol. Comput. **66**, 100944 (2021). https://doi.org/10.1016/j.swevo.2021.100944
3. Kazimipour, B., Li, X., Qin, A.K.: A review of population initialization techniques for evolutionary algorithms. In: 2014 IEEE Congress on Evolutionary Computation (CEC), pp. 2585–2592 (2014). https://doi.org/10.1109/CEC.2014.6900618
4. Kim, K., Shan, Y., Nguyen, X.H., McKay, R.I.: Probabilistic model building in genetic programming: a critical review. Genet. Program Evolvable Mach. **15**(2), 115–167 (2013). https://doi.org/10.1007/s10710-013-9205-x
5. Mei, Y., Nguyen, S., Xue, B., Zhang, M.: An efficient feature selection algorithm for evolving job shop scheduling rules with genetic programming. IEEE Trans. Emerg. Top. Comput. Intell **1**(5), 339–353 (2017). https://doi.org/10.1109/TETCI.2017.2743758
6. Nguyen, S., Mei, Y., Zhang, M.: Genetic programming for production scheduling: a survey with a unified framework. Complex Intell. Syst. **3**(1), 41–66 (2017). https://doi.org/10.1007/s40747-017-0036-x
7. Nguyen, S., Zhang, M., Tan, K.C.: Surrogate-assisted genetic programming with simplified models for automated design of dispatching rules. IEEE Trans. Cybern. **47**(9), 2951–2965 (2017). https://doi.org/10.1109/TCYB.2016.2562674
8. Pinedo, M.L.: Scheduling. Springer, USA (2012). https://doi.org/10.1007/978-1-4614-2361-4

9. Planinić, L., Đurasević, M., Jakobović, D.: On the application of ϵ-lexicase selection in the generation of dispatching rules. In: 2021 IEEE Congress on Evolutionary Computation (CEC), pp. 2125–2132 (2021). https://doi.org/10.1109/CEC45853. 2021.9504982

10. Poli, R., Langdon, W.B., McPhee, N.F.: A Field Guide to Genetic Programming. Lulu Enterprises Ltd., UK (2008)

11. Đurasević, M., Jakobović, D.: Comparison of ensemble learning methods for creating ensembles of dispatching rules for the unrelated machines environment. Genet. Program Evolvable Mach. **19**(1), 53–92 (2017). https://doi.org/10.1007/s10710-017-9302-3

12. Đurasević, M., Jakobović, D.: A survey of dispatching rules for the dynamic unrelated machines environment. Expert Syst. Appl. **113**, 555–569 (2018). https://doi.org/10.1016/j.eswa.2018.06.053

13. Đurasević, M., Jakobović, D.: Creating dispatching rules by simple ensemble combination. J. Heuristics **25**(6), 959–1013 (2019). https://doi.org/10.1007/s10732-019-09416-x

14. Đurasević, M., Jakobović, D.: Comparison of schedule generation schemes for designing dispatching rules with genetic programming in the unrelated machines environment. Appl. Soft Comput. **96**, 106637 (2020). https://doi.org/10.1016/j.asoc.2020.106637

15. Đurasević, M., Jakobović, D.: Heuristic and metaheuristic methods for the parallel unrelated machines scheduling problem: a survey. Artif. Intell. Rev. (2022). https://doi.org/10.1007/s10462-022-10247-9

16. Đurasević, M., Jakobović, D., Knežević, K.: Adaptive scheduling on unrelated machines with genetic programming. Appl. Soft Comput. **48**, 419–430 (2016). https://doi.org/10.1016/j.asoc.2016.07.025

17. Vlašić, I., Đurasević, M., Jakobović, D.: Improving genetic algorithm performance by population initialisation with dispatching rules. Comput. Ind. Eng. **137**, 106030 (2019). https://doi.org/10.1016/j.cie.2019.106030

18. Wu, L., Wang, S.: Exact and heuristic methods to solve the parallel machine scheduling problem with multi-processor tasks. Int. J. Prod. Econ. **201**, 26–40 (2018). https://doi.org/10.1016/j.ijpe.2018.04.013

19. Yu, L., Shih, H.M., Pfund, M., Carlyle, W.M., Fowler, J.W.: IIE Trans. **34**(11), 921–931 (2002). https://doi.org/10.1023/a:1016185412209

20. Zhang, F., Mei, Y., Nguyen, S., Zhang, M.: Guided subtree selection for genetic operators in genetic programming for dynamic flexible job shop scheduling. In: Hu, T., Lourenço, N., Medvet, E., Divina, F. (eds.) EuroGP 2020. LNCS, vol. 12101, pp. 262–278. Springer, Cham (2020). https://doi.org/10.1007/978-3-030-44094-7_17

21. Zhang, F., Mei, Y., Nguyen, S., Zhang, M.: Correlation coefficient-based recombinative guidance for genetic programming hyperheuristics in dynamic flexible job shop scheduling. IEEE Trans. Evol. Comput. **25**(3), 552–566 (2021). https://doi.org/10.1109/TEVC.2021.3056143

22. Zhang, F., Mei, Y., Nguyen, S., Zhang, M.: Evolving scheduling heuristics via genetic programming with feature selection in dynamic flexible job-shop scheduling. IEEE Trans. Cybern. **51**(4), 1797–1811 (2021). https://doi.org/10.1109/TCYB.2020.3024849

23. Zhang, F., Mei, Y., Zhang, M.: Surrogate-assisted genetic programming for dynamic flexible job shop scheduling. In: Mitrovic, T., Xue, B., Li, X. (eds.) AI 2018. LNCS (LNAI), vol. 11320, pp. 766–772. Springer, Cham (2018). https://doi.org/10.1007/978-3-030-03991-2_69

MTGP: Combining Metamorphic Testing and Genetic Programming

Dominik Sobania(✉) ⓘ, Martin Briesch ⓘ, Philipp Röchner ⓘ,
and Franz Rothlauf ⓘ

Johannes Gutenberg University, Mainz, Germany
{dsobania,briesch,proechne,rothlauf}@uni-mainz.de

Abstract. Genetic programming is an evolutionary approach known
for its performance in program synthesis. However, it is not yet mature
enough for a practical use in real-world software development, since usually many training cases are required to generate programs that generalize to unseen test cases. As in practice, the training cases have to
be expensively hand-labeled by the user, we need an approach to check
the program behavior with a lower number of training cases. Metamorphic testing needs no labeled input/output examples. Instead, the program is executed multiple times, first on a given (randomly generated)
input, followed by related inputs to check whether certain user-defined
relations between the observed outputs hold. In this work, we suggest
MTGP, which combines metamorphic testing and genetic programming
and study its performance and the generalizability of the generated programs. Further, we analyze how the generalizability depends on the number of given labeled training cases. We find that using metamorphic testing combined with labeled training cases leads to a higher generalization
rate than the use of labeled training cases alone in almost all studied
configurations. Consequently, we recommend researchers to use metamorphic testing in their systems if the labeling of the training data is
expensive.

Keywords: Program Synthesis · Metamorphic Testing · Genetic
Programming

1 Introduction

Genetic programming (GP) [7,21] is an evolutionary algorithm-based approach
to automatically generate programs in a given programming language that meet
user-defined requirements. In GP-based program synthesis, these specifications
are usually given as input/output examples, which define the expected output
from a generated program for a given input, and are used as training data during
the evolutionary search.

With the introduction of new selection methods [17,31], variation operators [16] and grammar design techniques [10], GP-based program synthesis has

© The Author(s), under exclusive license to Springer Nature Switzerland AG 2023
G. Pappa et al. (Eds.): EuroGP 2023, LNCS 13986, pp. 324–338, 2023.
https://doi.org/10.1007/978-3-031-29573-7_21

made great progress in the last years. Recently, it has been shown that GP-based approaches for program synthesis are even competitive in performance with state-of-the-art neural network-based approaches [26].

Unfortunately, GP-based program synthesis is not yet mature enough for a practical use in real-world software development, since many input/output examples are usually required (up to 200 are regularly used in the literature [13]) to generate programs that not only work on the training cases, but also generalize to previously unseen test cases. In practice, however, the training cases have to be labeled manually by the user which is very expensive and time consuming. So it is necessary to reduce the number of required training cases in order to minimize the user's manual effort. However, simply reducing the number of training cases is not sufficient, since a small training set can easily be overfitted which will on average lead to a poor generalization of the generated programs. Consequently, we need a supplementary approach to specify and check the desired program behavior without adding more manually defined training cases.

With metamorphic testing [6], we do not need additional hand-labeled training cases as we execute a program multiple times (starting with a random input, followed by related inputs) and check whether the relations between the observed outputs logically fit the user's domain knowledge. E.g., a function that assigns a grade based on the score achieved in an exam could be executed with a random score. If we then increase this score, the function must return an equal or better grade, otherwise the metamorphic relation is violated and the function must be incorrect. Such metamorphic relations, where labeling is not necessary, could be used together with a classical (but smaller) hand-labeled training set. We expect that these additional relations help to improve the generalization ability of GP-generated programs.

Therefore, in this work, we suggest an approach that combines metamorphic testing and genetic programming (MTGP) and study its performance and the generalization ability of the generated programs on a set of common program synthesis benchmark problems. To analyze how GP's generalization ability depends on the number of given training cases, we perform experiments for different (labeled) training set sizes.

MTGP is based on a grammar-guided GP approach which uses lexicase [31] for the selection of individuals during evolution. Since lexicase selection is not based on an aggregated fitness value, but considers the performance on individual cases, it is well suited to take hand-labeled training cases in combination with metamorphic relations into account during selection. To study this combination, we analyze different sizes of the hand-labeled training set and add further tests based on the metamorphic relations defined for the considered benchmark problem. More specific, the tests based on the metamorphic relations can be constructed automatically based on random inputs. A candidate program is executed first on the random input and then on a follow-up input (based on the random input). After that, the outputs of the candidate program on the random input and the follow-up input are compared regarding to a pre-defined metamorphic relation. If the relation holds, the test is passed, otherwise it is failed. So the outcome of a metamorphic test can therefore be treated in the same way as that of a test based on labeled input/output examples, with the difference that no expensive manual labeling is necessary for an automatically generated metamorphic test case. In our experiments, we find that incorporating metamorphic testing in combination with hand-labeled training cases leads to a higher generalization rate than the use of hand-labeled training cases alone (as usual in GP-based program synthesis) in almost all studied configurations.

Following this introduction, we present in Sect. 2 recent work related to GP-based program synthesis and work on metamorphic testing. In Sect. 3 we describe metamorphic testing as well as its integration into GP in detail. Furthermore, we present the used program synthesis benchmark problems together with their associated metamorphic relations. In Sect. 4 we describe our experimental setup and discuss the results before concluding the paper in Sect. 5.

2   Related Work

The main approaches in GP-based program synthesis are stack-based GP and grammar-guided GP [30]. These approaches differ primarily in their program representation and their techniques used to support different data types.

Stack-based GP approaches use different stacks for the separation of different data types [33]. During program execution, data is taken as input from the appropriate stacks and the results are pushed back to the associated stack. In current systems, the individual program instructions are also on their own stack, which allows changes to the program flow at runtime [32].

Grammar-guided GP approaches use a context-free grammar to represent the supported control structures and statements in their relationship to each other and to distinguish different data types [11,34]. In principle, this technique can be used to create programs in any programming language. In recent years, however, using grammar-guided GP approaches, mainly Python programs have been evolved [10,24,27].

Regardless of the GP approach used, the program synthesis results have been significantly improved in recent years, primarily through the use of lexicase selection and its variants [10,12,15,17,19]. In contrast to selection methods such as tournament selection, lexicase selection is not based on an aggregated fitness value (see evaluation bottleneck [22]), but selects on the basis of the results on individual training cases [31] which allows to include also the structure of the given training data.

Also independent of the used approach, mainly input/output examples are used for training in GP-based program synthesis. However, there exists also work which uses additional information, such as the textual description of the problem [20] or formal constraints [3] to improve the program synthesis performance.

Since the input/output examples given as training data are only an incomplete problem definition, it is important that the generated programs not only work on the training data, but also produce correct results on previously unseen inputs. In order to improve the generalization ability of the programs generated by GP, the literature knows approaches that generate smaller programs, either by post-simplification [14] or by a selection at the end of a run [25]. Another option is to use batch lexicase selection [1] to improve generalizability [29]. In addition to that, recently a method has been presented that can be used to predict whether the programs generated by GP will generalize to unseen data or not [28].

Metamorphic testing introduced by Chen et al. [6] is a method from software development that allows to check certain properties in the program under test without the need of explicitly specifying the expected output of a test (see Sect. 3.1 for a detailed description). In the field of evolutionary computation, metamorphic testing was used, e.g., for the genetic improvement of existing software [23].

However, to the best of our knowledge, no work so far studied the impact of combining metamorphic testing and GP on the program synthesis performance and the generalizability of the generated programs.

3   Methodology

In this section, we describe the basics of metamorphic testing and show how it works with some illustrative examples. Furthermore, we present the selected program synthesis benchmark problems together with their metamorphic relations. Lastly, we describe in detail how metamorphic testing and GP-based program synthesis can be combined.

3.1   Metamorphic Testing

Metamorphic testing [6] is a method from software development to check if certain logic properties hold in a given function f. These properties are defined by so-called metamorphic relations which describe the logic connection between a given (random) base input I with its observed output $f(I)$ and a further follow-up input I' with its corresponding output $f(I')$.[1] The key advantage of metamorphic testing compared to classical test methods is that we need no expensive labeled input/output examples as we are just interested if the metamorphic relation between the observed outputs $f(I)$ and $f(I')$ holds.

As a first intuitive example (based on the example given in [2]), we can think of a simple web search. E.g., as base input I we search for the exact term "Genetic Programming" without filters in a scientific search engine and find $f(I)$ results. As follow-up input I' we search for the same term "Genetic Programming" but limit the results to publications since 2022 and get $f(I')$ results. As metamorphic relation, we define that $f(I) \geq f(I')$, as additional filters should lead to an equal or lower number of results. Figure 1, shows screenshots from Google Scholar illustrating this example. We see that $f(I) = 246,000$ and $f(I') = 6,650$, so the metamorphic relation holds as $f(I) \geq f(I')$.

---

[1] More than one follow-up test is also possible, but in this work we focus on exactly one follow-up test.

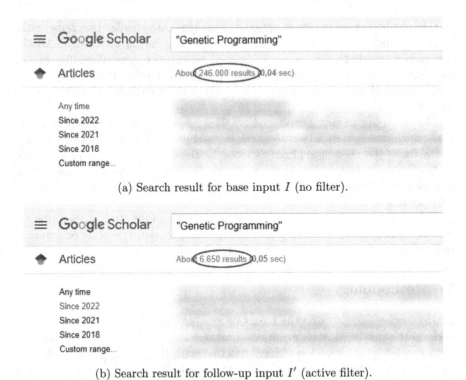

(a) Search result for base input I (no filter).

(b) Search result for follow-up input I' (active filter).

Fig. 1. Screenshots from Google Scholar illustrating a simple example of metamorphic testing. For the base input I (a), we see the corresponding output $f(I) = 246,000$ and for the follow-up input I' (b) the output is $f(I') = 6,650$, so the defined relation $f(I) \geq f(I')$ holds.

As a second, more technical, example, we choose the sine function, where we expect that it is 2π periodic. Consequently, for the inputs I and $I' = I + 2\pi$, a metamorphic relation could be defined as $f(I) = f(I')$ [4,5]. Figure 2 illustrates this example for $I = -3.7$. We see that constructed follow-up input $I' = -3.7 + 2\pi$ leads to the same result, so the relation $f(I) = f(I')$ holds.

3.2   Benchmark Problems and Metamorphic Relations

For the evaluation, we selected three problems from the program synthesis benchmark suite by Helmuth and Spector [18] which differ both in difficulty, according to a recent meta study [30], as well as in the data types used for input and output. Furthermore, we define two metamorphic relations for each of the benchmark problems for better comparison. However, more metamorphic relations are conceivable. We focused on relatively simple metamorphic relations which can be formulated by a user even with basic domain knowledge about the problem. The benchmark problems and metamorphic relations are defined as follows:

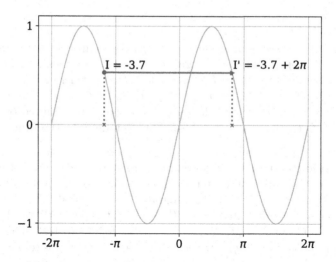

Fig. 2. An example of the sine function. We see that the result of the base input $I = -3.7$ and the follow-up input $I' = -3.7+2\pi$ is identical, so the defined metamorphic relation $f(I) = f(I')$ is satisfied.

Count Odds Problem:

Definition: A program should be generated that returns the number of odd values in a given list of integers. More formal, we search for a function f that maps the given vector of integers $I = (x_1, \ldots, x_n) \in \mathbb{Z}^n$ to the number of odds $f(I) = c \in \mathbb{N}_0$ as return value.

Metamorphic Relations: *1)* As first metamorphic relation, we require that the output of the program under test does not change when the input list is extended by an arbitrary number of even integers. For a given input $I = (x_1, \ldots, x_n)$ and the program f we calculate $f(I) = c_I$. As follow-up test, we create an extended input $I' = (x_1, \ldots, x_n, 2, 4, 6, \ldots, 2, 4, 6)$ by extending the input I with a random repetition of the vector $(2, 4, 6)$. The corresponding output of the manipulated input is $f(I') = c_{I'}$. This relation holds if $c_I = c_{I'}$. We chose 2, 4, and 6 as we expect that these values are known to be even numbers for users with basic domain knowledge about the problem (no knowledge about the `modulo` function needed). *2)* Analogously, we require for a second metamorphic relation that the output value increases if we extend the input by an arbitrary number of odd integers. Contrarily to the first relation, we extend the input $I' = (x_1, \ldots, x_n, 1, 3, 5, \ldots, 1, 3, 5)$ by a random number of repetitions of the vector $(1, 3, 5)$. Consequently, the corresponding output is $f(I') = c_{I'}$. This relation holds if $c_I < c_{I'}$. This time, we chose 1, 3, and 5 as trivial odd numbers.

Grade Problem:

Definition: We search a program that maps the numeric score achieved by a student to a discrete grade based on given thresholds. Specifically, we search for a function f that maps the input I to its grade $f(I) = g$, where $I = (t_1, t_2, t_3, t_4, s)$ with $t_i, s \in \mathbb{N}_0$, $t_i, s \leq 100$ for $i \in \{1, 2, 3, 4\}$ and $t_i < t_j$ for $i > j$ and $i, j \in \{1, 2, 3, 4\}$ where t_i are the thresholds and s is the score achieved by a student and $g = \{A, B, C, D, F\}$ is the corresponding grade with the order $A \succ B \succ C \succ D \succ F$.

Metamorphic Relations: *1)* As first relation, we require that a better numeric score leads to an equal or better discrete grade. For a given valid random input $I = (t_1, t_2, t_3, t_4, s)$ and the program f we calculate $f(I) = g_I$. After that, we create a manipulated input $I' = (t_1, t_2, t_3, t_4, s + k)$ based on I with $k \in \{0, \ldots, 100 - s\}$ and grade $f(I') = g_{I'}$ as follow-up test. The metamorphic relation holds if $g_{I'} \succeq g_I$. *2)* Second, we define the opposite relation which requires that a lower numeric score leads to an equal or worse discrete grade. We create a manipulated input $I' = (t_1, t_2, t_3, t_4, s - k)$ based on I with $k \in \{0, \ldots, s\}$ and grade $f(I') = g_{I'}$ as follow-up test. This second metamorphic relation holds if $g_{I'} \preceq g_I$.

Small or Large Problem:

Definition: The generated program should classify a given integer either as small, large, or in between. More precise, we search for a function f that maps a given integer $I = n \in \mathbb{Z}$ to its label $f(I) = l$, where $l = \{\texttt{"small"}, \texttt{""}, \texttt{"large"}\}$ with the order $\texttt{"small"} \prec \texttt{""} \prec \texttt{"large"}$. The function f should return $\texttt{"small"}$ if $n < 1,000$, $\texttt{"large"}$ if $n \geq 2,000$, and an empty string ($\texttt{""}$) otherwise.

Metamorphic Relations: *1)* First, we require that for an increased input the label also has to stay equal or increase according to the defined label ordering. For a given integer $I = n$ we compute the resulting label $f(I) = l_I$. Following this, we manipulate the input $I' = n + k$ with a random $k \in \mathbb{N}$ with the corresponding label $f(I') = l_{I'}$. The relation holds if $l_{I'} \succeq l_I$. *2)* Consequently, as second relation, we define that for an decreased input the resulting label has to stay equal or decrease according to the label ordering. For the given input $I = n$ we calculate the label $f(I) = l_I$. Then, we manipulate the input $I' = n - k$ with $k \in \mathbb{N}$ and $f(I') = l_{I'}$. If $l_{I'} \preceq l_I$, the relation is satisfied.

3.3   Incorporating Metamorphic Testing in GP

MTGP benefits from the use of lexicase selection, since lexicase considers the individual cases instead of an aggregated fitness value [31]. Consequently, different types of (training) cases can be used simultaneously for the selection. In MTGP, these are hand-labeled training cases for which we know the input and the corresponding output, as well as cases based on metamorphic relations.

For the hand-labeled training cases, both the inputs and the outputs are known. In order to check whether a candidate program solves a training case or not, we run the program with the given input I and check whether the generated output $f(I)$ matches the expected (in a real-world scenario hand-labeled) output O. If $f(I) = O$, then the test is passed, otherwise not.

To test the metamorphic relations defined for a benchmark problem, we provide random inputs as base inputs for which the outputs do not need to be known. Thus, any number of random entries can be generated at low cost. In our experiments, we have ensured that these random inputs do not already exist in the training or test set. In practice, they could be chosen freely. To check whether a candidate program satisfies a metamorphic relation or not, we execute it with a given base input I (one of the randomly generated inputs) and save the output $f(I)$. We then change the input randomly according to the manipulation rule of the metamorphic relation to create the follow-up input I' and run the program again. The test is passed if the observed outputs $f(I)$ and $f(I')$ satisfy the metamorphic relation, otherwise it is failed. Since we have defined two metamorphic relations for all benchmark problems considered in the experiments, we determine at random, with probability $p = 0.5$, which of the two relations is used each time a base input is selected.

One of the key advantages of using metamorphic testing this way is that during a GP run always many different metamorphic tests are executed, since the manipulation of the given basic input happens randomly. Our hope is, that this further supports the generalizability of the generated programs.

4   Experiments and Discussion

In this section, we study the performance of MTGP and analyze how well the generated programs generalize to previously unseen tests cases. To analyze how the generalization ability depends on the number of given training cases, we perform experiments for different labeled training set sizes. Below, we describe the experimental setup and discuss our results.

4.1   Experimental Setup

For the implementation of MTGP, we use the PonyGE2 framework [9]. Our used grammars are based on the program synthesis grammars provided by the PonyGE2 framework for the problems from the benchmark suite [18] and follow the principle suggested by Forstenlechner et al. [10] according to which the grammar of a problem is restricted to the data types (and dependent functions) that are defined for the input and output of the considered problem, in addition to required base data types (e.g., like `integer` and `Boolean`). This allows to keep the used grammars small and effective but still expressive.

We initialize a run with position independent grow [8] and use a population size of 1,000. The maximum initial tree depth is set to 10 and the maximum overall tree depth is limited to 17. As variation operators, we use sub-tree crossover and sub-tree mutation. For the sub-tree crossover we set the probability to 0.9 and for the sub-tree mutation we set the number of mutation steps to 1. As mentioned above, we use lexicase [31] as selection method. A GP run is stopped either after a program is found that solves all labeled training cases and all defined metamorphic tests or after 300 generations.

For every considered benchmark problem, we have 200 labeled training cases that we can use in the experiments. The test set consisting of 1,000 labeled cases is used to check if a candidate program also generalizes to unseen cases. Further we provide a large set of randomly generated inputs (800 for each benchmark problem) which we use as base inputs for checking if the defined metamorphic relations hold for a candidate program or not. In the experiments, we choose from these available training cases and randomly generated inputs depending on the considered configuration as specified below.

4.2   Results and Discussion

As we investigate the impact of using metamorphic testing in GP, we compare a standard GP approach, which only uses the training data, and the novel MTGP approach, which also includes metamorphic tests. In addition, we examine how

Table 1. Success rates on the training set s_{training} and the test set s_{test} achieved by standard GP and MTGP for all studied labeled training set sizes $|T_{\text{training}}|$ and benchmark problems. For MTGP, we use in addition $200 - |T_{\text{training}}|$ metamorphic tests. Best success rates achieved on the test set s_{test} are printed in **bold** font.

		Standard GP		MTGP	
Problem	$|T_{\text{training}}|$	s_{training}	s_{test}	s_{training}	s_{test}
Count Odds	25	43	**28**	32	20
	50	59	**46**	43	35
	100	70	**64**	61	59
	200	83	81	-	-
Grade	25	83	4	74	**14**
	50	86	12	76	**15**
	100	81	23	84	**33**
	200	93	45	-	-
Small or Large	25	94	2	58	**7**
	50	79	11	77	**25**
	100	91	**35**	86	29
	200	89	43	-	-

standard GP and MTGP perform on different training set sizes $|T_{\text{training}}|$. There-fore, we analyze for both approaches the labeled training set sizes 25, 50, 100, and 200. MTGP also uses $200 - |T_{\text{training}}|$ metamorphic tests so that MTGP considers exactly 200 tests during the training phase (e.g., for $|T_{\text{training}}| = 25$, MTGP uses 175 metamorphic tests).

As a first step, we study the achieved success rates on the training set s_{training} and the test set s_{test} as a performance indicator. The success rate on the training set s_{training} measures the percentage of runs in which a program is found that is successful on all given training cases (including metamorphic tests for MTGP). To determine the success rate on the test set s_{test}, we take from each run the candidate program that performs best on the training data and measure the percentage of candidate programs that solve all previously unseen test cases. For every studied configuration, we performed 100 runs.

Table 1 shows the success rates on the training set s_{training} and the test set s_{test} achieved by standard GP and MTGP for all studied labeled training set sizes $|T_{\text{training}}|$ (and corresponding metamorphic tests) and benchmark problems. Best success rates achieved on the test set s_{test} are printed in **bold** font.

For most configurations, we see that standard GP achieves a higher success rate on the training data s_{training} compared to MTGP. Our assumption is that this is because the additional metamorphic tests prevent an overfitting to the training data. On the test data, MTGP performs best for the Grade problem and for most of the configurations of the Small or Large problem (compared to the corresponding standard GP runs). For the Count Odds problem best results for s_{test} are achieved with standard GP.

More important than the pure success rates, however, is how well the pro-grams found on the training data generalize to unseen test cases. In practice, a program synthesis approach which is known for its high generalization can simply be executed again if no solution was found in the first run. If the gen-eralization is expected to be poor, it is unclear whether solutions found on the training data also work on previously unseen test cases, regardless of the suc-cess rate achieved on the training data. Large sets of additional test cases (to check for generalizability) are not available in a real-world scenario as manually labeling additional input/output examples is far too expensive. Therefore, in the second step, we analyze the generalization rate

$$G = \frac{s_{\text{test}}}{s_{\text{training}}} \cdot 100.$$

Table 2 shows the generalization rate G achieved by standard GP and MTGP for all studied labeled training set sizes $|T_{\text{training}}|$ (and corresponding metamor-phic tests) and benchmark problems. Again, best generalization rates G are printed in **bold** font.

We see that on average, best generalization rates G are achieved with MTGP, as MTGP performed best in 7 out of 9 configurations where we have results for standard GP as well as for MTGP. The differences are particularly obvious for the Grade and the Small or Large problem when only a small labeled training set ($|T_{\text{training}}| = 25$) is used. MTGP achieves for the Grade problem a generalization

Table 2. Generalization rate G achieved by standard GP and MTGP for all studied labeled training set sizes $|T_{\text{training}}|$ and benchmark problems. For MTGP, we use $200 - |T_{\text{training}}|$ metamorphic tests in addition to the considered labeled training cases. Best results are printed in **bold** font.

| Problem | $|T_{\text{training}}|$ | Generalization rate G | |
		Standard GP	MTGP
Count Odds	25	**65.116**	62.5
	50	77.966	**81.395**
	100	91.429	**96.721**
	200	97.59	-
Grade	25	4.819	**18.919**
	50	13.953	**19.737**
	100	28.395	**39.286**
	200	48.387	-
Small or Large	25	2.128	**12.069**
	50	13.924	**32.468**
	100	**38.462**	33.721
	200	48.315	-

rate G of 18.919 and for the Small or Large problem of 12.069 while standard GP achieves only 4.819 and 2.128, respectively.

So far we have only studied MTGP with a reduced labeled training set. But can the generalization rate even be increased compared to standard GP even if the complete labeled training set ($|T_{\text{training}}| = 200$) is used during the run? To answer this question, we run MTGP this time with 800 metamorphic tests.

Table 3. Generalization rate G achieved by standard GP and MTGP for all considered benchmark problems. All configurations use as labeled training set size $|T_{\text{training}}| = 200$. For MTGP, we use 800 metamorphic tests in addition to the considered labeled training cases. Best results are printed in **bold** font.

| Problem | $|T_{\text{training}}|$ | Generalization rate G | |
		StandardGP	MTGP
CountOdds	200	97.59	**100.0**
Grade	200	48.387	**61.538**
Small orLarge	200	48.315	**62.069**

Table 3 shows the achieved generalization rates G for standard GP and MTGP for all considered benchmark problems. As before, best generalization rates G are printed in **bold** font.

We see that even when the complete labeled training set $T_{training}$ is used during a run, using metamorphic testing can improve the generalization rate G. MTGP performs best on all studied benchmark problems. For the Count Odds problem, we even achieve a perfect generalization rate G of 100.

In summary, the generalization ability of the generated programs can be increased by using metamorphic testing. On average, best generalization rates are achieved with MTGP. The major advantage of metamorphic tests is that they do not require the expensive manual calculation of the expected outputs, since they work exclusively with random inputs which can be generated automatically.

5   Conclusion

GP [7, 21] is an evolutionary approach that is well known for its performance in automatic program synthesis. Even if GP is competitive in performance to other state-of-the-art program synthesis approaches [26], it is not yet mature enough for a practical use in real-world software development, as many input/output examples are usually required during the training process to generate programs that also generalize to unseen test cases. As in practice, the training cases have to be labeled manually by the user which is very expensive, we need a supplementary approach to check the program behavior with a lower number of manually defined training cases.

With metamorphic testing [6], we do not need labeled input/output examples. The program is executed multiple times, first on a given input (which can be generated randomly) and followed by related inputs to check whether certain user-defined metamorphic relations hold between the observed outputs.

Therefore, in this work we suggested MTGP, an approach that combines metamorphic testing and GP and studied its performance and the generalization ability of the generated programs on common program synthesis benchmark problems. Further, we analyzed how the generalization ability depends on the number of given training cases and performed experiments for different labeled training set sizes.

We found that incorporating metamorphic testing in combination with hand-labeled training cases leads to a higher generalization rate than the exclusive use of hand-labeled training cases in almost all configurations studied in our experiments, including those using smaller labeled training sets as usual in GP-based program synthesis. Even with the largest considered labeled training set, the generalization rate could be increased by a large margin on all studied benchmark problems with the use of metamorphic testing. Consequently, we recommend researchers to use metamorphic testing in their GP approaches if the labeling of the training data is an expensive process in the considered application domain.

In future work, we will study MTGP on additional program synthesis benchmark problems and further analyze the usage of the metamorphic tests as well as the given labeled training cases during a run to gain a deeper understanding of the implications of incorporating metamorphic testing in GP.

References

1. Aenugu, S., Spector, L.: Lexicase selection in learning classifier systems. In: Proceedings of the Genetic and Evolutionary Computation Conference, pp. 356–364 (2019)
2. Arrieta, A.: Multi-objective metamorphic follow-up test case selection for deep learning systems. In: Proceedings of the Genetic and Evolutionary Computation Conference, pp. 1327–1335 (2022)
3. Błądek, I., Krawiec, K., Swan, J.: Counterexample-driven genetic programming: heuristic program synthesis from formal specifications. Evol. Comput. **26**(3), 441–469 (2018)
4. Chen, T.Y., et al.: Metamorphic testing: a review of challenges and opportunities. ACM Comput. Surv. (CSUR) **51**(1), 1–27 (2018)
5. Chen, T.Y., Kuo, F.C., Liu, Y., Tang, A.: Metamorphic testing and testing with special values. In: SNPD, pp. 128–134 (2004)
6. Chen, T., Cheung, S., Yiu, S.: Metamorphic testing: a new approach for generating next test cases. Department of Computer Science, The Hong Kong University of Science and Technology, Technical report (1998)
7. Cramer, N.L.: A representation for the adaptive generation of simple sequential programs. In: Proceedings of the International Conference on Genetic Algorithms and the Applications, pp. 183–187 (1985)
8. Fagan, D., Fenton, M., O'Neill, M.: Exploring position independent initialisation in grammatical evolution. In: 2016 IEEE Congress on Evolutionary Computation (CEC), pp. 5060–5067. IEEE (2016)
9. Fenton, M., McDermott, J., Fagan, D., Forstenlechner, S., Hemberg, E., O'Neill, M.: PonyGE2: grammatical evolution in python. In: Proceedings of the Genetic and Evolutionary Computation Conference Companion, pp. 1194–1201 (2017)
10. Forstenlechner, S., Fagan, D., Nicolau, M., O'Neill, M.: A grammar design pattern for arbitrary program synthesis problems in genetic programming. In: McDermott, J., Castelli, M., Sekanina, L., Haasdijk, E., García-Sánchez, P. (eds.) EuroGP 2017. LNCS, vol. 10196, pp. 262–277. Springer, Cham (2017). https://doi.org/10.1007/978-3-319-55696-3_17
11. Forstenlechner, S., Nicolau, M., Fagan, D., O'Neill, M.: Grammar design for derivation tree based genetic programming systems. In: Heywood, M.I., McDermott, J., Castelli, M., Costa, E., Sim, K. (eds.) EuroGP 2016. LNCS, vol. 9594, pp. 199–214. Springer, Cham (2016). https://doi.org/10.1007/978-3-319-30668-1_13
12. Helmuth, T., Abdelhady, A.: Benchmarking parent selection for program synthesis by genetic programming. In: Proceedings of the 2020 Genetic and Evolutionary Computation Conference Companion, pp. 237–238 (2020)
13. Helmuth, T., Kelly, P.: PSB2: the second program synthesis benchmark suite. In: Proceedings of the Genetic and Evolutionary Computation Conference, pp. 785–794 (2021)
14. Helmuth, T., McPhee, N.F., Pantridge, E., Spector, L.: Improving generalization of evolved programs through automatic simplification. In: Proceedings of the Genetic and Evolutionary Computation Conference, pp. 937–944 (2017)

15. Helmuth, T., McPhee, N.F., Spector, L.: Lexicase selection for program synthesis: a diversity analysis. In: Riolo, R., Worzel, B., Kotanchek, M., Kordon, A. (eds.) Genetic Programming Theory and Practice XIII. GEC, pp. 151–167. Springer, Cham (2016). https://doi.org/10.1007/978-3-319-34223-8_9
16. Helmuth, T., McPhee, N.F., Spector, L.: Program synthesis using uniform mutation by addition and deletion. In: Proceedings of the Genetic and Evolutionary Computation Conference, pp. 1127–1134 (2018)
17. Helmuth, T., Pantridge, E., Spector, L.: On the importance of specialists for lexicase selection. Genet. Program. Evolvable Mach. 21(3), 349–373 (2020)
18. Helmuth, T., Spector, L.: General program synthesis benchmark suite. In: Proceedings of the 2015 Annual Conference on Genetic and Evolutionary Computation, pp. 1039–1046 (2015)
19. Helmuth, T., Spector, L.: Explaining and exploiting the advantages of down-sampled lexicase selection. In: ALIFE 2020: The 2020 Conference on Artificial Life, pp. 341–349. MIT Press (2020)
20. Hemberg, E., Kelly, J., O'Reilly, U.M.: On domain knowledge and novelty to improve program synthesis performance with grammatical evolution. In: Proceedings of the Genetic and Evolutionary Computation Conference, pp. 1039–1046 (2019)
21. Koza, J.R.: Genetic Programming: On the Programming of Computers by Means of Natural Selection, vol. 1. MIT Press, Cambridge (1992)
22. Krawiec, K.: Behavioral Program Synthesis with Genetic Programming, vol. 618. Springer, Cham (2016). https://doi.org/10.1007/978-3-319-27565-9
23. Langdon, W.B., Krauss, O.: Evolving sqrt into 1/x via software data maintenance. In: Proceedings of the 2020 Genetic and Evolutionary Computation Conference Companion, pp. 1928–1936 (2020)
24. Schweim, D., Sobania, D., Rothlauf, F.: Effects of the training set size: A comparison of standard and down-sampled lexicase selection in program synthesis. In: 2022 IEEE Congress on Evolutionary Computation (CEC), pp. 1–8. IEEE (2022)
25. Sobania, D.: On the generalizability of programs synthesized by grammar-guided genetic programming. In: Hu, T., Lourenço, N., Medvet, E. (eds.) EuroGP 2021. LNCS, vol. 12691, pp. 130–145. Springer, Cham (2021). https://doi.org/10.1007/978-3-030-72812-0_9
26. Sobania, D., Briesch, M., Rothlauf, F.: Choose your programming copilot: a comparison of the program synthesis performance of Github Copilot and genetic programming. In: Proceedings of the Genetic and Evolutionary Computation Conference, pp. 1019–1027 (2022)
27. Sobania, D., Rothlauf, F.: Challenges of program synthesis with grammatical evolution. In: Hu, T., Lourenço, N., Medvet, E., Divina, F. (eds.) EuroGP 2020. LNCS, vol. 12101, pp. 211–227. Springer, Cham (2020). https://doi.org/10.1007/978-3-030-44094-7_14
28. Sobania, D., Rothlauf, F.: A generalizability measure for program synthesis with genetic programming. In: Proceedings of the Genetic and Evolutionary Computation Conference, pp. 822–829 (2021)
29. Sobania, D., Rothlauf, F.: Program synthesis with genetic programming: the influence of batch sizes. In: Medvet, E., Pappa, G., Xue, B. (eds.) Genetic Programming. EuroGP 2022. LNCS, vol. 13223, pp. 118–129. Springer, Cham (2022). https://doi.org/10.1007/978-3-031-02056-8_8
30. Sobania, D., Schweim, D., Rothlauf, F.: A comprehensive survey on program synthesis with evolutionary algorithms. IEEE Trans. Evol. Comput. (2022)

31. Spector, L.: Assessment of problem modality by differential performance of lexicase selection in genetic programming: a preliminary report. In: Proceedings of the 14th Annual Conference Companion on Genetic and Evolutionary Computation, pp. 401–408 (2012)
32. Spector, L., Klein, J., Keijzer, M.: The Push3 execution stack and the evolution of control. In: Proceedings of the 7th Annual Conference on Genetic and Evolutionary Computation, pp. 1689–1696 (2005)
33. Spector, L., Robinson, A.: Genetic programming and autoconstructive evolution with the Push programming language. Genet. Program. Evolvable Mach. 3(1), 7–40 (2002)
34. Whigham, P.A., et al.: Grammatically-based genetic programming. In: Proceedings of the Workshop on Genetic Programming: From Theory to Real-world Applications, vol. 16, pp. 33–41. Citeseer (1995)

Interacting Robots in an Artificial Evolutionary Ecosystem

Matteo De Carlo[1]([✉]) [iD], Eliseo Ferrante[1,2] [iD], Jacintha Ellers[1] [iD],
Gerben Meynen[1] [iD], and A. E. Eiben[1] [iD]

[1] Vrije Universiteit Amsterdam, De Boelelaan 1111,
1081 HV Amsterdam, The Netherlands
m.decarlo@vu.nl

[2] Technology Innovation Institute, Abu Dhabi, United Arab Emirates

Abstract. In Evolutionary Robotics where both body and brain are malleable, it is common practice to evaluate individuals in isolated environments. With the objective of implementing a more naturally plausible system, we designed a single interactive ecosystem for robots to be evaluated in. In this ecosystem robots are physically present and can interact each other and we implemented decentralized rules for mate selection and reproduction. To study the effects of evaluating robots in an interactive ecosystem has on evolution, we compare the evolutionary process with a more traditional, oracle–based approach. In our analysis, we observe how the different approach has a substantial impact on the final behaviour and morphology of the robots, while maintaining decent fitness performance.

Keywords: Evolutionary Computing · Evolutionary Robotics · Robot Interaction · Artificial Ecosystem · Interactive Robot Ecosystem

1 Introduction

In evolutionary robotics, research is focused on how to evolve whole robotic ecosystems using artificial evolution [6]. Currently, the state of practice focuses on evolving the morphology (the "body") and the controller (the "brain") of individual robots performing simple tasks such as locomotion. Within this context, typically the robots (the "phenotypes") are created and evaluated one by one, to calculate the fitness values based on their behavior or other traits in a simulator. Thus, while the genotypes co-exist in the data set used as the population, the phenotypes only exist in isolation: they are never present in the same space at the same time, and they cannot interact with each other. This is in stark contrast with natural evolution. The main aim of this paper is to implement and investigate a more natural mechanism, where the robots that form the population co-exist and can physically interact. A similar scheme has been studied within the field of evolutionary robotics through the framework of embodied evolution, whereby a population of robots evolves their controller within the same environment, albeit the body is fixed and remains the same.

G. Pappa et al. (Eds.): EuroGP 2023, LNCS 13986, pp. 339–354, 2023.
https://doi.org/10.1007/978-3-031-29573-7_22

Our interactive evolutionary robotics system is based on a few design choices. First, the whole population is present in the given environment, where they can move around simultaneously. Second, the mate selection mechanism is decentralized: robots 'decide' themselves with whom they engage in producing offspring and this depends on whom they have encountered before the moment of reproduction. Third, the number of offspring of a robot depends on its fitness, but the population size is kept constant by a mechanism where the total number of offspring for all population members together equals the population size.

In this paper, we are interested in the implication of such a system on the evolutionary dynamics. More precisely, we would like to understand how the dynamics of the proposed interactive evolutionary robotics system compare to those of the standard one where reproduction happens offline and centralized. Additionally, we are interested to observe population dynamics, especially if the separation between different groups of robots occurs and if will be enough to have different species emerge in the environment.

To achieve the above objectives, we compare traditional oracle–based isolated evolution with a more natural setup with interactive robots and a parent selection algorithm running on the robot instance that we call an "Interactive Robot Ecosystem". On top of a distributed mate selection algorithm, what characterizes this new system is the physical presence of the robots in the same space, where the position of the robots is preserved across their lifetime, allowing the population of robots to slowly migrate to different areas. In addition, the robots can collide with each other.

To introduce the new system gradually, in this paper survival selection remained centralized: while mating selection happens locally for each individual, survival selection is rank based. The rank is constructed by a central entity. A distributed survival selection is theoretically possible, but it creates complications like the management of a variable population size, therefore the study of a possible implementation is left for future work.

In this work, our research objectives are:

1. prove that our simplified implementation of an "Interactive Robot Ecosystem" successfully evolve robots;
2. observe whether the population evolved in an interactive ecosystem differs from one evolved with a more traditional oracle–based approach;

The remaining of this paper is organized as follows: Sect. 2 is a Related Work section that compares our contribution with the work done by other researchers in the past. In Sect. 3 we explain in detail the various components of our experiments, including the Robot design space, the genotype we used, and the evolutionary algorithm we used. In Sect. 4 we detail all of the configuration details specific to our experiment, important for the repeatability of the experiment. In Sect. 5 we analyse the results of our data and in Sect. 6 we summarize our findings and draw final conclusions.

2   Related Work

Our work is inspired by the Evosphere, first introduced in [7]. The Evosphere is a system where evolution is not dictated by the algorithm design, but rather aims at creating a complex dynamic ecosystem of robots that adapt to the environment. Its purpose is two–fold: it is aimed at engineers to enable the evolution of robot designs that are adapted to specific environments, and it is aimed also at researchers as a platform to study the complex dynamics of evolution using robots instead of animals. While the aim is high, there are still many unknowns before a fully working implementation of an Evosphere is realised. Our work aims at reducing this gap, focusing on studying the effects that a decentralized evolutionary algorithm has on the behaviour and the morphology of evolving robots.

Some early work on simple simulated entities in dynamic ecosystems was already conducted in [5] by T. Buresh et al. In this work, we learn how a dynamic environment with simple rules can already show a great deal of interesting behaviours. It also teaches us how a variable population size can show very different patterns and could be quite difficult to stabilise in a complex system.

In a work from Nicolas Bredeche et al. [3] the researchers aimed at designing an algorithm, *mEDEA*, that was capable of adapting a population of physical robots in a dynamic unknown environment, using fixed morphology wheeled robots. They also manage to validate their algorithm on a real hardware platform. This work studied was especially interested in how certain behavioural dynamics that do not directly help with the assigned task, would be generated as a product of evolution. This is highly relevant to us, as we aim at studying similar effects on the evolutionary dynamics of our population. Although in our case the robots can demonstrate behavioural and morphological changes.

A study similar to ours was already conducted in a work by Berend Weel et al. [18]. In this paper, the authors aim at building an Artificial Life habitat. Our work is partially inspired by Weel's work, and it shares several similarities, namely the morphologically evolving robotic platform, the physical presence of multiple robots in the same environment, and basing the mate selection only between robots that are physically close. Despite these similarities, we identified some fundamental differences in relation to our work. The work from B. Weel et al. always spawns new robots in the center of the arena, shot in a random direction, not considering the impact that this choice has on population dynamics. It heavily uses a Reinforcement Learning algorithm to learn locomotion, which has a huge impact on the effective behaviour of the robots. Finally, our approach has the population size fixed and it is aimed at studying the effects on the morphologies and behaviour of the robots, while their approach has a variable population size and is more involved in designing a system that has stable population size dynamics without explicitly defining a fitness function.

Similar work from other groups includes studies from Takashi Ito et al. [10] of dynamic predator-prey systems of digital creatures with dynamic population size and both short and long-term cyclic patterns. We also include the work of Takaya Arita et al. [1] who study the feasibility of a more biologically realistic developmental model, eco-evo-devo, to evolve several virtual creatures across several different environments, including a predator-prey environment with dynamic population sizes. Both works show interest in interactive ecological systems, even if limited to digital–only creatures.

3   Method

In the following section, we are going to describe the evolutionary algorithm implementations, the robotic platform, and the genotype encodings we used.

3.1   Evolutionary Algorithm

In this work, we compare two different evolutionary approaches. The main difference between the oracle–based approach and the interactive ecosystem lies in the evolutionary algorithms used. In this section, we explain the two implementations of the two evolutionary algorithms used, with a variation on the second version.

Oracle–Based. As a baseline, we selected a state-of-the-art, simple evolutionary algorithm that implements a centralized reproduction mechanism. Every generation is composed of 100 robots, which are sampled to create 50 new offspring. The 50 new offspring are evaluated, a fitness value is calculated, and then the two populations are merged into one with a size of 150 individuals, which is shrunk down to the original size of 100 using a selection operator. This algorithm is usually referred to as a steady–state evolutionary algorithm, with a $100 + 50$ configuration. For both the mate selection and the survival selection we used tournament selection with a tournament size of 2. This particular configuration is been battle–tested in our lab and it's very robust and consistent at generating good robots.

Interactive Robot Ecosystem. The algorithm works as follows: it evaluates a population of robots in the same space, every robot with a physical location. The movement of all individuals is tracked and is used at the end of the simulation to compile a list of candidates per robot. Specifically, each robot becomes a mate candidate for another robot if the two robots ever get closer than the mating range threshold from each other. At this point, part of the population is killed to make space for new individuals. To simplify our study, we kill the lower 3/4 of the population based on fitness (speed-2) in an oracle–based manner; as previously mentioned, this is to reduce the complexity of our system and to introduce only a small change at the time with respect to the baseline. To replace the eliminated 3/4 of the population, each surviving robot is assigned a number of offspring to generate, proportional to his fitness. Each robot is guaranteed to get at least one offspring, with the best robot gets the generate the most number of children. In this paper, we assigned a value of fitness based on a linear scale, with the best robot getting extra robots to compensate for any rounding error. Each surviving robot generates the number of individuals assigned to them by mating with the best (sexual selection) among the candidate partners in the list assigned to that robot.

Every new offspring is placed in the environment; we implemented two different strategies to select the spawn location:

Spawn Closed to the Parent (Local). In this first variant, every new robot is spawned in a random location close to the main parent. If no location is found, the spawning radius is doubled. In theory, this should be beneficial to the genome pool, as robots that are genetically similar and more likely to produce working robots.

Spawn in a Random Position (Random). Instead of spawning the newborn robot in the vicinity of the parent robots, we spawn it in a random location inside a spawning area. This is to test if the spawning location of the robots has any influence on the evolutionary process.

3.2   The Robots

The experiments for this paper have been validated on the Revolve platform. Revolve is a modular robotic platform inspired by RoboGen [2] and similar other robotic platforms [4,8,16]. A big advantage of modular robotic platforms is that they often provide the possibility to craft our evolved robots in hardware, often at a reasonable cost, and challenge them with the harsh rules of reality.

Revolve itself is a classical modular robotic framework, using Isaac Gym for its physics simulation [13]. The robots are based on 3 types of modules: a Core module (Fig. 1a), a Brick module (Fig. 1b), and a Joint module (Fig. 1c). Every robot has to have one and only one Core module and any number and combination of Brick and Joint modules. Core and Brick have 4 attachment sides, Joint has 2. Brick and Joint modules can be attached at a rotation of 90°, to achieve a richer space. Another major difference with RoboGen is that there are no sensor or wheel modules in Revolve. In Fig. 2 we present an example of what a robot can look like.

For the controller architecture, for almost all of our research, we have used a Central Pattern Generator (CPG) as described in [11,12], with no sensory input. A CPG network is an interconnected network that outputs an oscillating rhythmic signal. This signal is then wired into the joint servos as a position signal. While not perfect, we have proven this architecture to be quite successful with our joint–based robots. An example of a CPG network for a spider robot is in Fig. 3.

Traits. To observe quantifiable changes in our populations, we defined some measurable morphological and behavioural traits; most traits are from [14,15].

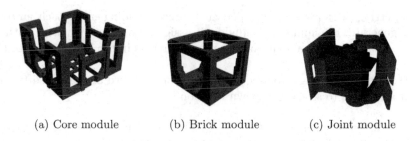

(a) Core module   (b) Brick module   (c) Joint module

Fig. 1. Robots are built using three types of modules. Starting with only one Core module, the robot grows by connecting Brick and Joint modules to the Core or other already connected modules.

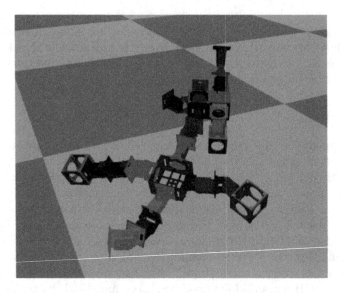

Fig. 2. Example of an evolved robot randomly chosen from one of the runs.

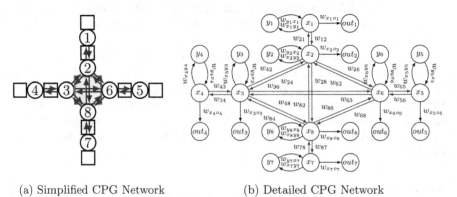

(a) Simplified CPG Network   (b) Detailed CPG Network

Fig. 3. Example of CPG network for a Spider robot.

Morphological traits are measurable properties of the bodies of the robots.

- *Branching modules count*: the number of brick modules that have all the attachment slots occupied.
- *Limbs*: considering the robot as a tree of modules, it is the number of leaf modules. The value is normalized per robot by the number of all possible limbs available.
- *Length of limbs*: measures the average number of modules in a limb.
- *Coverage*: considering the 3D bounding box that encompasses the robot, this trait is the ratio between the area that is occupied by modules and the total area of the rectangle.
- *Joints*: the number of active joints (motors) in the body normalized by the number of modules in a robot.
- *Hinge count*: the number of active joints (motors) in the body.
- *Brick count*: the number of brick modules in the body.
- *Proportion*: considering the 3D bounding box that encompasses the robot when viewed from above, this trait is the ratio between the two sides of this rectangle.
- *Z-depth*: The height of the robots (z-axis is perpendicular to the floor).
- *Robot size*: the number of modules in the body.
- *Symmetry*: the relative symmetry of the body around the head on the x and y-axis.
- *Vertical simmetry*: the relative symmetry of the body around the head on the z-axis.
- *Height to base ratio*: Ratio between the size of the base of the 3D bounding box that encompasses the robot and its height.
- *Base density*: similar to the measurement of coverage, measures the ratio between modules and empty spots, but it only looks at the vertical slice of the modules at the same height of the head.

Behavioural traits are measurable properties of the behaviour of the robots, i.e. the observable properties of the lifetime of the robots. Any data collected during the first 5 s of evaluation is discarded, using only 120 s of data.

- *Speed-1*: describes the average robot speed (cm/s), and is calculated by dividing the length of the robot's path during its lifetime by the robot's lifetime (120 s). This measurement is highly dependent on the sample rate, which in our case 8 Hz. The speed measure is formally defined in Eq. 1.

$$v = \frac{\sum_{t=1}^{T} |s_t - s_{t-1}|}{\Delta t} \tag{1}$$

- *Speed-2*: also referred to as *Displacement Speed*, it describes the average robot speed (cm/s), and is calculated as if the robot took the shortest path from the start position s_0 to the end position s_t. We use *Speed-2* as *fitness*. It is formally defined by Eq. 2.

$$v_{disp} = \frac{s_T - s_{t_0}}{\Delta t} \tag{2}$$

– *Balance*: we use the rotation of the head in the x–y plane to define the balance of the robot. We describe the rotation of the robot with three dimensions: roll ϕ, pitch θ, and yaw ψ. Thus, we consider the pitch and roll of the robot head, expressed between $0°$ and $180°$ (because we are not interested in whether the rotation is clockwise or anti-clockwise). Perfect balance corresponds to $\theta = \phi = 0°$, so that the higher balance, the less rotated the head is. Formally, balance is defined by Eq. 3.

$$b = 1 - \frac{\sum_{t=0}^{T} |\phi_t| + |\theta_t|}{180 \cdot 2 \cdot \Delta t} \tag{3}$$

3.3   Genotype Encodings

A considerable advantage of using a modular robotic platform is that a sequence of modules and rules on they should be attached together can be easily encoded in a digital genotype.

To encode our robots' genotype, we mainly used two different methods combined together: a Tree-based direct representation for the body and a CPPN-NEAT indirect representation for the brain.

Tree-Based Direct Representation. In a tree-based representation [9], the genotype is a tree data structure in which each node represents a module of the robot. The Core module is always the root of the tree and it can only be present once in the entire genotype. Brick and Joint modules can be attached and form a tree of modules, with the Core as the root of the tree.

CPPN-NEAT Indirect Representation. While the structure of the brain is determined by its body structure, to develop the parameters of the brain we use an encoding based on Hypercube-based Neuro Evolution of Augmenting Topologies (HyperNEAT) [17] that encodes the weights of the CPG controller network. This ensures that controller configurations can be inherited when an offspring has a different number of joints.

4   Experimental Setup

For all versions of our experiments, we use the same robot modules and the same genotype. Every robot has a maximum size of 25 modules and it's evaluated in the same infinite open arena with no obstacles. All individuals are evaluated at the same time, and collisions between robots are disabled in the oracle–based baseline and enabled in both Interactive Robot Environment setups. All robots are simulated for 125 simulated seconds, and all data of the first 5 s is discarded, to avoid registering weird behaviour at spawning. This gives us data for 120 s of simulated time. The data is sampled 8 times a second, which is also the update rate of the controller. During the 120 s of lifetime, we collect and save the robots'

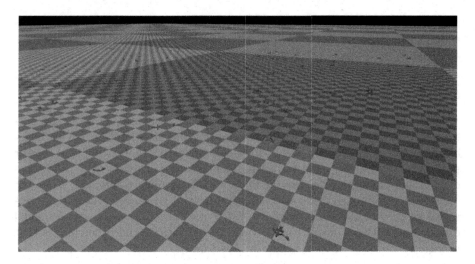

Fig. 4. Screen capture of the simulator running an evaluation of an entire population of interactive robots.

positions and orientations. These will later be used to calculate behavioural traits and the fitness of the robots. The fitness function used to evaluate all robots is the "speed-2" trait (displacement speed).

All configurations of the evolutionary algorithms have a population size of 100 individuals and all evolutionary runs last 100 generations.

In the interactive ecosystem, we spawn the first generation in a square of 5×5 meters, equally distanced in a grid of 10×10 robots. The mating range, the range within which two robots exchange their genome, is 2 m. In the first variation (local), offspring robots are spawned within a ring-shaped area around the first parent with a radius between 2 and 3 m. In the second variation (random), offspring robots are spawned in the same square area used to spawn the first generation. For both configurations, all offspring are spawned at a location that is at least 2 m away from every other robot already present. If no location is found, the spawn area is increased in size. In the local variant, we double the maximum radius of the ring. In the random variant, we double both sides of the square. The process of doubling is repeated every time no location is found until all offspring are spawned. At the end of the 120 s, every robot that survives keeps its position in the next generation. For statistical significance, each configuration is repeated 10 times, for a total of 30 independent evolutionary runs.

5   Analysis

In our analysis, we compare the three different setups analysing the differences in the performance of the robots and investigating any morphological and behavioural differences. The three different setups are the oracle–based setup that acts as our baseline and the interactive robot ecosystem setup, in the two variants that differ on how offspring are placed (either *locally* around the parent or *randomly* in a square spawn area).

5.1   Fitness Analysis

When looking at Fig. 5, the median fitness of our baseline shows a distinct better performance with values easily over-passing the mark of 2 cm/s. Our interactive ecosystem approach seems to show still decent but significantly lower performance, until we dissect the final generation and reveal that the fitness distribution plot (Fig. 6) across the two setups is extremely different. While the oracle–based approach develops uniformly distributed fitness values, the interactive ecosystem approach seems to distribute fitness values in a multimodal distribution with an obvious mode close to zero fitness and other modes that average to values even higher than the baseline.

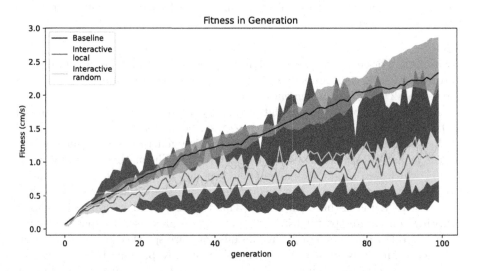

Fig. 5. Median fitness over generations. The lower and upper bounds are the first and third quantiles of the medians of the different runs.

The distribution shape of the speed-1 values (Fig. 7) sheds some light on what is happening. Speed-1 values are directly proportional to respectively the path length while Speed-2 values are directly proportional to the distance between starting and end positions of a robot. What we observe is that in the interactive ecosystems there are a substantial amount of robots that travel a long path length (speed-1) but this doesn't transform into high fitnesses (speed-2). This can be interpreted by imagining a substantial amount of robots in our interactive setup that prefer to travel in circles instead of straight lines. This is likely an indirect effect of having robots share their genome with other robots in proximity.

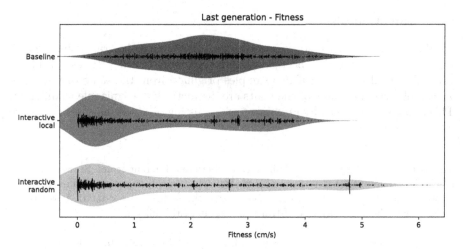

Fig. 6. Violin and Swarm plots representing fitness values (speed-2) distribution in the last generation. Data is aggregated across all runs.

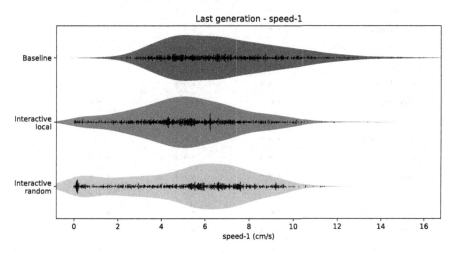

Fig. 7. Violin and Swarm plots representing average speed values (speed-1) distribution in the last generation. Data is aggregated across all runs.

Instead, robots that travel in straight lines end up outside of the group of robots and get fewer chances of sharing their genetic code with other individuals.

One notable observation: the oracle–based approach has not converged after 100 generations and might need more generations to achieve the robots' full potential. Both interactive ecosystems seem to have converged instead.

5.2   Trait Analysis

We did a comprehensive statistical comparison between the trait values, which were previously defined in Sect. 3.2. Using the Mann-Whitney U test (Table 1), which is aimed at testing if two samples originate from the same distribution, we can observe how most of the traits are distinctively recognizable as different. This effect seems especially prominent when testing our baseline against the two interactive setups.

Table 1. Table listing statistical significance between different traits. Stars mark how low is the probability that two samples are from the same distribution (Mann–Whitney U test). One star $\star < 0.05$, two stars $\star\star < 0.01$, three stars $\star\star\star < 0.001$. It is easy to observe how the vast majority of traits are significantly different.

Morphological Traits	Baseline Inter. local	Baseline Inter. random	Inter. local Inter. random
Speed-2 (fitness)	$\star\star\star$	$\star\star\star$	-
Speed-1	$\star\star\star$	$\star\star\star$	-
Head balance	$\star\star\star$	$\star\star\star$	$\star\star\star$
Branching modules count	-	$\star$	$\star\star\star$
Limbs	$\star\star\star$	$\star\star\star$	-
Length of limbs	$\star\star\star$	$\star\star\star$	-
Coverage	$\star\star$	$\star\star\star$	$\star\star\star$
Joints	$\star\star\star$	$\star\star\star$	$\star\star\star$
Hinge count	$\star\star\star$	$\star\star\star$	-
Brick count	$\star\star\star$	$\star\star\star$	$\star\star\star$
Proportion	$\star\star\star$	$\star\star\star$	$\star\star\star$
Z depth	-	$\star\star\star$	$\star\star\star$
Absolute size	$\star\star$	$\star\star\star$	$\star\star\star$
Symmetry	$\star\star\star$	$\star\star\star$	$\star$
Vertical symmetry	$\star\star$	$\star\star$	$\star\star\star$
Height base ratio	-	$\star\star\star$	$\star\star\star$
Base density	$\star$	$\star\star\star$	$\star\star\star$

When observing the evolution of the traits and their final distribution, we observe how the generated robots differ. The major difference seems to be in the modality of the data, while in our baseline most data is unimodal, in an interactive ecosystem robots evolve with a great spread and along multiple modes. The multimodality is also apparent within the data of single runs, from which we can deduce that this is not just a by-product of the high variance between different runs, but it is apparent within the same environment. We sampled two significant examples, size and number of bricks, in Figs. 8 and 9.

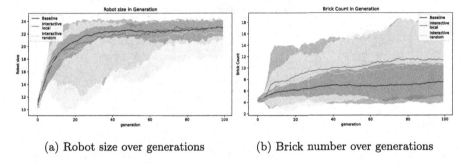

(a) Robot size over generations   (b) Brick number over generations

Fig. 8. Evolution of two representative morphological traits over the generations. Data is sampled as the average of individual runs. The data is bounded between the 0.5% and 95% of the average values (variance between runs).

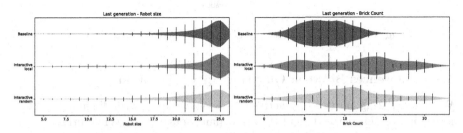

(a) Number of modules (size) distribution in the last generation   (b) Number of Brick modules distribution in the last generation

Fig. 9. Violin and Swarm plots representing the distribution of two sample morphological measurements in the last generation. Data is aggregated across all runs.

Finally, we take a look at the difference in Age within the populations of the different setups (Fig. 10). Here we can observe a strong difference between our baseline and our interactive ecosystem approach. In the baseline population, elite individuals seem to linger much longer inside the populations while in the interactive environment, they get replaced much more quickly. We believe this is partially influenced by differences in survival selection mechanism, which is stronger in the interactive ecosystem approach, even if it should also be extremely favourable to elitist robots. It is likely also influenced by the difference in evaluation: while in the baseline each robot is evaluated only once, in an interactive environment robots are re-evaluated for each generation making it more likely that the fitness will change over the course of different generations.

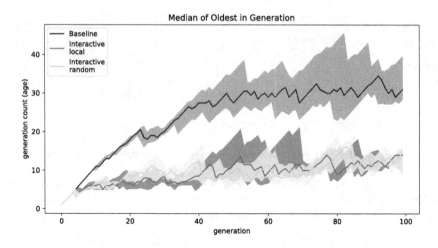

Fig. 10. Oldest individual per generation, median across 10 runs.

5.3   Differences Between Offspring Spawn Location

When comparing our two interactive ecosystem variants, with different spawning location strategies, it is difficult to draw many conclusions. While fitness results are very similar, we can observe statistical significance between many different traits and we could not clearly determine the cause of it. What we can observe is a greater variance in robot morphologies when the spawn location was random, which could imply a greater exploration of the fitness landscape. But this does not translate into higher fitness values. What we were hoping to observe is greater fitness values in the local variant, under the assumption that having similar robots in proximity would produce fewer bad combinations, improving the fitness landscape exploration.

6   Conclusion

From observing fitness plots, behavioural traits and morphological traits we can observe a clear difference in the results of the three different configurations. The oracle algorithm seems to perform better, and that was expected from us. Years of experience and tuning are playing a role here, while the interactive ecosystem setup is just in its infancy. Nevertheless, the robots that evolved in the interactive ecosystem seem to be capable and there is a consistent portion of the population with fitness values comparable to the baseline approach.

What we hoped and achieved to observe is that just introducing a distributed mating approach with a physical presence in the same ecosystem, making it an interactive ecosystem, would generate new dynamics in the evolutionary algorithm. It is likely that physical contact did not play a major role in our system, but it opens up many possibilities for future work, e.g. predator–prey dynamics and fighting for fertile land.

In further studies, we aim at extending our interactive ecosystem by introducing a distributed mechanism of survival selection with variable population size. In addition, we are interested in removing the fitness function entirely and only measuring a fitness value a posteriori, similar to how fitness is measured in the field of Biology.

References

1. Arita, T., Joachimczak, M., Ito, T., Asakura, A., Suzuki, R.: ALife approach to eco-evo-devo using evolution of virtual creatures. Artif. Life Robot. **21**(2), 141–148 (2016)
2. Auerbach, J., et al.: RoboGen: robot generation through artificial evolution. In: Artificial Life 14: Proceedings of the Fourteenth International Conference on the Synthesis and Simulation of Living Systems, pp. 136–137. MIT Press (2014)
3. Bredeche, N., Montanier, J.M., Liu, W., Winfield, A.F.: Environment-driven distributed evolutionary adaptation in a population of autonomous robotic agents. Math. Comput. Model. Dyn. Syst. **18**(1), 101–129 (2012)
4. Brodbeck, L., Hauser, S., Iida, F.: Morphological evolution of physical robots through model-free phenotype development. PLoS ONE **10**(6), e0128444 (2015)
5. Buresch, T., Eiben, A.E., Nitschke, G., Schut, M.: Effects of evolutionary and lifetime learning on minds and bodies in an artificial society. In: 2005 IEEE Congress on Evolutionary Computation, Edinburgh, Scotland, UK, vol. 2, pp. 1448–1454. IEEE (2005)
6. Doncieux, S., Bredeche, N., Mouret, J.B., (Gusz) Eiben, A.E.: Evolutionary robotics: what, why, and where to. Front. Robot. AI **2**(MAR), 1–18 (2015)
7. Eiben, A.E.: EvoSphere: The world of robot evolution. Lecture Notes in Computer Science (including subseries Lecture Notes in Artificial Intelligence and Lecture Notes in Bioinformatics) 9477, 3–19 (2015), iSBN: 9783319268408
8. Hale, M.F., et al.: Hardware design for autonomous robot evolution. In: 2020 IEEE Symposium Series on Computational Intelligence (SSCI), Canberra, ACT, Australia, pp. 2140–2147. IEEE, December 2020
9. Hupkes, E., Jelisavcic, M., Eiben, A.E.: Revolve: a versatile simulator for online robot evolution. In: Sim, K., Kaufmann, P. (eds.) EvoApplications 2018. LNCS, vol. 10784, pp. 687–702. Springer, Cham (2018). https://doi.org/10.1007/978-3-319-77538-8_46
10. Ito, T., Pilat, M.L., Suzuki, R., Arita, T.: Population and evolutionary dynamics based on predator-prey relationships in a 3D physical simulation. Artif. Life **22**(2), 226–240 (2016)
11. Lan, G., Carlo, M.D., Diggelen, F.V., Tomczak, J.M., Roijers, D.M., Eiben, A.E.: Learning directed locomotion in modular robots with evolvable morphologies (2020). arXiv: 2001.07804. Publisher: arXiv
12. Lan, G., Jelisavcic, M., Roijers, D.M., Haasdijk, E., Eiben, A.E.: Directed locomotion for modular robots with evolvable morphologies. In: Auger, A., Fonseca, C.M., Lourenço, N., Machado, P., Paquete, L., Whitley, D. (eds.) PPSN 2018. LNCS, vol. 11101, pp. 476–487. Springer, Cham (2018). https://doi.org/10.1007/978-3-319-99253-2_38
13. Makoviychuk, V., et al.: Isaac gym: high performance GPU-based physics simulation for robot learning, August 2021. http://arxiv.org/abs/2108.10470. arXiv:2108.10470 [cs]

14. Miras, K., Ferrante, E., Eiben, A.E.: Environmental influences on evolvable robots. PLoS ONE **15**(5), e0233848 (2020)
15. Miras, K., Haasdijk, E., Glette, K., Eiben, A.E.: Search space analysis of evolvable robot morphologies. In: Sim, K., Kaufmann, P. (eds.) EvoApplications 2018. LNCS, vol. 10784, pp. 703–718. Springer, Cham (2018). https://doi.org/10.1007/978-3-319-77538-8_47
16. Moreno, R., Liu, C., Faina, A., Hernandez, H., Gomez, J.: The EMeRGE modular robot, an open platform for quick testing of evolved robot morphologies. In: Proceedings of the Genetic and Evolutionary Computation Conference Companion, Berlin, Germany, pp. 71–72. ACM, July 2017
17. Stanley, K.O., D'Ambrosio, D.B., Gauci, J.: A hypercube-based encoding for evolving large-scale neural networks. Artif. Life **15**(2), 185–212 (2009)
18. Weel, B., Crosato, E., Heinerman, J., Haasdijk, E., Eiben, A.E.: A robotic ecosystem with evolvable minds and bodies. In: 2014 IEEE International Conference on Evolvable Systems, Orlando, FL, USA, pp. 165–172. IEEE, December 2014

Author Index

G. Pappa et al. (Eds.): EuroGP 2023, LNCS 13986, pp. 355–356, 2023.
https://doi.org/10.1007/978-3-031-29573-7

Printed in the United States
by Baker & Taylor Publisher Services